A POLITICAL THEOLOGY OF CLIMATE CHANGE

A Political Theology of Climate Change

Michael S. Northcott

William B. Eerdmans Publishing Company
Grand Rapids, Michigan / Cambridge, U.K.

Published 2013 by
Wm. B. Eerdmans Publishing Co.
2140 Oak Industrial Drive N.E., Grand Rapids, Michigan 49505 /
P.O. Box 163, Cambridge CB3 9PU U.K.

Printed in the United States of America

19 18 17 16 15 14 13 7 6 5 4 3 2 1

Library of Congress Cataloging-in-Publication Data

Northcott, Michael S., author.
A political theology of climate change / Michael S. Northcott.
pages cm
Includes bibliographical references and index.
ISBN 978-0-8028-7098-8 (pbk.: alk. paper)
1. Global warming — Political aspects.
2. Climatic changes — Political aspects.
3. Geopolitics. 4. Environmental ethics. I. Title.

QC981.8.G56N675 2013
363.738'7401 — dc23

2013030577

www.eerdmans.com

For Jill

Contents

Acknowledgements

I am grateful to the School of Divinity in the University of Edinburgh for the periods of research leave in 2008 and 2011 that made the research for this book possible, and to Dartmouth College whose wonderful library, set in the midst of the Appalachian Mountains and above the broad and kayak-friendly Connecticut River, provided a productive writing space in the summers of 2011 and 2012 for pulling it together. I am thankful to colleagues and graduate students at Boston College; Dartmouth College; the Universities of Edinburgh, Exeter, and Glasgow; the 2010 meeting of the Society for the Study of Christian Ethics in Cambridge; and the 2013 conference on Religious and Spiritual Approaches to Climate Engineering at the Institute for Advanced Studies in Sustainability in Potsdam, where I presented earlier versions of material in this book. Some material in chapters 1 and 2 also appeared in the journal *Studies in Christian Ethics,* and I am grateful to the editor, Dr Susan Parsons, for permission to reproduce it here.

This book reflects, and is intended to undergird, my teaching in the University of Edinburgh, and I am grateful to successive cohorts of students who continue to inspire me. The book also reflects conversations with graduate students in ecology and religion, and I am grateful to Jeremy Kidwell, Daniel Miller, and Paul Peterson who read and commented on early drafts of this book. I am also thankful to colleagues who have kindly read and commented on parts or the whole of this book: Nick Adams, Robin Gill, David Grumett, and Jolyon Mitchell. Bruno Latour gave a memorable series of Gifford Lectures in the University of Edinburgh in 2013 as I worked on late drafts, and his lectures, and my conversations with him, were a great stimulus while I edited the penultimate draft. Finally, Bruno Latour and Noah Toly generously read that draft and made very helpful corrections and suggestions.

Scripture quotations are from the King James Version, minimally adapted by the author to modern English style. I make no apologies for relying on this old English version; despite the 'scientific' inaccuracies of the underlying texts, the cadences, metaphors, and timbre of this version, like Shakespeare's plays, have passed down into the English language, definitively shaping contemporary spoken and written English, and my use of it here is a reminder that the past is a chain of memory without which we cannot really know the present or imagine the future.

1. The Geopolitics of a Slow Catastrophe

In the summer of 2012 a larger extent of the Arctic Ocean was open sea than at any time in the 200,000-year history of *Homo sapiens*. The extent of ice loss from 2007 is far above predictions from climate models, including those of the Intergovernmental Panel on Climate Change. So extensive is the melting trend that some scientists predict the imminent disappearance of multiyear ice formation, and the disappearance of summer ice altogether between 2020 and 2035.[1] Ice core evidence indicates that neither of the ice caps of the planet have melted completely, even in summer, in the last two million years.[2] The effects of this much open water on the Northern Hemisphere are already unfolding. The temperature of the North Atlantic was two degrees Celsius above historic norms in 2012, which contributed to an extensive drought in North America that year. Warmer oceans sustain stronger storms. Hurricane Sandy was the largest storm system ever recorded in the North Atlantic. High pressure over Greenland, in a summer when satellites revealed that 97 percent of the surface area of Greenland's ice was melting, pushed a late tropical storm inland toward the northeastern United States, causing a fifteen-foot tidal surge which destroyed thousands of homes and businesses, taking out electricity and transportation systems in much of New York City, coastal New Jersey, and Delaware. In the same year northern Europe had an exceptionally wet summer, while Iceland and Greenland basked in unprecedented summer heat.

While the earth as a whole had warmed only 0.8 degrees Celsius from the

1. James E. Overland and Muyin Wang, 'When Will the Summer Arctic Be Nearly Sea Ice Free?' *Geophysical Research Letters* (2013), DOI: 10.1002/grl.50316.

2. Wieslaw Maslowski, Jaclyn Clement Kinney, Matthew Higgins, and Andrew Roberts, 'The Future of Arctic Sea Ice', *Annual Review of Earth and Planetary Sciences* 40 (2012): 625-54.

pre-industrial era to 2013, a comprehensive study of archaeological climate 'proxies', which include ice cores and fossilised tree rings, combined with contemporary observations from satellites and ocean- and land-based thermometers, indicates that the pace of climate warming in the twentieth century is unprecedented since the end of the last ice age.[3] At the same time the Mauna Loa record of CO_2 in the atmosphere revealed an unprecedented annual jump in atmospheric CO_2 of 2.97 parts per million (ppm) in 2012. While the period from 1750 to 1950 saw a rise in atmospheric CO_2 from 270 to 310 ppm, from 1950 to 2013 atmospheric concentrations had risen to 400 ppm. This rapid rise coincides with the consumer revolution in the Northern Hemisphere and represents what Will Steffen and colleagues call the 'great acceleration.'[4] The geophysical consequences of the great acceleration have given humanity an unprecedented material influence over the earth. In this book I argue that this influence, while unprecedented in earth history, has analogies with prescientific beliefs about the influence human beings believed they had over the climate, and nature, before the Copernican revolution. Indeed, such beliefs are still in evidence until the Enlightenment; the 1755 Lisbon earthquake provoked speculation on whether it was an instance of divine judgement on human activities in general or on the Portuguese.[5] Arguably, then, climate change is taking human culture back into familiar cultural territory, even as it is taking the earth's physical state into 'new climatic territory'.[6]

Tangible evidence of a new climate state is most notable to contemporary Europeans and North Americans in increasingly dramatic changes in weather in the Northern Hemisphere, which natural scientists believe are a result of the influence of melting Arctic ice. Melting ice is affecting ocean currents, and especially the Gulf Stream, which brings heat from the tropics into northern Europe, and the related high atmosphere jet stream. This is producing 'stuck' weather patterns and a growing number of intense precip-

3. Shaun A. Marcott, Jeremy D. Shakun, et al., 'A Reconstruction of Regional and Global Temperature for the Past 11,300 Years', *Science* 339 (2013): 1198-1201.

4. Will Steffen, Paul J. Crutzen, and John R. McNeill, 'The Anthropocene: Are Humans Now Overwhelming the Great Forces of Nature?' *Ambio* 36 (2007): 614-21.

5. Susan Nieman, *Evil in Modern Thought: An Alternative History of Philosophy* (Princeton, NJ: Princeton University Press, 2002), 4.

6. Tim Flannery, Australian Chief Climate Commissioner, 'Australia Enters New Climatic Territory', Australian Broadcasting Corporation, 4 March 2013, at http://www.abc.net.au/7.30/content/2013/s3703207.htm (accessed 9 March 2013); Julienne C. Stroeve, Mark C. Serreze, et al., 'The Arctic's Rapidly Shrinking Sea Ice Cover: A Research Synthesis', *Climatic Change* 110 (2012): 1005-27.

itation events, stronger storms, including snow storms, as well as heat waves and droughts.[7] It is also provoking the release of quantities of subterranean methane beneath the Arctic Ocean and from subarctic lands and oceans. Methane has a shorter life in the atmosphere than CO_2, but it has a warming potential seventy-two times that of CO_2 for two decades after release. Weather balloons and satellites above the open ocean northeast of Norway and eastern Siberia in 2013 recorded substantial methane release from the ocean floor was under way in this area, sustained by the growing melt of surface ice.[8] Speleological investigations of caves in Siberia reveal that climate-changing quantities of both carbon and methane were last released from Siberian permafrost 500,000 years ago; that event triggered global temperature change 1.5 degrees Celsius above the pre-industrial global surface temperature average.[9] Annual releases of carbon from frozen trees and soils in subarctic tundra would potentially exceed annual emissions of greenhouse gases emitted from human activities. So methane and carbon release from a melting Arctic and subarctic region potentially presage a catastrophic and far more sudden increase in global temperatures than the gradual existing warming from 1750 to the present of 0.8 degrees Celsius.

Another amplifying effect from anthropogenic greenhouse gas emissions is that warmer atmospheric temperatures create more surface water condensation, which increases atmospheric water vapour. Atmospheric water vapour is presently rising at 1 percent *annually,* and this is promoting more extreme precipitation and storm events.[10] At the same time rising land temperatures create more extreme heat events, enduring droughts, drying of soils and forests, and stronger wildfires. Together these events constitute a daily pattern of weather across the globe that already shows more extremes and marked effects on species productivity as well as human communities. Temperature extremes in the summer of 2012 on the American and Eurasian continents saw significant declines in agricultural production. In the U.S. a six-month drought reduced wheat, soy, and corn production by 20 percent. In northern Europe production of cereals, fruit,

7. Jennifer A. Francis and Stephen J. Vavrus, 'Evidence Linking Arctic Amplification to Extreme Weather in Mid-latitudes', *Geophysical Research Letters* 39 (2012): L06801.

8. *Arctic News,* 1 February 2013, at http://arctic-news.blogspot.co.uk/2013/02/dramatic-increase-in-methane-in-the-arctic-in-january-2013.html (accessed 4 March 2013).

9. A. Vaks, O. S. Gutavera, et al., 'Speleothems Reveal 500,000-year History of Siberian Permafrost', *Science* 1228729 (2013): 1-6.

10. K. H. Rosenlof, S. J. Oltmans, D. Kley, et al., 'Stratospheric Water Vapor Increases over the Past Half-Century', *Geophysical Research Letters* 28 (2001): 1195-98.

and vegetables was down because of summer-long rains and lack of sunshine. Poor summer weather not only affects the appearance and quantity but also the quality of fruit and vegetables because, as agricultural scientist Mike Gooding observes, 'the nutrients available to the plant might well be reduced. We do know that rainfall, for example, will often cause leaching and loss of nutrients from the soil, and at certain times that will certainly reduce the amount of protein that ends up in the produce.'[11]

It was once thought that climate change would make temperate agriculture more productive, since elevated atmospheric CO_2 and warmer temperatures would increase crop productivity. Instead, growing weather extremes are already making farming more challenging, in temperate as well as semi-temperate and tropical zones. Consequently, although human food production was at record high levels in 2011, the Food and Agriculture Organisation in 2012 warned of an ongoing rise in global food prices, which had already doubled since 2000, and suggested food production risks failing to meet rising consumption.[12]

A Slow Catastrophe

The connection between temperature rise and rising anthropogenic emissions of carbon dioxide was first accurately modelled by New York glaciologist Wallace Broecker in 1975, and there are now four Global Circulation Models run by supercomputers belonging to United States and United Kingdom government agencies that are increasingly accurate in tracking observed climate changes and in matching predictions to observations.[13] According to these models, present rates of rising greenhouse gas emissions will see global temperatures rise by 4 to 7 degrees Celsius by 2100 compared

11. Professor Mike Gooding, Interview, *Farming Today*, BBC Radio 4, 9 January 2013.

12. Food and Agriculture Organisation of the United Nations, *World Food Situation: Food Price Index* (Geneva: FAO, 2012), at http://www.fao.org/worldfoodsituation/wfs-home/foodpricesindex/en/ (accessed 11 October 2012); for a contrary view see Johan Swinnen and Pasquamaria Squicciarinni, 'Mixed Messages on Prices and Food Security', *Science* 335 (2012): 405-6.

13. Wallace S. Broecker, 'Are We on the Brink of a Pronounced Global Warming?' *Science* 189 (1975): 460-63; Broecker's model overestimated warming for 2005-10 while the IPCCs models, begun in 1982, have proven more accurate since 2005; for a detailed account of the development of reliable climate models see Paul N. Edwards, *A Vast Machine: Computer Models, Climate Data, and the Politics of Global Warming* (Cambridge, MA: MIT Press, 2010).

to pre-industrial temperatures.[14] But there remain substantial uncertainties as to how 'sensitive' the climate is to greenhouse gas emissions, and much depends upon the potential of feedbacks — such as the melting of frozen methane — to drag the system into a warmer episode.

A one-third increase in atmospheric CO_2 from pre-industrial levels of 270 parts per million (ppm) to 400 ppm in 2013 has provoked a globally averaged warming of 0.8 degrees Celsius and significantly elevated warming in the Arctic region and in North Africa. This slow historic increase in temperature is at first sight comforting. But it is occurring at a faster rate than study of climate 'proxies' such as ice cores and tree rings indicate has occurred before in planetary history. Nonetheless, the relatively modest rate of warming seems to indicate that the atmosphere and oceans have so far performed well in soaking up greenhouse gas emissions. But earth responses to atmospheric pollution are accelerating because the *rate* of greenhouse gas emissions growth has risen sharply since 1950, with an annual rise of 2 percent to 2000, 2.7 percent from 2000 to 2010, and 3 percent from 2011.[15] Whereas atmospheric CO_2 levels rose at 1.5 ppm per decade from 1750 to 1950, from 1950 to 2010 they rose at an average of *14 ppm per decade.*[16] Though only a trace gas, the proportion of CO_2 in the upper atmosphere closely corresponds to changes in earth's temperature, and the recent growth trajectory of atmospheric CO_2 has produced a decadal temperature rise of 0.2 degrees Celsius since 1960, whereas the rate of warming was below 0.02 degrees Celsius from 1750 to 1960.[17] According to the World Bank and the International Energy Authority, three degrees of warming looks increasingly likely by mid-century, four degrees by 2080, and even six degrees of warming by century's end with ongoing unrestrained growth in greenhouse gas emissions.[18]

NASA's Chief Scientist James Hansen argues that the measures or

14. Richard A. Bettis, Matthew Collins, et al., 'When Could Global Warming Reach 4° C?' *Philosophical Transactions of the Royal Society* 369 (2011): 67-84.

15. Jos G. L. Olivier, Greet Janssens-Maenhout, et al., *Trends in Global CO_2 Emissions: 2012 Report* (The Hague: PBL Netherlands Environmental Assessment Agency).

16. The rising level of atmospheric CO_2 is recorded at the Mauna Loa observatory and the data are continuously updated at the National Oceanographic and Atmospheric Administration website: http://co2now.org/Current-CO2/CO2-Now/noaa-mauna-loa-co2-data.html (accessed 17 May 2013).

17. James Hansen, Makiko Sato, et al., 'Global Temperature Change', *Proceedings of the National Academy of Sciences* 103 (2006): 14288-93.

18. World Bank, *Turn Down the Heat: Why a 4° C Warmer World Must Be Avoided* (Washington, DC: World Bank, 2012); International Energy Authority, *2011 World Energy*

'proxies' of past climates, such as ice cores and fossilised tree rings, reveal that just two degrees of warming correlated to a largely ice free Arctic and a much reduced ice mass on Greenland. Greenland if melted would represent seven metres of sea level rise, inundating the first and second floors of many buildings in London, New York, Singapore, and Tokyo, as well as the Nile Delta, most of Bangladesh, Louisiana, Florida, parts of eastern England, and the Netherlands.[19] At six degrees of warming the Antarctic ice sheet will also begin to collapse. With much polar ice gone, the flood would rise two hundred feet, or a sixth of the way up the newly constructed Shard office block in London, standing as it does just a few feet above the Thames River. On the current trajectory of greenhouse gas emissions growth, by the end of the present century, or within the lifetime of my grandson Jacob, the planet will be a 'new creation', but not of the making of God or evolution.

While warmer conditions can provoke much more rapid sea level rise of one metre every twenty years, as they have in past deglaciation events, present sea level rise of only 3 millimetres a decade is hardly perceptible across the lifetime of a non-scientific observer. Hansen uses the analogy of a Christmas tree light to explain the perception problem of climate change. The contribution of human greenhouse gas emissions to climate heating is equivalent to the heat from two one-watt tungsten filament Christmas tree lights per square metre of the earth's surface. The imperceptible quantity of extra heat per square metre sets up a contrast 'between the awesome forces of nature and the tiny light bulbs. Surely their feeble heating could not command the wind and waves'?[20] Climate science models drawing together historic proxies and real time observations of temperature change are confirming that this small heat load per square metre is translating into a large-scale but slow rate of global warming. But human beings have yet to modify their behaviours significantly in response. Other species, oceans, and land areas are, however, already showing observable responses. Satellite photography and spectography reveal a marked increase in the growing season in the

---

Outlook (Paris: IEA, 2011). The authoritative IPCC's Sixth Scientific Assessment Report was not available at time of writing.

19. James Hansen, 'Dangerous Anthropogenic Interference: A Discussion of Humanity's Faustian Climate Bargain and the Payments Coming Due', Distinguished Public Lecture series, Department of Physics, University of Iowa, 26 October 2004, and at http://www.columbia.edu/~jeh1/2004/dai_complete_20041026.pdf (accessed 15 March 2013).

20. James Hansen, 'Can We Defuse the Global Warming Time Bomb?' Working Paper (2003), NASA Goddard Institute for Space Studies, and Columbia University Earth Institute, http://pubs.giss.nasa.gov/docs/2003/2003_Hansen.pdf (accessed 15 March 2013).

subarctic region in the last thirty years, which shows up as a greening of the northern tundra in Siberia and Canada in spring and autumn as plants respond to increased CO_2 and warmer temperatures.[21] Scientists also observe that species are gradually migrating in many places on the planet.[22]

Carbon Wars

While the Arctic region has experienced the most warming in the last fifty years, it is a largely uninhabited region, and hence the effects of warming in the region are more visible to satellites and shipping than to significant numbers of human beings. The changing topography of the ice affects the hunting practices of indigenous human communities in the Arctic Circle and the behaviours of polar bears, but there is so far no evidence of increased human suffering from Arctic ice melt.[23] By contrast, drought and raised temperatures in North and East Africa are already seriously impacting human life. Enduring drought in the Horn of Africa is the cause of growing instability in the region, while declines in crop productivity are, along with food speculation by investment bankers and others, causing significant food price rises. Climate change was strongly implicated in the riots which sparked the 'Arab Spring' and subsequent civil conflicts, or wars, in many Middle Eastern and North African countries, including Bahrain, Egypt, Libya, Mali, Syria, and Tunisia.[24] Many governments in the region were corrupt and lacked democratic accountability, and their replacement with more accountable governments might in principle be said to be a benefit from climate-induced social upheaval. However, there has been great loss of life, large numbers of refugees have fled conflict zones to neighbour-

21. L. Xul, R. B. Myneni, et al., 'Temperature and Vegetation Seasonality Diminishment over Northern Lands', *Nature: Climate Change*, advance publication, online 10 March 2013, DOI: 10.1038/NCLIMATE1836.

22. Delphine Nicolas and Aure'lie Chaalali, 'Impact of Global Warming on European Tidal Estuaries: Some Evidence of Northward Migration of Estuarine Fish Species', *Regional Environmental Change* 11 (2011): 639-49; Wayen Hsuing and Cass R. Sunstein, 'Climate Change and Animals', *University of Pennsylvania Law Review* 155 (2007): 1696-1740.

23. I discuss the effects of floating ice loss and melting permafrost on indigenous Arctic communities in Michael S. Northcott, *A Moral Climate: The Ethics of Global Warming* (London: Darton, Longman and Todd, 2007), 35-43.

24. Sarah Johnstone and Jeffrey Mazo, 'Global Warming and the Arab Spring', *Survival: Global Politics and Strategy* 53 (2011): 11-17; also Caitlin E. Werrell and Francesco Femia, eds., *The Arab Spring and Climate Change* (Washington, DC: Center for American Progress, 2013).

ing countries, including one million from Syria alone, and many of these countries have not yet moved on toward civil peace.

Weather-related problems around the Horn of Africa long preceded the Arab Spring, with enduring drought in the region and especially in Ethiopia and Eritrea, Somalia, and Yemen since 1996, contributing to the collapse of lawful rule and provoking millions of people to flee to neighbouring countries which are also experiencing climate-related difficulties.[25] In 2013 a breakdown of orderly government in Mali, and a murderous terrorist attack on a natural gas facility in Algeria, prompted external intervention by France and other Western agencies. Though the Malian adventure is described as a new military 'intervention', in reality the West is already intervening in Mali, not only in its ongoing quest for energy resources, but also because Western-originated greenhouse gas emissions are a cross-border infraction which are raising temperatures and reducing food and water availability in North Africa and the Middle East.

The science of climate change presents a political problem which is unique. The science indicates that the earth is warming faster than at any previous point in human or earth history. But although the speed of change is unprecedented in the intergenerational history of humankind, the rate of change from the point of view of one human generation, and even more from the point of view of the four- or five-year term of a democratically elected government, is imperceptibly slow. Furthermore, one or even two degrees of warming averaged across the planet's surface does not sound catastrophic in its implications. Temperatures fluctuate by far more than this on a daily basis, from day to night, and from day to day. Slow change leading to catastrophic outcomes is therefore counterintuitive. But for non-scientists to realise this, the rise in conflicts in North Africa and the Middle East needs to be presented in ways that make the connections between climate and culture. Modern political scientists tend, however, to decontextualise politics from geography, and culture from nature, and hence are more likely to read signs of growing conflict in Islamic terrains, and the overflow of terrorism and other problems into the West, as evidence of a 'clash of civilisations' rather than as evidence of climate change. As I argue in what follows, political theology offers an alternative perspective because it situates culture in creation, and politics in the geography of the nations.

The relationship between climate change and civil conflict is under-

25. Clionadh Raleigh, 'Political Marginalization, Climate Change, and Conflict in African Sahel States', *International Studies Review* 12 (2010): 69-86.

written in a study of the relationship between civil war and warming in the Tropics associated with the El Niño weather pattern between 1950 and 2004. Solomon Hsiang and his colleagues found that there was a doubling of the likelihood of civil conflict in tropical countries in El Niño years.[26] Another study reveals a contiguity between depleting ground water and declining precipitation in the Mediterranean region and countries affected by civil unrest and economic crisis in the period from 2000 to 2009.[27] The growing association between climate change and civil conflict generates a growing tendency to frame climate change as a threat to national security. Defence establishment and intelligence agencies increasingly describe climate change as analogous to other global conflicts, including the Cold War and the so-called 'war on terror'. As U.S. Admiral Thomas J. Lopez, a former NATO commander, puts it:

> Climate change will provide the conditions that will extend the war on terror. You have very real changes in natural systems that are most likely to happen in regions of the world that are already fertile ground for extremism. Droughts, violent weather, ruined agricultural lands — those are the kinds of stresses we'll see more of under climate change. The result of such changes will be more poverty, more forced migrations, higher unemployment. These conditions are ripe for extremists and terrorists.[28]

A study commissioned by the Central Intelligence Agency argues that the U.S. should anticipate that climate change 'will produce consequences that exceed the capacity of the affected societies or global systems to manage, and therefore will have global security implications serious enough to compel international response'.[29] A U.S. Defence task force found that, while

26. Solomon M. Hsiang, Kyle C. Meng, et al., 'Civil Conflicts Associated with the Global Climate', *Nature* 476 (2011): 438-41.

27. Katalyn A. Voss, James S. Famiglietti, et al., 'Groundwater Depletion in the Middle East from GRACE with Implications for Transboundary Water Management in the Tigris-Euphrates-Western Iran Region', *Water Resources Research* 49 (2013): 904-13.

28. Admiral Thomas J. Lopez, former Commander in Chief, U.S. Naval Forces in Europe, cited in Gordon R. Sullivan, Frank Bowman, Lawrence P. Farrell, et al., *National Security and the Threat of Climate Change* (Alexandria, VA: CNA Corporation 2007), 17, note 28; see also Arija Flowers, 'National Security in the 21st Century: How the National Security Council Can Solve the President's Climate Change Problem', *Sustainable Development Law & Policy* 11 (2011): 50-55, 90-94.

29. John D. Steinburger, Paul C. Stern, et al., *Climate and Social Stress: Implications for Security Analysis* (Washington, DC: National Academies Press, 2013), 2.

climate change may in the long term affect low-lying and drought-prone areas of the United States, its most immediate effects 'will come from the most vulnerable regions of the world where the United States obtains vital fuel and strategic mineral resources and combats terrorism'.[30] In the national security perspective, empire is also an underlying theme. The United States views climate change primarily as an external threat to its ability to continue to garner sufficient energy from the Middle East and other regions already being destabilised by extreme weather in order to sustain its heavily fossil fuel–dependent consumption and infrastructure. Analogously, British sociologist Anthony Giddens argues that 'energy security' is the biggest challenge that modern societies face as drought and food insecurity destabilise energy-rich developing countries, while melting permafrost, rising oceans, and increasing heat threaten the highly complex and relatively fragile international energy extraction and supply network.[31]

A climate challenged planet without significant mitigation of fossil fuel use will see a growth in civil conflict and mass migration as populations — particularly in developing countries — experience growing energy, food, and water insecurity. These challenges will likely be so grave that billions of people will be forced to migrate from their ancestral lands. The largest number of people seriously affected by three degrees of warming live in Asia. More than two billion people depend for food and water on the three rivers of the Indus, the Ganges, and the Brahmaputra, all of which rise in the Himalayan ice mass, the largest terrestrial ice mass on earth outside of the Arctic and Antarctica.[32] As with the polar regions, air temperatures are rising twice as fast over the Himalayan ice mass than average global temperatures, and the glaciers are already showing accelerated retreat. With three degrees of global warming these three great rivers will first flood as the ice melts in unprecedented quantities and will then reduce to a trickle. This will radically cut food and water supplies, while also reducing the production of hydroelectricity on which China and India are particularly reliant. China and India also have access to nuclear weapons. Conflict at their borders is already endemic. As Mark Lynas argues, deglaciation in the Himalayas may well provoke conflicts 'between these two nuclear-armed

30. *Report of the Defense Science Board Task Force on Trends and Implications of Climate Change for National and International Security* (Washington, DC: Office of the Under Secretary of Defense, 2011), xii-xiii.

31. Anthony Giddens, *The Politics of Climate Change* (Cambridge: Polity, 2009).

32. Mark Lynas, *Six Degrees: Our Future on a Hotter Planet* (London: Harper Perennial 2008), 138-39.

countries as water supplies dwindle and political leaders quarrel over how much can be stored behind dams in upstream reservoirs.'[33]

To highlight the impacts of dramatic temperature rises later in the twenty-first century, the UK Meteorological Office published a map of a four degree warmer planet in 2009. It shows that three quarters of the earth's land area will be unsuitable for food growing or secure human settlement in a four degree warmer world, including most of Africa, the Americas, and Asia.[34] In the Northern Hemisphere the areas that will remain fertile and habitable are North America above the forty-ninth parallel, northern Europe, and northern Russia; in the Southern Hemisphere habitable land will include the southern cone of South America, South Africa, New Zealand, and, possibly, the northern territories in Australia. The remaining land areas will either be desert, flooded, or subject to such climatic extremes as to make agriculture and human settlement infeasible.

Harald Welzer argues that the 'carbon wars' of the present century will dwarf even the scale of killing of the world wars of the twentieth century.[35] The German Environment Ministry argues that climate change will 'contribute to an increased potential for conflict', and that the spread of such conflicts, and the waves of migration they will provoke, is the principal threat to national security.[36] The British Ministry of Defence estimates that by 2036

> nearly two thirds of the world's population will live in areas of water stress, while environmental degradation, the intensification of agriculture, and pace of urbanisation may reduce the fertility of, and access to, arable land. There will be constant pressure on fish stocks, which are likely to require careful husbanding if major species are not to become depleted or extinct. Food and water insecurity will drive mass migration from some worst-affected areas and the effects will be felt in more affluent regions.[37]

33. Lynas, *Six Degrees,* 140.

34. Met Office, 'The Impact of a Global Temperature Rise of 4° C (7° F)', at http://web archive.nationalarchives.gov.uk/20100623194820/http:/www.actoncopenhagen.decc.gov.uk/content/en/embeds/flash/4-degrees-large-map-final (accessed 9 October 2012).

35. Harald Welzer, *Climate Wars: What People Will Be Killed for in the 21st Century* (Cambridge: Polity Press, 2011).

36. German Environment Ministry, *Climate Change and Conflict: Environmental Policy* (Bonn: Federal Ministry of the Environment, 2002), cited in Ragnhild Nordas and Nils Peter Gleditsch, 'Climate Change and Conflict', *Political Geography* 26 (2007): 627-38.

37. *The DCDC Global Strategic Trends Programme 2007-2036* (London: Ministry of De-

With less than one degree of warming the number of conflicts in the developing world is already rising, while the United Nations estimates that there are 100 million people who are already climate refugees. As states fail in their basic responsibilities to ensure that people can live securely and obtain food and water for their families, large-scale societal collapse seems the only realistic prospect in the more populous regions of Africa, Asia, and the Americas, where extremes of climate are already challenging the capacity of farmers to grow enough food.

Climate Apocalypse

The unfolding of the geopolitical consequences of climate change is a gradual process which mirrors the slow response of the climate system to gradually accumulating industrial greenhouse gas emissions. The slow pace of such change is likely to continue, short of tipping points, such as the release of large quantities of methane into the atmosphere. There is a political as well as a perception difficulty with this slow pace. Mitigating climate change requires dramatic, large-scale political interventions in fossil fuel extraction and marketing, and hence in the energy systems and behaviours that these fuels sustain. But these systems and behaviours are so intrinsic to industrial civilisation and modern consumerism that radical reform without a real and present climate catastrophe lacks popular support, and hence influential advocacy, in most political domains. Both in popular culture and in science writing there is therefore a growing tendency to display climate change events in terms of a more sudden and imminent collapse in order to promote cultural support for larger and more timely political change. Roland Emmerich's film *The Day After Tomorrow* depicts a catastrophic and extremely sudden cooling of the Northeast Atlantic as the Gulf Stream, which draws heat from the Tropics into the North and East Atlantic, ceases to function.[38] The sudden cooling of the North Atlantic produces an extreme winter storm which engulfs the eastern seaboard of the United States in glacial conditions with twenty feet of snow. The film ends with shots of vast numbers of people on highways attempting to flee New York

fence, 2007) and at http://cuttingthroughthematrix.com/articles/strat_trends_23jan07.pdf (accessed 4 November 2011).

38. Roland Emmerich, *The Day After Tomorrow* (Hollywood, CA: 20th Century Fox, 2004).

City. The sudden and apocalyptic collapse of technological systems that the film depicts is not that distant from the effects of Hurricane Sandy in 2012: some apartment blocks were without water mains, functioning sewerage, heat, and power for three months after the hurricane hit. While the overnight shift to an extreme winter state is scientifically unrealistic, the film highlights a potential climate change scenario in which the extent of fresh meltwater from melting Arctic ice shuts off the thermohaline ocean currents which presently bring heat from the Tropics, in the form of the Gulf Stream, into northern latitudes, reducing the severity of winters through North America and northern Europe and warming the summers.

A second film, Franny Armstrong's *Age of Stupid,* depicts a man, played by Pete Postlethwaite, in old age and ensconced on a technological Noah's Ark somewhere in the ice-free Arctic Ocean.[39] Through a series of filmic flashbacks, which he manipulates on a transparent digital screen, Postlethwaite recounts the history of the end of history. Climate protest marches, failed international conferences, wars for the remaining fossil fuels under the earth's crust, and growing extreme weather events all precede the eventual melting of polar ice. Finally the sea engulfs the world's principal cities and low-lying residential areas and farmlands, leaving remnant human groups who eke out a living on high lands and in technological lifeboats of the kind in which Postlethwaite is depicted.

A number of recent examples of science writing by established climate scientists take an equally apocalyptic turn. James Lovelock, who discovered the ozone hole over the Antarctic and formulated the Gaia hypothesis, envisages that, as an interconnected set of living systems, the earth, or Gaia, will compensate for the ecological excesses of too many human beings by ridding herself of most of them. In *The Revenge of Gaia* he uses grim apocalyptic language to describe the gradual collapse of the planetary processes that have evolved to produce an atmosphere congenial to human and mammalian life.[40] This collapse will involve a massive cull of the human and other keystone species. 'Gaia', the mythological name Lovelock gives to his now widely accepted scientific theory of an interactive living planet, takes on the visage of an implacable and vengeful goddess who gets her own back on the humans that have ravaged her. Gaia as goddess is indifferent to human suffering in enacting her 'revenge' on heedless industrial humanity.

In *Requiem for a Species* Australian scientist Clive Hamilton argues that

39. Franny Armstrong, *Age of Stupid* (London: Spanner Films, 2009).
40. James Lovelock, *The Revenge of Gaia* (London: Allen Lane, 2006).

unmitigated climate change will bring the end of the human species within two hundred years.[41] Australia has seen some of the most extreme climate change–related weather events, which helps explain Hamilton's particularly bleak outlook. Temperatures in 2012 reached an unprecedented 49 degrees Celsius north of Adelaide, and weather forecasters deployed a new colour — deep purple — to display temperatures above 40 degrees. Wildfires took hold in many areas close to major cities, including Melbourne and Sydney, in 2011 and 2012, causing a number of deaths as well as extensive destruction of property and bushfires. In both years the wildfires and extreme heat were rapidly followed by extreme flooding which in 2011 inundated much of Queensland and significant parts of the city of Brisbane. But Australia is also home to a deep vein of climate denialism. This might be said to reflect the extremes of heat and drought that settlers have endured for more than a hundred years. The widely expressed, if complacent, 'Aussie' view is that no matter what settlers throw at the land, 'she'll be alright'. The extent of Australian denialism also reflects the considerable influence of Rupert Murdoch's ownership of a number of climate-denying Australian newspapers and broadcasting outlets.[42]

In *Storms of My Grandchildren* James Hansen gives a less bleak but still apocalyptic account of the near and middle future that anthropogenic climate change will likely visit on some of those alive in the present and on future generations. Planet earth is in 'imminent peril', and the continued use of fossil fuels 'threatens not only the other millions of species on the planet but also the survival of humanity itself — and the timetable is shorter than we thought'.[43] Hansen believes humanity has at most fifteen years to turn the presently rising consumption of fossil fuels downwards before it sets in train a series of climate feedbacks which will be irreversible, committing the planet to a post-glacial era unprecedented in the last half million years.

The increasingly frequent resort of natural scientists to the language of catastrophe indicates that those who follow the science most closely, and

41. Clive Hamilton, *Requiem for a Species: Why We Resist the Truth about Climate Change* (London: Earthscan, 2010).

42. I use the word 'denialism' since the scientific method that has produced strong consensus on the human influence on the climate is premised on evidential and theoretical scepticism, whereas denialists are mostly non-scientists and resist contrary evidence to the strongly held belief that human beings are *not* influencing the climate: Michael Shermer, 'I Am a Sceptic, but I'm Not a Denier', *New Scientist* 206.2760 (2010): 36-37.

43. James E. Hansen, *Storms of My Grandchildren: The Truth about the Coming Climate Catastrophe and Our Last Chance to Save Humanity* (London: Bloomsbury, 2009).

understand its implications, are *more afraid* than the non-scientists.[44] But the resort to climate catastrophism of scientists, as well as more popular and activist climate change discourses, also indicates a belief that *fear* of extreme consequences from inaction may motivate rapid and effective political change and so mitigate the causes of climate change. The connection of an apocalyptic register to a politics of crisis and revolution is long established in political and religious discourse, and it has particular associations with the industrial epoch. Marx and Engels predicted that industrialism would eventually collapse from its own internal contradictions as the process of creative destruction on which industrial capitalism relies, in which 'all that is solid melts into air', provoked social revolution and revolutionary change toward more equitable social conditions. However, the age of revolutions from 1789 to 1989 did not produce cultures and societies that were more sensitive to ecological limits. Indeed, communist societies were more prone to ignore natural signs of ecological disaster than capitalist ones because their modes of government did not rely on popular consent. Nonetheless, those on the left of the political spectrum tend to embrace the science of climate change because it seems to present a natural as well as cultural crisis in industrial capitalism which will force a radical reform of the tendencies of industrial societies toward inequality in power and property between social classes. Thus for Ulrich Beck the risks from ecological crises are the 'unintended side-effects' of modernity. They reveal the inherent contradictions in the 'self-evident truths' of modernity, including the equation of increasing welfare and material progress, and its 'abstraction from environmental consequences and hazards'.[45]

Shooting the Messenger

Access to and use of energy comprise one of the main drivers of inequality in industrial societies, as Ivan Illich first argued.[46] The availability of tools and lifestyles dependent on fossil fuels shapes a cultural enslavement to these fuels, which becomes an increasingly dominant, if hidden, driver and sustainer of political institutions and ideologies. Hence the potential for cli-

44. Bruno Latour, 'Waiting for Gaia: Composing the Common World through Arts and Politics', Lecture at French Institute, London, November 2011, and at http://www.bruno-latour.fr/sites/default/files/124-GAIA-LONDON-SPEAP_0.pdf (accessed 19 February 2013).

45. Ulrich Beck, *World at Risk* (Cambridge: Polity Press, 2009), 109.

46. Ivan Illich, *Energy and Equity* (New York: Harper and Row, 1974).

mate catastrophe presents an apparent political necessity to tackle inequality, both within and between nations, as it is manifest in the excess use of energy by the rich, who use it in multiples of hundreds more than the earth's poorest inhabitants.[47] If access to fossil fuels for flying, driving, heating and cooling large homes and offices, meat eating, and other energy-intense activities has to be rationed in order to prevent climate catastrophe, then the rich, as well as the poor, will be forced to live more equitably because they will be forced to live and to meet their material needs more locally. It is not hard to see, then, why climate science is embraced by the political left and resisted by those on the right. If untrammelled capitalism is the means to advance human flourishing, as the right believes, then climate science and climate scientists are the enemy. Conversely, if inequality and excessive cultural focus on material consumption are the causes of political as well as moral declension in late modern capitalist societies, as the left believes, then climate science offers a geophysical rationale for promoting the politics of equality and dematerialisation.

Predictions of climate catastrophe therefore represent a *politics* because climate science indicates that, absent a *levelling* of unequal uses of fossil fuels between rich and poor and between developed and developing countries, the earth itself will enforce a levelling on the presently disequalising tendencies of fossil-fuelled industrial capitalism through climate catastrophe. Unmitigated climate change by the end of the century will flood the rich cities of the powerful and disrupt their global resource extraction and wealth accumulation systems, as well as turning the lands of the poor into deserts. Climate apocalyptic therefore represents a politics, even a political theology, and not just a natural scientific theory. It is a political theology because, like the apocalyptic of the New Testament, it indicates the imminence of a moment of judgment on the present form of civilisation, and the end of an era in which humans expanded their influence over the earth without regard to planetary limits and without apparent consequences. The climate crisis will also make visible the relationship which was formerly hidden between the foundation and structure of the earth and human history; and this unveiling of what was hidden will bring a levelling on human society.

That climate science, as well as climate apocalyptic, represents a politics, and a geophysical boundary or end-point to the current trajectory of global capitalism, is why climate *science* is rejected very extensively by those

47. M. Korbetis and D. Reay, 'New Directions: Rich in CO_2', *Atmospheric Environment* 40 (2006): 3219-20.

on the right of the political spectrum who have not in the past tended to reject mainstream natural science as contrary to their worldview. The seeds, in other words, of either acceptance of or resistance to climate science, and hence of climate apocalyptic, lie not in the science itself but in the politico-theological worldviews of those who hear and respond to the science, with most of those who resist climate science on the right of the political spectrum and those who accept it to the middle or left.[48] Hence even prominent advocates of environmental conservation on the right, such as Roger Scruton, resist the conclusions of mainstream climate scientists that the earth is warming because of greenhouse gas emissions, and that regulation of these emissions is therefore necessary to ecological conservation.[49]

The extent of resistance to climate science on the political right is particularly marked in Anglo-Saxon domains, including Australia and New Zealand, Britain, the United States, and Canada. Influential media outlets in these domains, including Fox News, the *Wall Street Journal,* and the UK's *Daily Telegraph,* regularly represent climate sceptic views. Climate denialists such as the UK's Nigel Lawson and A. W. Montford and the United States' Fred Singer and Steve McIntyre have a considerable following.[50] Coal, oil, and gas corporations have poured vast funds into anti–climate science lobby groups and think tanks. The rationale for this lobbying effort is not hard to discern. The capital and stock values of fossil fuel companies, which are the most powerful of modern economic corporations, rest in part on their claimed reserves of fossil fuels. If, as climate scientists claim, the burning of these reserves will destabilise the climate, extinguish myriad species, inundate coastal cities, and lead to the desertification or flooding of much farmland, then the value of these reserves is contentious.

Climate science is not the first kind of environmental science to be resisted because of its implications for industry and commerce. Rachel Carson's book *Silent Spring* uses an extensive apocalyptic rhetoric, from the title

48. James Painter and Teresa Ashe, 'Cross-national Comparison of the Presence of Climate Scepticism in the Print Media in Six Countries, 2007-10', *Environmental Research Letters* 7 (2012): 044005.

49. Roger Scruton, *Green Philosophy: How to Think Seriously about the Planet* (London: Atlantic Books, 2012).

50. Nigel Lawson, *An Appeal to Reason: A Cool Look at Global Warming* (London: Duckworth, 2009); A. W. Montford, *The Hockey Stick Illusion: Climategate and the Corruption of Science* (London: Stacey International, 2010); S. Fred Singer and Dennis T. Avery, *Unstoppable Global Warming: Every 1500 Years* (Lanham, MD: Rowman and Littlefield, 2007); Steve McIntyre's influential climate sceptic website is at http://climateaudit.org (accessed 14 March 2013).

itself, which suggests that pesticides would kill all bird life and hence bird song in the spring, to the claim that the chemicalisation of agriculture and the human food chain threaten a range of cancers and hormonal changes which endanger human health.[51] Carson's apocalyptic rhetoric also recalls the origin of agrochemicals in the development of chemical weapons first deployed in the Great War. As Peter Sloterdijk argues, the military creation of toxic air for disabling and killing people was the first time the possibility of the turning of the atmosphere against humanity through scientific intervention arose into popular consciousness.[52] The most apocalyptic science of the twentieth century was not mustard gas and agrochemicals but nuclear weapons technology and the related development of nuclear-powered electricity generators. The prospect that much life — including much human life — might be extinguished through an all-out nuclear war, and the nuclear winter of an atmospherically driven cooling of the earth, was the nuclear horizon of my own childhood and came close to being realised in the Cuban missile crisis.

As with climate science, scepticism about cancers from agrochemicals or the prospect of nuclear winter was strongest on the political right, while it was liberal progressives who took up the warnings of *Silent Spring* and acknowledged the threat of nuclear winter, and hence questioned the unregulated use of agrochemicals and the growing deployment of nuclear weapons.

While there are precedents, that climate science has been treated as a politics rather than a science has surprised the scientists themselves. Climate scientists have been accused of anti-capitalism, corruption, fraud, perfidy, and selfishness in promoting the claim that anthropogenic greenhouse gas emissions are changing the biophysiology of the earth. The claimed fraudulence of climate science was given a considerable boost by the theft, and publication in news media, of a tranche of emails from the Climate Research Unit (CRU) of the University of East Anglia during the Copenhagen Climate Summit of 2009. Journalists claimed the emails showed scientists had manipulated data, suppressed results which did not fit the theory of anthropogenic climate change, and used their role as peer reviewers for journals to suppress contrary findings.[53]

51. Rachel Carson, *Silent Spring* (New York: Fawcett, 1962).

52. Peter Sloterdijk, *Terror from the Air* (Los Angeles: Semiotext[e]; Cambridge, MA: MIT Press, 2009).

53. James Delingpole, 'Climategate: The Final Nail in the Coffin of Anthropogenic Global Warming', *Daily Telegraph*, 20 November 2009; and Fred Pearce, *The Climate Files: The Battle for the Truth about Global Warming* (London: Random House, 2010). For an overview from

Subsequent enquiries by other scientists, including one by the Royal Society, found that there was no truth to the allegations against the CRU and that its scientists had neither distorted their data nor used undue influence to suppress the research of others.[54] However, media reporting of the allegation of fraud was widespread and occasioned a significant decline in public trust of climate scientists and increased scepticism about the theory of anthropogenic climate change; in a 2010 poll 48 percent of the American public agreed with the claim that the problem of climate change had been 'greatly exaggerated', and fewer than one third believed that climate change represented a serious problem. Prior to the email theft in 2006, by contrast, 70 percent thought the problem was either serious or 'somewhat serious'.[55] The extreme weather events of 2012 dampened climate sceptic discourse and publicity, and there was a shift back toward the majority view that climate change is happening, even if most Americans continue to believe there is nothing they can do about it.

Some argue that climate scepticism reflects poor media reporting of the science by giving equal weight to mainstream scientific views and to the opinions of climate denialists.[56] But scientists associated with the 'sceptic' position are mainly funded by the fossil fuel industries.[57] Like tobacco companies in the 1960s, instead of admitting that use of their product poses scientifically verifiable risks to planetary, and hence human, well-being, fossil fuel companies have gone after the scientists and fostered a culture of blame and victimisation in which prominent scientists, such as Michael Mann and Phil Jones, have been subjected to hate campaigns and multiple death threats; in Mann's case an attorney general even attempted to sue him for fraud in the Virginia State court.[58]

Widespread support for climate scepticism is not just because of elite and corporate promotion of anti-science messages. There is a very large class of

---

a science studies perspective see Brigitte Nerlich, '"Climategate": Paradoxical Metaphors and Political Paralysis', *Environmental Values* 19 (2010): 419-42.

54. Sheila Jasanoff, 'Testing Time for Climate Science', *Science* 328 (2010): 695-96.

55. Joane Nagel, 'Climate Change, Public Opinion, and the Military Security Complex', *The Sociological Quarterly* 52 (2011): 203-10.

56. Maxwell T. Boykoff, *Who Speaks for the Climate? Making Sense of Media Reporting on Climate Change* (Cambridge: Cambridge University Press, 2011), 130-35.

57. Naomi Oreskes and Erik M. Conway, *Merchants of Doubt: How a Handful of Scientists Obscured the Truth on Issues from Tobacco Smoke to Global Warming* (New York: Bloomsbury, 2011).

58. Michael Mann, *The Hockey Stick and the Climate Wars: Dispatches from the Front Lines* (New York: Columbia University Press, 2012).

individuals and groups who perceive their beliefs and interests to be threatened by government action and regulation, and in particular government restraint in the use of fossil fuels. The 'three dollar gallon' in the United States, taxes on flights in Europe, car and home energy efficiency standards, carbon taxes — all are represented as unfair impositions by the state on the freedoms of individuals and businesses to consume as they will and to pursue their own goods by their own lights. In his populist climate novel *State of Fear,* Michael Crichton portrays climate science and climate apocalyptic as a ruse by the nation-state to continue to grow its power and influence over the lives of citizens and to restrain their freedoms.[59] On this account ordinary citizens and businesses are the victims of climate science; their freedoms to pursue their chosen model of the good life are threatened by the claim of climate scientists that human consumption activities visit harms on present and future people and thus require preventive action by lawful authority. Climate science is therefore seen as a cultural politics, and not 'objective' science, since it indicates that to live within planetary constraints governments and communities should deliberate together on ways to limit consumption, and that the human good is relative not only to individual preferences but also to natural order.

Climate scientists do not appreciate that their science represents a politics. They are surprised, as Bruno Latour notes, to be called lobbyists by climate denialists when they are simply reporting on climate models and data.[60] They believe that their climate change theory, observations, and data are unmediated by social or political worldviews or belief systems; for them climate change is physics and not politics. That it is treated by climate denialists as *politics* therefore comes as a surprise. But it is precisely because of the political claims implicit in climate science — that individuals are not autonomous, that there are limits to the capacity of the climate to absorb their pollution, that this pollution harms distant others in time and space — that political conservatives resist climate science.

The difficulty for political conservatives is that mitigating the extreme effects which climate catastrophists, and many climate scientists, argue will be produced by continuing growth in greenhouse gas emissions requires rapid and global-scale governmental and international efforts to restrain the use, and extraction, of fossil fuels. This seems to require governmen-

59. Michael Crichton, *State of Fear* (San Francisco: HarperCollins, 2004).

60. Bruno Latour, 'Anthropocene and Destruction', Lecture 4 of 'Facing Gaia: A New Enquiry concerning Natural Religion', The Gifford Lectures, University of Edinburgh, February 2013, at http://www.ed.ac.uk/schools-departments/humanities-soc-sci/news-events/lectures/gifford-lectures/series-2012-2013/lecture-four (accessed 13 March 2013).

tal regulation of everyday life so extensive that it will impact the design and heating and cooling of buildings, affect flying and driving behaviours, and restrain the consumption of carbon-intense products including meat, household mechanical and entertainment devices, long-distance foodstuffs, and even industrially made clothing. Resistance to climate change science is therefore strongest amongst individuals, groups, and organisations who identify most strongly with the core political claim that individual choice and the pursuit of self-interest in consumption and production activities are the sources of the human good, of freedom and flourishing.[61] On the other hand, those who are inclined to believe that the human good is advanced by a greater degree of collective action, including coordination by the state and civil society, are more likely both to accept that climate change is humanly caused and that it is remediable through collective action to address its causes, including restraint in the extraction and use of fossil fuels.

Climate Apocalyptic and the Anthropocene

There is growing concern among climatologists that their message is not being heeded. To highlight the significance of the crisis humanity faces, some natural scientists argue that humanity has embarked on a new planetary era. Since Charles Lyell, geologists have divided up the age of the earth according to their interpretations of the layers of rock and sediment in the earth's crust. They named the Holocene after the Greek word for the sun, *helios,* as an era in which the earth's climate has been stable and enduringly warm, as compared to previous eras. This stability and relative warmth made possible the development of agriculture, the resultant growth in human numbers in the last 8,000 years, and the rise of enduring human civilisations.

Paul Crutzen, whose early scientific work was on the possible atmospheric and climatic effects of a nuclear war, and who won a Nobel prize for his work on the ozone layer, argues that the great acceleration in greenhouse gas emissions since 1950, combined with a rapid expansion in deforestation, mining, and deep ocean fishing, marks the end of the Holocene, and the dawn of a new era he calls the 'Anthropocene'.[62] The term indicates

61. Andrew J. Hoffman, 'The Culture and Discourse of Climate Skepticism', *Strategic Organization* 9 (2011): 77-84.

62. P. J. Crutzen and E. F. Stoermer, 'The "Anthropocene"', *Global Change Newsletter* 41 (2000): 16-18.

that anthropogenic greenhouse gas emissions have taken over from variations in solar radiation as the dominant influence on the climate and hence on the biology and physiology of the earth. Scientists from the Royal Society have examined the 'stratigraphic' evidence for the new era and argue that higher concentrations of carbon and methane gases frozen in polar ice begin occurring at the industrial revolution in the eighteenth century.[63] On this account, the 'golden spike' which indicates the shift to the Anthropocene is the dawn of the industrial era and the great acceleration in coal burning which began after James Watt's invention of the steam engine.[64]

The term 'Anthropocene' indicates not only that the stable Holocene climate era is ending but also that humanity has a new power over, and hence collective responsibility for, the state of the earth and her future. Hence, according to the Potsdam climate impact modeller H. J. Schellnhuber, the dawn of the Anthropocene represents a second Copernican revolution.[65] The first revolution, inaugurated by the use of optics, which I discuss below, revealed that the earth, and hence humans, were not at the centre of the universe but instead that the earth revolved around the sun. This displacement of humanity from the earth system had cosmological and cultural implications which unfolded over succeeding centuries, as I chart below. But if the Copernican turn decentred humanity from the cosmos and reduced the perception of human influence over the earth and the skies, the Anthropocene is a second Copernican turn because it puts humanity back into planetary history as its most influential shaper. When the philosopher Hans Jonas highlighted this increased sense of human agency over the biosphere, he intended to promote a new moral sensitivity to humanity's relationship with life on earth and the conditions for its persistence.[66] However, when Schellnhuber explains this heightened human agency, he argues that it indicates a new agential *control* over the earth system. The questions the Anthropocene raises are these: First, what kind of world do we have? Second, what kind of world do we want? Third, what must we do to get there? The suggestion is that humanity needs to form clear intentions about *desired* planetary and climate states, and then use the instruments of earth sys-

63. Jan Zalasiewiz, Mark Williams, et al., 'The New World of the Anthropocene', *Environmental Science and Technology* 44 (2010): 2228-31.

64. Ignacio Ayestaran, 'The Second Copernican Revolution in the Anthropocene: An Overview', *Revista Internacional Sostenibilidad, Tecnología y Humanismo* 3 (2008): 146-57.

65. H. J. Schellnhuber, '"Earth System" Analysis and the Second Copernican Revolution', *Nature* 402 Supplement (1999): C19-C23.

66. Hans Jonas, *The Imperative of Responsibility: In Search of an Ethics for the Technological Age* (Chicago: University of Chicago Press, 1984).

tem engineering, climate modelling, and global meteorological governance to bring these about. On this account, climate science is a new political theology which recalls some of the cosmological assumptions of an earlier Copernican age, but recasts them in a hybrid of a human-centred cosmology and a Baconian, and post-Copernican, aspiration to technological control over the earth. As I discuss below, earlier agricultural peoples had a sense that in their relations with gods, species, and skies they had a role in shaping climates that gave either good harvests or famines, and they sacrificed to the gods to give them favourable weather. But the second Copernican revolution suggests that human beings are now *as* gods: they not only desire favourable weather, but they have the technical means to bring it about.

While one reading of the Anthropocene suggests that it involves a recovery and enhancement of human agency over the earth, another even more apocalyptic reading suggests that the Anthropocene, far from enhancing human intentionality and agential interaction with the earth, threatens to reduce it and hence to undermine the modern scientific imaginary of the human control of nature. If rising sea levels inundate cities and ports, and droughts destroy much presently viable cropland, the Anthropocene will potentially be an era in which human power over nature is greatly reduced. In these circumstances nature will wrest back control over the boundary between earth and sea from human sea defences, and over agricultural lands from the irrigation schemes, terracing, and crop rotations of farmers.

Geological studies indicate that the last time in which the atmosphere contained 400 parts per million of CO_2, which proportion the Mauna Loa Observatory recorded as this book went to press, was the mid-Miocene, around fourteen million years ago. At that time there was no permanent ice on earth, sea levels were seventy feet above present levels, and temperatures were three to six degrees above the present.[67] Climate inertia, and in particular the vast amount of terrestrial and oceanic ice now present on earth, will hold back Miocene sea levels and temperatures for some hundreds of years. But many scientists believe that by the end of the present century, without imminent and dramatic reductions in fossil fuel use and deforestation, sea level will be rising ten to twenty millimetres a decade instead of three millimetres as at present. In the centuries to come, the coastline of the earth will move to a new level seventy feet above the present. This will

67. Aradhna K. Tripati, Christopher D. Roberts and Robert A. Eagle, 'Coupling of CO_2 and Ice Sheet Stability over Major Climate Transitions of the Last 20 Million Years', *Science* 326 (2009): 1394-97.

mean the abandonment of all present coastal settlements and cities below that level, including the majority of cities and deep ports around the world. The Miocene was also much warmer than the Holocene, and temperatures raised far beyond the present will see widespread drought, which will render nonviable many current large agricultural areas, including the prairies of the United States and Russia, most of southern Europe, and much agricultural land in Asia, North Africa, and sub-Saharan Africa.

When seen in these terms, the climate crisis reads as a truly apocalyptic event, albeit one that may unfold relatively slowly and across generations. Many, however, interpret the evidence of future substantial change as indicating a sudden and imminent catastrophe which mirrors, if not in speed, then in effect, the cataclysms of apocalyptic films such as *The Day After Tomorrow* or science fiction novels such as J. G. Ballard's *The Drowned World* and Edward Abbey's ironically titled *Good News*. In these genres, the priority for those who are not going to be caught out by rising seas and crop failures is *survival*. Climate activist Bill McKibben imagines a kind of 'Mad Max' response to the large-scale breakdown of law and order extreme climate change will likely precipitate:

> If you think about the cramped future long enough, for instance, you can end up convinced you'll be standing over your vegetable patch with your shotgun, warding off the marauding gang that's after your carrots. . . . The Marines aren't going to be much help there — they're not geared for Mad Max — but your neighbours might be. Imagining local life in a difficult world means imagining taking more responsibility not only for your food but for your defense.[68]

Some survivalists and extreme climate activists are already preparing for this imagined future by setting up home on plots of good and remote land well above sea level, taking their household energy supplies off grid, and preparing to defend themselves from city migrants scavenging for food. An analogous survivalist mentality is also reflected in the militaristic and security responses of a 'fortress world' and an 'armed lifeboat' evident in official climate crisis literature.[69]

68. Bill McKibben, *Eaarth: Making a Life on a Tough New Planet* (New York: Henry Holt and Co., 2010), 145-46.

69. Christian Parenti, *Tropic of Chaos: Climate Change and the New Geography of Violence* (New York: Month Books, 2011), 216.

Another response to the threat of a climatic apocalypse is of humanity in exile from the earth. James Lovelock reports with a certain kind of satisfaction that Gaia may ultimately slough off the human species, as a body may slough off a virus: the earth will endure once Gaia has dealt with a species which has overextended its biological niche.[70] In *The World without Us* Alan Weisman envisages a planet in which extreme climate change has extinguished human life and the human influence over the planet thus gradually comes to an end.[71] Stephen Hawking in an interview a few years ago envisaged a time when humanity would have used up the resources of earth and would need to expand into other earths and galaxies, and to use the insights and technologies of advanced physics to enable this expansion. Science fiction films of the settlement of other planets, such as *Avatar,* indicate that such colonial expansion out from our earth will be as hubristic and violent as the responses to the present climate crisis threaten to be.

There is an incipient nihilism to the wilder fringes of climate catastrophism that some call 'climate porn'. There is also evidence that catastrophism, far from promoting a greater sense of shared responsibility for the state of the earth, instead encourages the spread of climate denialism, or passive acceptance of a coming climate disaster that political and personal change can do nothing to assuage.[72] In Lars Von Trier's film *Melancholia* a wedding party is held in a grand country house in upstate New York. The bride and groom and their families and friends are mostly drawn from the elite of Manhattan's advertising and banking circles. But as the post-wedding limousine is approaching the house along a winding narrow track — with some diverting difficulty — we see that scientists are tracking the path of a giant meteor which, if it impacts earth, will extinguish most human life, most life, on earth. As the wedding party unfolds, the imminent threat hangs over events: the bride retreats to the bath, eventually falls out with the groom, and severs her relationship with her boss. She stays on in the house with a few close relatives and friends and their children, and they track the path of the meteor online and with their own telescope. As the meteor grows in size in the sky, members of the party start to fall apart. One of the men, who had all along denied that the meteor would hit, goes

70. James Lovelock, *The Revenge of Gaia: Why the Earth Is Fighting Back and How We Can Still Save Humanity* (London: Penguin Books, 2007).

71. Alan Weisman, *The Earth without Us* (London: Random House, 2008).

72. Thomas D. Lowe, *Is This Climate Porn? How Does Climate Change Communication Affect Our Perceptions and Behaviour?* Working Paper 98, Tyndall Centre for Climate Research, 2008.

to the stable block and shoots himself. Another member of the party takes to her four-wheel-drive vehicle and drives wildly with two children around the estate. The bride meanwhile makes brunch for everyone and creates a small ritual space on a rise in the lawn with branches arranged as an open pyramid. As the time approaches she takes the children and her sister by the hand and they sit together in the ritual space and await impact. The film's message seems to be that the one who at the end endures in love for those around her is the one who faces the crisis and refuses to deny it. The crisis is an *apocalypse* which reveals her character and the characters of those around her. As a metaphor for the climate crisis, and the competing imaginaries of denial, control, and care that it evokes, *Melancholia* is superb.

Agrarianism and Apocalypse

Climate apocalyptic serves many of the same purposes as New Testament apocalyptic. First, it is an unveiling: it reveals that humanity's influence over the planet has become so large in scale that it is reaching a limit point which puts humanity's enduring tenure on earth at risk. Second, it heralds judgment: the earth will punish the heedless affluent generations who continue to spew fossil fuels into the atmosphere regardless of the limits of the earth and their duties to future generations and species. Third, it is a call to moral and political transformation. Preventing the evil of extreme climate change, and preventing an age of extremes from descending into war, requires radical political and personal change — from fossil fuel dependence to reliance on the renewable energies of the earth, from a global to a local economy, from a wasteful culture of excess to a culture of conservation, care, and reuse, and from a culture which encourages the rich to pile up wealth regardless of growing inequality to communities where rich and poor live closer together and share more equitably the capacities of the atmosphere and ocean to absorb greenhouse gas emissions so all can meet their needs but not their greed.

There is, however, a crucial contrast between climate apocalyptic and the Apocalypse of John of Patmos. John reads ecological and civilisational collapse not as the *end* of all things but as the end of a new *beginning* for the earth and its creatures which is already unfolding in the appearance on earth of the *logos,* the inner *ratio* of created life, in the person of Jesus Christ. There is no schadenfreude, no rejoicing in the tribulations that bring down the powerful from their thrones at the end. Instead, John's aim is to encourage the persecuted and powerless Christians, whom Rome threatened to overwhelm under the emperor Nero, that, provided they remain faithful in loving God and neighbour, the cataclysm to come will see them vindicated in the end. For John of Patmos, this vindication of the poor in history is revealed in a new messianic time, which the revelation of the *logos* as cosmic servant, the 'lamb of God' who becomes a victim, inaugurates on the public stage of empire. In this new time, neglect of divinely ordained limits to human power and wealth by the rich and the strong will at last be restrained, and the suffering of the poor and the weak will come to an end. Hence John's apocalyptic end of history is not only the end of one kind of imperious civilisation, but the inauguration of a new and transformed era, both for humanity and for the earth. This era will reveal a 'new heaven and a new earth' in which 'God shall wipe away all tears from their eyes; and there shall be no more death, neither sorrow, nor crying, neither shall there be any more pain: for the former things are passed away' (Revelation 21:4).

John Zizioulas argues that the Book of Revelation represents a new literary genre of 'cosmological prophecy' in which Christians are 'called to think of the Kingdom of God not only in terms of the salvation of the human being, but also in terms of the survival and wellbeing of the entire creation.'[73] This new beginning will not be without struggle, and Revelation describes an almighty struggle between good and evil in which catastrophe threatens to overwhelm the good. But the unveiling represented by the symbolic language of John is that, while the evil which is focused in Babylon (Rome) threatens catastrophe, evil will not finally prevail. John calls those who suspect that catastrophe will overwhelm them to persevere in their witness against the beast, for the hidden truth John declares is that at the very moment of the seeming escalation of evil's power over the good, evil will overreach itself and bring down catastrophe on itself, and the good will finally prevail.

Long before Lovelock's Gaia hypothesis, in the cosmology of Revelation is found the central claim that human life on earth is relationally interconnected to the 'heavens', or the sphere above the earth, and shares in the same destiny. In chapters 4 and 5 the throne of God is set in the midst of the heavens, and all living creatures gather around the throne of God and join in cosmic praise: 'And every creature which is in heaven, and on the earth, and

73. John of Pergamon, 'The Book of Revelation and the Natural Environment', in S. Hobson and J. Lubchenco et al., eds., *Revelation and the Environment AD 95-1995: Eastern Mediterranean* (Singapore: World Scientific Publ., 1997), 17-21.

under the earth, and such as are in the sea, and all that are in them, heard I saying, Blessing, and honour, and glory, and power, be unto him that sits upon the throne, and unto the Lamb for ever and ever' (Revelation 5:13). For John of Patmos the destiny of the cosmos, as well as human history, is revealed in the divine humanity of the Lamb of God, who has redeemed by his blood people of 'every kindred, and tongue, and people, and nation: and has made of them a kingdom and priests to God, and they shall reign on the earth' (Revelation 5:9-10). Messianic communities who worship the Lamb as Lord of heaven and earth anticipate a new cosmological era in which evil empires — and especially the empire of Rome, which already in John's time was persecuting the Christians — will at the last be restrained, and the good prevail. The growing power of the Roman Empire and its persecuting rage against the Christians are not denied by John. But he predicts that those who remain faithful to the Lamb of God, who put love before fear, and who refuse to return evil for evil, even toward their persecutors, will ultimately be vindicated. There will be suffering, even for the faithful, but those who endure through this suffering will ultimately overcome the power of Rome/Babylon and help to bring history to an ultimate end in which all empires and nations will have to recognise the authority of the Lamb of God.

What of the implication of natural events in these human struggles? Christ in the 'little apocalypse' of Mark's Gospel (Mark 13) envisages the apocalyptic fall of Jerusalem as being accompanied by natural catastrophes. The Book of Revelation envisages that nature will be subjected to suffering and destruction in a global conflagration before the new heaven and new earth come about. Some conservative Christians interpret these texts to indicate that the earth in its present form will be burned up and destroyed before the 'new heaven and new earth' announced by John of Patmos in Revelation 21.[74] This is indeed the plot of the best-selling *Left Behind* novels of Timothy LaHaye, which describe the exploits of a group of individuals who are left on earth after the saints are 'raptured', and who then promptly convert to Christianity and get caught up in the full rigours of a divinely visited punishment on humanity involving natural catastrophes of all kinds.[75] Some even interpret global warming as a portent of this earthly de-

74. See further Michael S. Northcott, 'Apocalyptic Christians and the Theology of Dominion', in Kyle Van Houtan and Michael Northcott, eds., *Diversity and Dominion: Dialogues in Ecology, Ethics and Theology* (Eugene, OR: Cascade Books, 2009), 236-70; and John Agnew, 'Religion and Geopolitics', *Geopolitics* 11 (2006): 183-91.

75. Michael S. Northcott, 'Earth Left Behind? Ecological Readings of the Apocalypse of

struction.[76] But the reading of Apocalypse as the foretelling of the ultimate destruction of the earth is deeply at odds with the way in which the early Christians understood the implications of the Christ events for creation. The triumph of Christ over earthly powers presages a new peace between humanity and creation. This is the import of Saint Paul's teaching about the implication of creation in the redemptive history of the children of God:

> For the earnest expectation of the creation waits for the manifestation of the sons of God. For the creation was made subject to vanity, not willingly, but by reason of him who hath subjected the same in hope, because the creation itself also shall be delivered from the bondage of corruption into the glorious liberty of the children of God. (Romans 8:19-21)

For the apostle Paul the salvation of humanity in the Incarnation represents the potential renewal of creation. Alienation, corruption, and death first entered the world through the original sin of Adam and Eve; and as the Second Adam, Christ therefore overcomes the cosmic effects of sin, including death, and hence, 'as in Adam all die even so in Christ are all made alive' (1 Corinthians 15:22). The Book of Genesis describes the sin of Adam in analogous terms, but the narrative includes a context that Paul does not — and this is that in Eden, Adam and Eve are forest-dwelling gatherers of fruits and herbs, whereas the exile from Eden requires them to raise their food through agriculture. In this ancient myth of origin, agriculture is the source of alienation between human beings that leads to the original murder, of Abel by Cain, using an agricultural implement, and to the rise of hierarchical civilisations in which men and women visit violence on one another and reduce one another to slavery. It is also the beginning of a mark of humanity on the earth which changes the destiny of all creatures and the earth, and not only the destiny of humanity. Hence, if the Anthropocene names the era in which humanity becomes central to the destiny of creation, theologically this era begins not with greenhouse gas emissions but with the neolithic 'fall into agriculture'.[77]

John in Contemporary America', in John Lyon and Jorunn Okland, eds., *The Way the World Ends? Revelation and Its Reception* (Sheffield: Phoenix Press, 2009), 112-31.

76. Conrad Ostwalt, 'Armageddon at the Millennial Dawn', *The Journal of Religion and Film* 4 (2000), at http://www.unomaha.edu/jrf/armagedd.htm (accessed 15 March 2013).

77. R. J. Berry, *God and Evolution* (London: Hodder and Stoughton), 68-71; and Denis Alexander, *Creation or Evolution: Do We Have to Choose?* (Oxford: Monarch Books, 2008), 241-43.

Although agrarianism is the source of civilisation and its rise reflects the new civilisational possibilities opened up for humanity by a stable climate era, the Old Testament describes the rise of agrarianism in ancient Mesopotamia as the *descent* and not the ascent of humanity. Before agriculture human beings were few in number and were able to hunt and gather other species in abundant forests teeming with life. These are the conditions which Adam and Eve are said to enjoy in the lush and forested Garden of Eden. In the Garden they are described as being made in the 'divine image', which in ancient Mesopotamia was a description conventionally conferred only on kings and rulers. The eviction of humanity from the Garden, the 'fall into agriculture', is mythically recast as the consequence of a primordial infraction of a divine command not to eat from the tree of the knowledge of good and evil. The forbidden knowledge is often interpreted as the knowledge of how God made the world and the inner laws of its workings; today we would call this science.[78] In the ancient world this science began when humanity worked out how to take seeds and plant and harvest them and so increase the availability of human food apart from what could be gathered from the wild. So with this knowledge the first humans can dwell outside the forest of Eden. But it carries with it the ambiguity that after their eviction, they must eat bread which they must produce 'by the sweat of their brow', in contest with weeds, and through the arduous work of hand milling. Within one generation their descendants — Cain and Abel — descend into fratricidal violence, the one killing the other with an agricultural implement. This new form of civilisation is described in Genesis 6–8 as descending into large-scale violence so that 'the earth is filled with violence' (Genesis 6:13). The Creator is said to bring about the destruction of most of humanity in order to end the violence, save only for the righteous man Noah, his family, and the animals, by a primordial flood.

While, as we have seen, some chart the perceptible rise of human influence over the planet to the industrial revolution, others argue that the true beginning of anthropic influence over the planet coincides with the transition to agriculture from hunter-gathering, of which we hear echoes in Genesis, around eight thousand years ago.[79] William Runciman argues that carbon and methane from deforestation, domestic animals, and the large-scale

78. Dexter J. Callender, *Adam in Myth and History: Ancient Israelite Perspectives on the Primal Human* (Winona Lake, IN: Eisenbrauns, 2000), 68-69.

79. G. Certini and R. Scalenghe, 'Anthropogenic Soils Are the Golden Spikes for the Anthropocene', *The Holocene* 21 (2011): 1269-74.

growing of rice likely was the reason for the modest increase of CO_2 in the atmosphere from 240 ppm to 280 ppm from 6,000 BC to 1850.[80] Humans, in other words, have been influencing the climate for most of recorded history. The proposal that agriculture marks the true beginning of the Anthropocene has considerable historic and theological resonance, since it was humanity's transition to agriculture eight thousand years ago which drove the first large-scale clearing of forests and made possible food surpluses, the rise of cities, and the emergence of large-scale and enduring cultural developments, including the religions of the Ancient Near East, China, and India. The contiguity between a stable climate and the rise of agriculture explains why a move to an unstable climate has such catastrophic implications for contemporary humans. More than any predecessor culture, *Homo industrialis* has overspread the earth and daily consumes food, energy, and other resources procured through complex and fragile networks of planes, ships, trucks, roads, pipelines, wires, and radio and satellite communications. These networks, even more than agriculture, are highly dependent upon the stable climate that human beings have enjoyed in the course of recorded history.

The dating of the Anthropocene to the beginning of agriculture brings into view the theological frame of the human relation to the earth in the language of creation, fall, restoration, and redemption. The Old Testament, from the rise of the first empire of Babel (Babylon), describes the rise of hierarchical and large-scale civilisations that agriculture made possible as evidence of human hubris and will to power which stand under the judgment of God.[81] In Egypt, the first and grandest of these civilisations, the ancestors of Israel languished, lured there by one of their own, Joseph the Egyptian prince, and the promise of grain and land, only to be enslaved by subsequent pharaohs.

Israel is called out of Egypt as a 'chosen people' to practise agriculture, but the Law of Moses establishes that the way Israel does agriculture contrasts with the agrarian practices of the empires that surround her, and hence the Torah is, as Jacob Taubes puts it, profoundly 'anti-Egyptian.'[82] The first hint of this difference is in Genesis 1, where Adam is described

80. W. F. Runciman, *Ploughs, Plagues and Petroleum: How Humans Took Control of Climate* (Princeton, NJ: Princeton University Press, 2005).

81. J. Severino Croatto, 'A Reading of the Story of the Tower of Babel from the Perspective of Non-identity', in Fernando F. Segovia and Mary Ann Tolbert, eds., *Teaching the Bible: The Discourses and Politics of Biblical Pedagogy* (Maryknoll, NY: Orbis Books, 1998), 203-23.

82. Jacob Taubes, *The Political Theology of Paul,* trans. Dana Hollander (Stanford, CA:

both as made from the soil and as made in the divine image (Genesis 1:26). In this original political theology of the Bible, the forest dweller who becomes a farmer images the King of Heaven. In Eden a peasant walks with the Creator, and his wife shares equally in this divine-human communion. Similarly, in Eden it is humanity as a whole — and not only the King — who represents God as vice-regent on earth and exercises 'dominion' over all creatures and the earth. By contrast, in Babylon and Egypt kings alone commune with the gods and act as their image and representative.[83] Here we find the root of Jewish and Christian experimentation with democracy and egalitarianism, and their joint critiques of the will to power. Analogously, in the Mosaic covenant every household is to be the possessor of an equal portion of the land that God gives to create a new non-hierarchical and post-slavery agrarianism.[84] The curse of agriculture, and its potential to sustain powerful kings and empires, is transformed by a cosmic covenant between God, the Hebrews, and the earth in which the land and its peasant farmers and its species all have rights, all must do justice, and all are purified by abstaining from work, and so honouring God, with the sacrifice of the Sabbath.

Israel's God is different from the gods of the other nations because Yahweh 'made the heavens and the earth' and the other gods did not (Jeremiah 10:9). Yahweh is the governor of the forces that govern the relations between heaven and earth, and hence Yahweh controls the climate: 'It is he who made the earth by his power' and 'whose voice provokes tumult in heaven' and 'mists to rise from the ends of the earth' (Jeremiah 10:13). This divine power over the elements, first revealed in the plagues in Egypt, the pillar of fire, and the rain of manna in the wilderness, is crucial for Israel in the land of Canaan since Canaan is not watered by a large river like the Nile or the Euphrates and Tigris, which watered respectively the empires of Egypt and Babylon. The God of Israel is therefore more directly implicated in the ability of the people to bring forth good harvests from the land than the gods of the other nations. Israel would eat in this dry land only when the rains came, and the rains over Canaan were fickle, and hence Canaanite

Stanford University Press, 2004), 31. See further Ellen Davis, *Scripture, Culture and Agriculture: An Agrarian Reading of the Bible* (Cambridge: Cambridge University Press, 2009).

83. Edward Mason Curtis, 'Man as the Image of God in Genesis in the Light of Ancient Near Eastern Parallels' (DPhil Dissertation, University of Pennsylvania, 1984), at http://repository.upenn.edu/dissertations/AAI8422896.

84. Clive S. Beed and Cara Beed, 'Economic Egalitarianism in Pre-monarchical Israel', *Faith and Economics* 54 (2009): 83-106.

farmers stored rain in underground cisterns to water their crops in the dry times.[85] But 'hewn cisterns' are seen by Hebrew theologians as a pagan heritage which Israel would do well not to rely on, since in so relying she may forget her dependence on God for the gift of a well-watered land. They also carry with them the implication of idolatry for they are 'made' by human hand and by those who worship gods which are also made by human hand.

In the religion of Israel, then, a God who controls the sky and what comes forth from the soil takes centre stage. Hence Israel's poets and prophets read Israel's relationship with God *through* her experience of climate. So for the psalmist, when the weather is good, and the rains fall gently and regularly on the fields, giving harvests which sustain the poor (Psalm 72), this is because the Israelites worship the true God and refrain from idolatry, are ruled by just rulers, observe the Sabbath, and give a tenth of the fruit of the land to God. But when the weather is bad and the fields give only a tenth of their usual produce, this is because Israel has trusted in other gods and in the works of her own hands, including cisterns and idols, and not in the God who made heaven and earth:

> Hath a nation changed their gods, which are yet no gods? But my people have changed their glory for that which doth not profit. Be astonished, O ye heavens, at this, and be horribly afraid, be ye very desolate, saith the LORD. For my people have committed two evils; they have forsaken me the fountain of living waters, and hewed them out cisterns, broken cisterns, that can hold no water. Is Israel a servant? Is he a homeborn slave? Why is he spoiled? (Jeremiah 2:11-14)

For Jeremiah the climate is political, and worship and agriculture are geopolitical. When Israel worships idols, and not the God who gives good land and makes good weather, Israel descends into slavery again, her rulers defeated by surrounding empires, and her farms desolated by drought and their walls broken down. When Israel returns to the worship of God, her people are freed from bondage, exile, and political power, and the land is again watered and gives good harvests. Idols do not save, whereas the God who made the world rescues the poor and the needy in their distress.

85. Archaeologists once thought Israelite farms could be distinguished from Canaanite by the presence of hewn cisterns, but it is now recognised that Canaanites as well as Israelites used cisterns for irrigation: Mark S. Smith, *The Early History of God: Yahweh and the Other Deities in Ancient Israel* (Grand Rapids: Wm. B. Eerdmans, 2004), 22.

At the heart of Judaism is a God who is encountered through nature and events rather than in words or texts.[86] Christianity, by contrast, and then Islam, is a form of religion that is less implicated in the weather, climate, and political power and more invested in words and texts. For John of Patmos, Rome's persecution of the Christians is not because they have *not* been faithful to God but because they *have*. Similarly, portents from the sky are signs that God is closing the present age because of its wickedness; they do not indicate that the Christians have lost divine favour. Heaven and earth may pass away, but this is not a judgment on the Christians. It is rather a reward of God's 'servants, the prophets and saints and all who fear God's name, both small and great' while it is a judgment which destroys 'those who destroy the earth' (Revelation 11:18). But there is a judgment for those who fail to honour the messenger and the Word: 'if any one takes away from the words of the book of this prophecy, God will take away his share of the tree of life and in the holy city' (Revelation 22:19). Revelation is an unveiling of something new, in which the Spirit renews the face of the earth, and the earth herself destroys those, and especially Babylon/Rome, who would destroy her. This fall of Rome, which may seem to drag the Christians down with it, is a new thing, not like the former things. But it is also an unveiling of the plan that was hidden since the foundation of the world: the reversal of the will to power that empires from Babel to Rome have shown, a reversal in which the weak and poor of the world shame the powerful and those who worship the Lamb shame the wise.

The Common Era as the Christocene

In a deep sense Christianity is a religion of freedom: from the power of nature *and* the power of politics. In Adam and after Eden, dominion over nature turns quickly into a curse, and Adam is turned from son of the soil to its servant, toiling in the fields to raise crops and put bread on his table in contest with weeds. After Adam, dominion becomes will to power, empires rise, and the earth as well as humans are subjected to violence and destruction. Not even Israel recovers peaceful dominion, though she is called from political bondage to a priesthood of the soil, to offer up sacrifices from her

86. Moses Mendelssohn, *Jerusalem, or Religious Power and Judaism* (1783), cited in Jan Assmann, *Of God and Gods: Egypt, Israel, and the Rise of Monotheism* (Madison: University of Wisconsin Press, 2008), 140.

tilling of the earth to the earth's Creator, and so to restrain, or slough off onto the scapegoat, the ancient curse. Christians for Saint Paul are liberated from the mark of Cain, and for John of Patmos from the politics of Babylon. They are promised the way back to Paradise and citizenship of the Kingdom of God. But these promises involve neither *rule* nor *agriculture.* The early Christians had no land, and few had positions of civic or political power; many were slaves, most were urbanites, few were landowners, fewer still were rulers or aspired to be.[87] As Peter Brown argues, the ethic of the early Christians was essentially person to person; they made no grand claims about political structures because Christians did not have access to them.[88] So the political theology of the New Testament and the early Church involved no geopolitical claims other than apocalyptic ones about the end of the world and the revelation of who is really Lord instead of the emperor. But it could not end there. As the Christians overspread the Mediterranean world, they invented a new politics, and even a new agriculture. How could it not be so when a sect turned into a Church, and converted two empires — Rome and Constantinople — in little more than four hundred years?

The success of the Christian mission as well as the delay of the *parousia* — the final revelation — encouraged the development of a more settled approach to agriculture and politics. Christ might one day return on the cloud. But he did not come in one generation, or even a dozen, after those who watched him ascend. Instead, the Christians began to envisage it likely that Christ would not return again until the gospel had been preached and the Church had converted *all* the nations (Matthew 28:19).

If the world was to be Christianised, it was inevitable that this would involve also a new cosmology, a new politics, a new spirituality of earth, its depths, and its heavenly realms. Hence the creeds of the second and third century affirm the incipient New Testament references to Christ descending, between the Crucifixion and the Resurrection, to the depths of Hades where he preached to the dead souls, enlivening them, and so brought the Resurrection into the history of the dead who had been waiting for it in the limbo of Sheol. Then Christ rises to earth, not as a resuscitated corpse but as a new kind of ethereal being who passes through walls and materialises in varying geographic locations with no evident means of transport. Finally,

87. There is one New Testament story of a ruler being converted, that of the Ethiopian prince in Acts; and intriguingly Ethiopia is the first nation in Christian history to announce its nationhood, and its borders have largely remained fixed since the third century.

88. Peter Brown, *Through the Eye of a Needle: Wealth, the Fall of Rome, and the Making of Christianity in the West, 350-550* AD (Princeton, NJ: Princeton University Press, 2013), 56.

Christ rises into the clouds to be seated on the right hand of God, as ruler of the sublunary world beneath the heavens.

The time of the Church inaugurates a new geospiritual relation of souls to bodies, of heaven to earth, and it is first revealed in the imaginary of the communion of saints, the 'cloud of witnesses' (Hebrews 12:1) that surrounds the worship of Christians. The same imaginary is evident on the walls of the catacombs where the Christians gathered for worship in the presence of the remains of the dead who had preceded them in the resurrection. And when the Christians move their worshipping spaces from caverns and domestic houses to purposely built structures for worship, they paint Christ on the roofs of their basilicas as the revealed Pantocrator — the geopolitical ruler over the heavens and over emperors and princes, who is also present on earth in the bread and the wine, and who is worshipped by beasts as well as humans even before he comes again. In the catacombs the Christians paint animals worshipping Christ, and in the desert the wild animals come to Anthony and Jerome to be healed by them, recognising in them the presence of the same divinity that the wild beasts recognised when they were 'with Christ' in the wilderness while 'the angels ministered to him' (Mark 1:13). So between the first and the second coming there is a new mediation between the earth and the heavens; it is revealed in the history of the nations in the success of Christian mission in overtaking the Mediterranean which is the middle of the world, and it is mediated in the mysteries of worship and the enduring presence of the divine Word and the divine Spirit in the Christians who, as Saint Paul has it, are the first 'new creation' before the 'new heavens and new earth' which come later.

The Christian era is a new era, which we might call the Christocene, in which the divine image in humanity is already being cleaned up and restored, and the world is already being repaired. The origin of this restoration and repair is the *logos* who is Christ, whose rule extends from the subterranean to the sublunary and from the beginning to the end of human history. The mediatorial presence of the *logos* on earth after the Ascension is centred in eucharistic worship and in the sacrifices of praise, and in moral struggle against evil and for the virtues in which the saints engage. Hence humanity in worship becomes the priestly mediator, the microcosm to the macrocosm, as the sixth-century Byzantine theologian Maximus the Confessor would put it.[89] In the priesthood of eucharistic worship, humanity

89. The key source for Maximus's cosmological Christology is *Ambiguum* 7 and 8, excerpted in Maximus the Confessor, *On the Cosmic Mystery of Jesus Christ: Selected Writings*

stands as Christ once did between earth and heaven, restoring the cosmic covenant which Israel's descent into empire and idolatry had broken.

In the Lord's Prayer, the most widely used prayer throughout the Common Era, Christians pray that the divine will 'be done on earth as in heaven' (Matthew 6:10). In Christian history until the Enlightenment, this petition was understood in climatic and meteorological terms since it is followed by the petition 'give us this day our daily bread' and the petition that 'debts be forgiven'. Rain and sunshine, clouds and temperature have been the determinants of agricultural produce since the original Levantine turn to agriculture in the eighth millennium before Christ. If they arrive in 'due season', the petitions in the Lord's Prayer are answered; if not, the crops fail, the people of God go hungry, they fall into debt, and they and their children lose their land.

The people of God in the time of Christ were often hungry. Many of Christ's contemporaries were losing their lands to urban elites as a result of the imperial burden of taxation imposed by the Jewish authorities as proxy rulers of the Roman empire.[90] Consequently, they had ceased to be full members of the people of Israel while also being enslaved to coercive wage labour, debt, and life-threatening hunger and poverty. Christ's parabolic acts of healing and his moral and political teachings therefore challenged the social and political arrangements which rendered the people of God landless. They also challenged the Jewish religion and its moral and political authority since the treasury of the Temple in Jerusalem had become the tax collection centre of the Herodian rulers of Palestine in the time of Christ. Hence the principal crime of which Christ is accused at his trial is blaspheming against the Temple.

It is, then, entirely appropriate that the rite the Christians adopted after the Ascension to bring down and make present the heavenly will on earth should be a meal of bread and wine. The Romans ruled an empire through violence and bread; they subsidised the mass production of bread in the cities of the empire even as their legionnaires and their taxes violently took the land away from the peasants who grew it and turned them into slaves. When the early Christians broke bread and distributed it equitably to all who were present, they became a political body in which every member had equal standing

---

from St. Maximus the Confessor, trans. P. M. Bowers and R. L. Wilken (Crestwood, NY: St. Vladimir's Seminary Press, 2003).

90. John Dominic Crossan, *The Historical Jesus: The Life of a Mediterranean Jewish Peasant* (San Francisco: Harper Collins, 1976), 341.

and whose head was Christ. At the same time they rejected the claim that Caesar was Lord or that they need rely on Roman bread for their sustenance; they were sustained by the bread which had 'come down from heaven', and not up from Babylon. The eucharist was the supreme cosmopolitical act of the early Church, in which heaven and earth are reconciled, empires cast down, and princes dethroned. At the eucharist slaves ate with free citizens, masters with their servants, and hence it was not long before slaves such as Onesimus would claim to have been freed by Christ.[91]

The restoration of created order which is first revealed in the risen body of Christ is experienced politically in the social body of Christians. Hence in the Christian household mutual subordination and love reverse the conventional hierarchy of masters and slaves, husbands and wives. Paul writes to the churches of Ephesus and Colossae not that slaves should demand freedom but that in their free service to those who are over them in the world they show a mutual love that undoes coercive power and economic claims to ownership. It was nine or ten centuries before this belief would result in the freeing of slaves in Christendom. But that it did eventually have this result should not be discounted. As Hilaire Belloc argues, Christianity is above all the religion of freedom from coercive power.[92] The gradual dissolution of the institution of slavery in the first millennium of the Christian era and the recovery by peasant farmers in the Middle Ages of usage rights over land are the most visible signs in the history of Europe of the political and economic fruit of eucharistic politics.

As well as subverting established political order, the eucharist challenges cosmological order since Christ as the heavenly Lord is also in time worshipped as the 'Lord of the harvest'. Hence not only did the Christians display Christ ruling from the heavens on the ceilings of their churches; they also read the climate, and the clouds which watered the crops, as governed by Christ. In the art of the catacombs and of the first three centuries of the Christian era, Christ is already a heavenly farmer, the good shepherd who goes in search of the lost sheep.[93] His presence is celebrated with sheaves of wheat, vine fruits, and loaves and fishes, in an echo of the Gospel stories of Christ 'feeding the crowds'. One of the key moments in the eucharistic rite is the re-creation of the offering of tithes, which were the tenth of the fruits of

91. Oliver O'Donovan, *The Desire of the Nations: Rediscovering the Roots of Political Theology* (Cambridge: Cambridge University Press, 1996), 185.

92. Hilaire Belloc, *The Servile State* (London: N. T. Foulis, 1902).

93. See further Susan Power Bratton, *Environmental Values in Christian Art* (Albany: State University of New York Press, 2003).

human labour on the land commanded of Israel to be given to Yahweh, and to the Levites, in honour of the covenant between God, the people, and the land. In the eucharist this ritual finds new form in the offertory procession when the people bring to the altar, along with the bread and wine, their alms, in the form of money to be distributed to the poor and to sustain the ministers of the Church who serve the poor and the faithful.

In time, the life of Christ and the climate-driven seasonality of the agricultural year were liturgically mapped onto each other so that the mirroring of earth and heaven of the Lord's Prayer finds a climatic as well as a political shape. Christ's birth is celebrated on December 25th, which is the date of the winter solstice in the Northern Hemisphere when solar insolation is at its lowest on the surface of the earth, and plant growth and photosynthesis are at a minimum. The gestation of Christ's earthly ministry in his journey of forty days in the wilderness is enacted in the Lenten fast of the European spring, when the soils are gradually receiving more sunlight and plants are beginning to respond. Christ's crucifixion and resurrection at Easter are celebrated as spring sunshine begins to warm the soils and the soils are prepared for the growing season. The Ascension of Christ is celebrated four weeks following Easter, when the crops have been sown in the field, and preceded by 'Rogation Sundays', when the priests go into the fields to bless the crops and the people pray for a good harvest. The season of Trinity commences after Ascension Day, and the earthly ministry of Christ in rural Palestine is rehearsed over the late spring and early summer until the harvest season, which, until the Reformation in northern Europe, began on August 1st, when the first of the wheat crop was gathered and turned into bread; hence August 1st is known as Lammas or 'loaf mass'. The liturgical year then is intricately tied to the sun-driven cycle of planting, growing, and harvesting of farmers and gardeners in the Northern Hemisphere.

Christ, Climate, and the Council of Nicaea

As if to underline the climatic significance of the dating of the liturgical calendar, the dating of Easter is not set on a fixed date in the Roman calendar; instead, at the Council of Nicaea, it was determined as the Sunday of the second week of the paschal moon following the spring equinox, a crucial date in agricultural planting. Christians in different parts of the Roman Empire — in Syria, Antioch, and Alexandria — had been celebrating Easter on different days, some following precisely the Jewish *pascha,* and some even celebrating Easter on a weekday, while others departed more or less from that date. The Council of Nicaea declared that there was to be one universal date of Easter, and from then until now in Western Christendom this date is set by the spring equinox and the paschal moon. The council was therefore in a way the first geopolitical climate conference, since, at the behest of the emperor, it drew up a new international agreement on the relationship of the worship of Christians and their calendar to the seasonal shifts in the earth's climate from the declining and rising hours of sunlight in the Northern Hemisphere. Mapping the life of Christ onto the earth's relationship to the sun, the Council of Nicaea affirmed and underwrote the Christocene era in which Christ's Lordship had put down all earthly powers under his feet and reigned from the sky as the Lord of heaven and earth.

Until the fifteenth century Christians believed, with Aristotle and Plato, that the sun travelled around the earth in a geocentric universe. Hence their maps of the cosmopolitical realm showed the earth at the centre of the universe; it was therefore entirely logical also that the sun and the Cross should be superimposed onto each other in Celtic northern Europe. Christ's rule over the earth was mirrored in the daily dying and rising of the sun, and in the effects of its diminishing and increasing light and warmth on the fields and forests.

Christians in the patristic and medieval eras thought themselves at the centre of a cosmos ordered and governed by Christ at its zenith. They thought, moreover, there was a direct relation between divine will and what happened on earth. It therefore seemed obvious to them, as it no longer is to us, that God had overarching responsibility for the weather and the climate, and that by worshipping God through their tithes, alms, good work, and craft they partnered with the Creator in caring for the world. In particular, and as is evident from a detailed study of medieval sacred architecture, in their church building and liturgies they helped to *maintain* the world, as the Israelites had once done in the worship of the Temple.[94] Hence, when the 'medieval warm period' produced remarkable crop surpluses on which city burghers and friars grew fat, the city guilds and merchants devoted a proportion of these to the great cathedral building enterprise. The development of Gothic sacred architecture — in which master builders used the

94. Margaret Barker, *Creation: A Biblical Vision for the Environment* (London: Continuum, 2010), 20. On the world-construction inherent in medieval sacred architecture see further Gordon Strachan, *Chartres: Sacred Geometry, Sacred Space* (Edinburgh: Floris Books, 2003).

arch first adopted in the Muslim world, as well as the flying buttress which was their own invention, to build to new heights — was itself therefore a product of varying insolation.

The relation of Christ to climate and cosmos in the medieval world-view is depicted in the small 1250 illumination known as the *Psalter Mappa Mundi.*[95] Christ stands above the earth, holding an orb in his left hand and giving a sign of blessing with two raised fingers with his right. Around his nimbus-encircled head are two pots, hanging in midair, attached to strings which are held by two angels who, like puppeteers, upturn the pots to give rain on the earth and so water the crops. In medieval cartography, the map is an icon, and each element in the iconography corresponds both to Christian belief about the relation of Christ, cosmos, and Christians and to physical properties, for in the medieval mind there is no split between spirit and matter, nature and culture. It therefore seemed obvious to the medieval mind that when the crops grew tall in two centuries of long and good summers, which are now called the 'Medieval Warm Period', God looked with favour on the prayers and virtues of the citizens and clergy of Catholic Europe. Equally, when the rivers of northern Europe turned to ice in the 'Little Ice Age' which began around 1490 and endured until 1650, it was widely believed that it was the transgressions and wars of kings and princes that had brought it on. As the crops failed in the fields, rivers froze for months in the winter, and Alpine glaciers began to advance down the mountainsides, it seemed obvious to radical preachers and would-be Reformers that God was displeased with the moral turpitude and wealth of the late medieval Church. Most believed, like Joachim of Fiore, that these climatic changes indicated the end of an age in which the Church had grown too powerful and too wealthy and that this presaged the apocalypse or second coming of Christ.[96] It was therefore an era of radical and millennialist preachers, in which long-settled partnerships between Church and state were overthrown in northern Europe, and the four horsemen of the apocalypse — in the form of plague, war, famine, and death — stalked Europe on a scale unknown since the fall of the Roman Empire. War and conflict grew to a crescendo of violence which, though sometimes referred to as the 'wars of religion', might more appropriately be called the first climate wars.

95. A digital reproduction of *Psalter Mappa Mundi* is available at http://minmaxsunt.files.wordpress.com/2012/08/413px-psalter_world_map_c_1265.jpg (visited 24 July 2013).

96. Andrew Cunningham and Ole Peter Grell, *The Four Horsemen of the Apocalypse: Religion, War and Famine in Reformation Europe* (Cambridge: Cambridge University Press, 2000).

It was during the Little Ice Age that theologians began to doubt the correlation between the will of the heavens and life on earth that was central to the microcosm-macrocosm relation affirmed by Christian Platonism and Orthodox theologians in East and West until this time. Nominalism has its origins in this time, and the leading nominalist philosopher William of Ockham developed a new linguistics and *theologic,* according to which there exists no sensible and perceptible correlation between divine or heavenly will and the unfolding of events on earth. Instead, the earth is governed by a divine Creator whose will is ultimately unknowable and hence experienced on earth as arbitrary, even though it is in itself an originally good will.[97] The names which are ascribed to earthly and sensible objects signify those objects univocally, or with certainty and specificity, rather than drawing their meaning equally from their representation of a larger theological grammar of being, as the medievals had formerly believed. So Ockham's nominalism both underwrites a new sense of distance between heaven and earth and gives greater powers to language to identify with some certainty, and without reference to a supernatural theologic, the fixed and peculiar properties of named objects and species in nature.

As George Lindbeck argues, nominalism sets up a strong division between two radically opposed perspectives: on the one hand, the absolute power of God's omnipotent and arbitrary will; and on the other, what God has 'in fact ordained'. This split emphasises the 'radically individualistic isolation of man before a God of absolute and arbitrary power'.[98] This in turn sustained a new politics, first of the 'divine right' of kings apart from the consent of the governed, and then in the modern era of the social contract according to which the individual is autonomous of the body politic into which she is born until she *contracts* some of her autonomous power by an act of will. This split also gave rise to a new religion in which the individual soul exists as an independent entity within the body, drawn by piety to a life of Puritan self-denial and a related quest for the inner feeling of divine presence. And the split gave rise to a new science in which the body of the earth and the human body become available for investigation and reorder-

97. Louis Dupré, *Passage to Modernity: An Essay in the Hermeneutics of Nature and Culture* (New Haven, CT: Yale University Press, 1993).

98. George Lindbeck, 'Nominalism and the Problem of Meaning as Illustrated by Pierre d'Ailly on Predestination and Justification', *Harvard Theological Review* 52 (1959): 43-60; and see also Michael S. Northcott, *The Environment and Christian Ethics* (Cambridge: Cambridge University Press, 1996).

ing by empirical science, freed through nominalist logic from the theologic symbolism of medieval cosmology.

Copernican Cosmology and the Climate Cosmopolis

The theological revolution of late medieval nominalism gave birth to two ideas which had not conventionally been thought separately before, these being 'nature' and 'culture'. Copernicus and Galileo discovered that humans were not — at least in natural scientific terms — at the centre of the universe. Instead, their bodies walked on a planet that circulated a rather ordinary star on the edge of a universe whose size far exceeded the medieval imaginary of the hands of Christ stretched around the earth in another of the world icons of the Mappa Mundi. This decentring was resisted by the Church not because it dethroned humanity but because Galileo's observations and theorisations of the movements of heavenly bodies revealed something more troubling to the medieval mind: matter moves according to mathematical logic, the laws of inertia and mobility, and not because it is moved by a divine Mover.[99] But soon the Church resolved the problem through its own nominalist theology. If there were two realms of truth and not one, and two tests of truth, one for 'nature' and the other for 'culture', then the hidden and arbitrary omnipotence of nominalist imagination could be confined as the *deus ex machina* of nature, as the divinity who governs imperceptibly and from a distance. It would therefore remain possible for the Church to say that God impregnated Mary and generated the Incarnate *logos;* God *moved* Mary. But it would no longer be necessary to say that God moved the earth or the clouds. And as the Church ceded the power of truth telling over the mathematical relations of heavenly bodies, it also gradually ceded the power of truth telling over the mathematical relations of earthly commerce and ended the religious ban on usury. Scientific measure rose inexorably as the guiding star of the new age, in the dissection of bodies, in the movement of the stars, and in the commerce between traders and companies and states.

The rise of the science of mathematics as the arbiter of relations between heavenly and earthly bodies had profound implications for political bodies. If in the Middle Ages humanity, and above all the body of the king and the liturgy of the Church, were the microcosm to the macrocosm, as

99. Louis Dupré, *Passage to Modernity,* 68-70.

Maximus has it, then in the new 'state of nature' science had unleashed, atoms and bodies moved and collided and were at risk of falling apart. Hence for Thomas Hobbes, in the midst of the vicious English civil war, where subjects no longer owed their duties to king or Church, only a social contract, an agreement of the atomised bodies to join themselves to one another by giving their consent to civil authority, could construct civil order. In a culture governed by calculations of interest, and where men read the Bible for themselves, there was no more a world soul, no higher power to which men might appeal, beyond the control of the sovereign.[100] Analogously, in the newly objectified relations of nature, as revealed in telescopes and as tested in the forge and the laboratory, nature's mechanical and mathematical relations are also subjected to human control; Copernicus, Galileo, Bacon, and Newton created the experimental facts of the new world of nature and put nature to the test to prove the veracity and repeatability of their facts.

The modern constitution, premised on the new governability of culture apart from nature, originated in Bacon and Newton's science of mechanism and Hobbes's and Locke's realm of contract and commerce. The common people may still have believed it rained for decades in England because of the divorces of a king, but by the end of the English Revolution, the constitution and the cosmos were also divorced. The king might have been restored to the throne after the Revolution, but his body was no longer sacred; instead, it was a symbol freely chosen by the people to underwrite their new powers over their own destiny. Henceforth, civil power and commerce, not the magical powers of kings and prelates, would construct the world of culture. Analogously, Bacon's new science of mechanism and technology would create and refashion the world of nature. After the Renaissance and the Reformation, humanity would arise as the new artificer of politics and of the machine; the world would be made new, as Bacon would claim, not in millennialist hopes of the end of history but in the power which science and the state would draw together in birthing the 'new atlantis' and the *novum organum*.

As Bruno Latour argues, the modern political constitution is uniquely premised on this new division between nature and culture, which is the outcome of the Cartesian division between subject and object and the separation of matter and bodies from a transcendent, increasingly 'wholly other' and distant, God. Science constructs the modern 'cult of facts' in experi-

100. Bruno Latour, *We Have Never Been Modern*, trans. Catherine Porter (Cambridge, MA: Harvard University Press, 1993), 19.

ments and laboratories whose new objects enable engineers, industrialists, and medics to create bridges, steam engines, and antibiotics, and so control and direct the energies of nature into the fabrications of an industrial society.[101] Nature is made new by being turned into scientific facts, and so brought under the power of knowledge through the mediation of experimental physics, chemistry, and biology. Culture is made new as the sphere of moral and political fabrication, unfolding from the mediation of the consciences and the consent of atomised persons who by contract and commerce create history anew. To *be* modern is to affirm this new separation of powers. Air is no longer a medium of spirits or souls; air is conditional on human thoughts only when air pumps and condensers are made that can manipulate the temperatures of enclosed spaces, or planes seed clouds with manufactured particles. Political sovereignty is not the gift of nature or the gods; government does not arise from a covenant between a divine representative and the cosmos. Instead, it issues from will, the will to power, and the consent of the people to cede that will to the civil powers who rule over them for the purposes of civil peace.

These new separations of nature and culture, of bodies and souls, of objective and subjective realms are the source of the tremendous productive and technological powers released by the modern rise of human power over nature and cultures. But at the same time the unfolding of the engineering powers of modern science and the societal construction of the modern constitution involve the constant production of *hybrids* of nature and culture which do not respect these new separations. As Latour puts it, scientists go 'on and on constructing Nature artificially and stating that they are discovering it', while political theorists go on and on 'constructing the Leviathan by dint of calculation and social force'. Consequently, 'on the one hand, the transcendence of Nature will not prevent its social immanence; on the other, the immanence of the social will not prevent the Leviathan from remaining transcendent'. Hence it is necessary that added to the forced separation of nature from culture, of science from politics, is the forced exclusion of a Creator who as the Incarnate God constructs nature and culture. The separation could only be made to work by the scientists 'ridding Nature of any divine presence', and the politicians 'ridding Society of any divine origin'.[102]

101. Bruno Latour, *The Modern Cult of the Factish Gods* (Durham, NC: Duke University Press, 2010).

102. Latour, *We Have Never Been Modern*, 31-33.

To be modern, then, is to deny that the weather is political, or that politics influences the climate. To be modern is to deny that there is a God who is the author of nature *and* culture, for in the separation of nature from culture the moderns feel that they are invincible. Sweeping away the prejudices and superstitions of old, the newly revealed laws of nature would liberate Europe and thence the world from a prescientific past and give the humans unprecedented power over the ordering of nature. At the same time, in the human sciences society also would be refashioned, remade by social facts and positive law; the 'iron laws' of society and economics would become the new physics of human relations.

Into this newly ordered and humanly controlled world, so all powerful that it even begins to show up in the gases of the atmosphere and in gases trapped in Arctic ice, the claims of climate scientists are an incomprehensible hybrid; they seem more like the superstitions of forest dwellers than the utterings of Enlightened scientists. A fossil fuelled economy is changing the weather. How can this be? Enlightened rulers no longer claim to influence the weather; if their divorces no longer influence the weather, why should their coal mines?

Even if the models of climate scientists do demonstrate, to them, a human influence, what about the models of political economists? Is it imaginable that the productive powers, the 'wealth of nations', unleashed by the new laws of commerce and competitive interest could show up in the sky? How can climate models, constructed in computers and from proxies like ice cores, challenge the hard commercial realities of producer supply and consumer demand? Climate science seems to require a cosmopolis — a politics which is cosmic and not merely rational. But meteorology since Newton is not a human science. Moderns don't participate in the weather except when they watch the weather report or walk in the rain. They don't *make* rain, except in an experiment. Nor of course does God. As Peter Sloterdijk puts it, beyond animism and deism, modern weather has no author other than itself. The weather is self-made; the weather *performs* for forecasters to comment on and people to gossip about. Natural science and the modern constitution have

> exonerated every still possible God of any responsibility for weather phenomena and elevated him to super-climatic zones. Zeus and Jupiter may well have hurled lightning bolts, the God of modern Europeans is a *deus otiosus* and *eo ipso* inactive in the climate domain. The modern weather report can therefore present itself as a discipline of regional ontology in

> which one speaks of causes, but not of instigators. It speaks of that which happens by itself, and of that which, in the best of cases, gets 'reflected' as objective quality data in a subjective medium.[103]

In the new world of science and the social contract, nature is the objective backdrop for the Enlightened renaissance of culture, knowledge, and wealth. The Enlightenment cosmopolis is the product of the governing wills of reasonable minds; natural boundaries, organic relations, ought to play no part in the cosmopolitan constitutions of the moderns.

It is no wonder that the modern nations, and their newly constructed cosmopolitical United Nations, have failed to find a political 'solution' to the dissolution of solid carbon into atmospheric gases. For Latour and Sloterdijk, science as well as politics is a realm of mediation. There are no 'facts' that are not human constructs, whether natural or social. In this perspective the real problem with climate talks is that they are predicated on the claimed *objectivity* of natural science rather than the new *subjectivity* of the climate of the Anthropocene. The earth ceases to be an objective background to human culture when she calls time on the modern project to extend technological control from the earth's depths to the heavens.

Political Theology in the Anthropocene

Although the first human mark of civilisation on sedimentary layers may be traced by future archaeologists to the layers of soil tilled by neolithic farmers — the first Adams — I argued above for a more important date in the anthropomorphising of the earth, which is the birth of Christ. But perhaps it would be more theologically appropriate to date the beginning of the Christocene era to the rise and fall of Adam, since in the first Adam 'all die' and in the second Adam, Christ, 'all are made alive' (1 Corinthians 15:22). Christ is revealed in the New Testament as Lord of the cosmos from Adam to the new era of the Church, and we may reasonably call this era the Christocene. The natural and cultural mark of Mesopotamian and then Christian culture gradually extended from the Middle East to northern Europe, with the contiguous advance of agrarian pastoralism and Christian worship in the Christian era.

103. Peter Sloterdijk, *Terror from the Air*, trans. Amy Patton and Steve Corcoran (Los Angeles: Semiotext[e]; Cambridge, MA: MIT Press, 2009), 87.

But when priests no longer walked the fields of post-Reformation Europe, and peasants in the industrial revolution were forced from small farms into mines and factories, and to breathe the coal-fouled air of industrial cities, another era, the Anthropocene, may be said to have begun. The dating of the Anthropocene at the onset of the industrial revolution is then indeed the most appropriate from a theological point of view, since it is with the rise of coal, optics, and commerce that the sense of co-agency between Christ, Church, and cosmos is lost in post-Reformation Europe. In its place Cartesianism and Baconianism promote scientific enquiry and power over nature as the dominant forces brought together by the state in shaping a New World. But *anthropogenic* climate change is nonetheless a strange outcome of this newly secular politics and natural science. Climate science is, in Latour's terminology, a hybrid of nature and politics. But such hybrids, like the 'witches' who were persecuted with ever-increasing ferocity after the Reformation, are supposed to be forbidden by the new separation of nature from culture.

The failure of political institutions, including national governments and the United Nations, to act prudentially in mitigating climate change, so that even as I write these words a melting Arctic drives winter storms south in a European 'spring' and rattles the window panes, reflects the modern constitution of the nation-state as a cultural and secular, rather than created and providential, agency in human and earth history. The failure is ultimately a politico-theological failure. The world in the Enlightenment is not *alive*. It has no *soul;* it is *not* the Spirit-breathed creature of a divine Creator, nor even a Gaian goddess. And the nations, although gatherings of individual bodies, bear no intrinsic ensouled relation to the mechanical laws which govern the world of heavenly and earthly atoms and organisms. That the earth interacts relationally with humans, that humans are becoming again the microcosm to the macrocosm, is forbidden not only by Newtonian physics but by the modern constitution and the laws of commerce.

In the chapters which follow I offer a genealogy of the modern science-informed cosmopolis in which weather is not authored by humans, and of how Bacon, Descartes, Newton, Locke, and Kant *decentred* the human and the political from the earth at the same time as they released cultural and scientific powers that gave modern humans powers sufficient to become the dominant biochemical influence over the climate and soils and oceans of the earth. In dialogue with the writings of Alfred North Whitehead, Giambattista Vico, Carl Schmitt, Alasdair MacIntyre, Bruno Latour, and Jacob Taubes, I attempt to construct a new political theology of climate change in which I recover the relation of natural and human history and the role of ecosystem boundaries and ecological limits in the constitution of the nations. I also chart a course to find again the theological as well as cosmopolitical roots of the nations — in the divine and providential ordering of history from the first covenant between God and Israel, in the political responsibilities of the nations to sustain rule over limited terrains and to guard a just and fair distribution of the fruits of the earth within the ecological limits of these terrains without damaging the fruitfulness of the earth for future generations. A due acknowledgement of the *ecological* as well as political duties of the nations to their own citizens and their own lands is an essential prerequisite for the unfolding of forms of rule which will see the nations prepared to reduce fossil fuel dependence, and to rule as unusable most of their remaining reserves of fossil fuels.

The current refusal of the nations to acknowledge their ecological responsibilities to promote economic and political practices within limits also promotes a growing tendency to scapegoat the poor and the destitute both within and beyond national borders. In the midst of the ongoing failure of the nations to restrain fossil fuel extraction, I will argue that Christians have particular duties to witness to the nations of the divine and providential importance of *ascesis* and restraint. As climate change produces growing extreme weather events and reduces the availability of food and water in many terrains, Christians also have duties to remind the nations of their responsibilities to the growing number of migrants who will appear at their borders as a result of climate change. In short, only nations which acknowledge their ecologically and politically situated character will be able to treat migrants as friends and not enemies.[104]

104. John Macmurray, *The Clue to History* (London: SCM Press, 1938), 118-19.

2. Coal, Cosmos, and Creation

Coal is the substance above all others that originated a new relationship between humanity and the planet in which human beings became a greater influence on the atmosphere of the planet than the tilt of the earth in relation to the sun.[1] Before humans began mining coal in large quantities they harvested energy for cooking and heating from plants — mostly trees and scrub — supplemented by rudimentary forms of wind and water power. The release of modern human beings from the limits of plant-derived energy created possibilities for technological progress unknown in previous eras. Coal-fuelled steam engines were the first vehicles to move human beings faster than the gallop of a horse or a wind-driven boat, and hence opened up the modern era of extreme speed and synchronous global time. Millions of years before its large-scale use kicked off the Anthropocene, coal was seminal to humanity in a more fundamental sense. Coal is stored sunlight. Burning coal releases the prehistoric carbon stored in the earth's crust into the atmosphere.[2] This stored carbon represents the activities of millions of plants and other creatures which in the Carboniferous geological era gradually sank into the earth's crust. As they did so, they locked down sufficient quantities of CO_2 to change the balance of carbon and oxy-

1. Milutin Milankovitch first identified periodic and fractional changes in the tilt of the earth towards the sun as the cause of previous major climate shifts in M. Milankovitch, *Kanon der Erdebestrahlung und seine anwendung auf das eiszeitenproblem* (Belgrade: Königlich Serbische Akademie, 1941); see also J. D. Hays, J. Imbrie, and N. J. Shackleton, 'Variations in the Earth's Orbit: Pacemaker of the Ice Ages', *Science* 194 (1976): 1123-31.

2. See further Rolf Peter Sieferle, *The Subterranean Forest: Energy Systems and the Industrial Revolution,* trans. Michael P. Osman (Cambridge: White Horse Press, 2001); and J. S. Dukes, 'Burning Buried Sunshine: Human Consumption of Ancient Solar Energy', *Climatic Change* 61 (2003): 31-44.

gen in the atmosphere, thus creating the climatic conditions which enabled the evolutionary rise of mammals, including latterly *Homo sapiens,* and the demise of reptiles as the dominant genus of species on the planet. Without coal there would not only not be modern people; there would be no people.

Coal plays a seminal role in the genealogy of the climate crisis because of its peculiarly carbon-intense characteristics. Coal also played a *material* role in shaping the culture of early modern Europe for the industrial age. Fifteenth- and sixteenth-century coal mining in England and Germany shaped a new attitude of dominion and control over nature and fostered the other principal economic and technical processes and institutions that characterise the industrial age.[3] The earth-invasive techniques and extensive tailings associated with coal mining fostered the cultural belief in the earth as spiritually 'dead', which played out seminally in European philosophy. At the same time new mathematical accounts of nature's wealth acquired cultural prominence, including double-entry bookkeeping and the mathematisation of the 'laws of nature'. These promoted the mechanistic cosmology that fostered the rise, and gradual detachment, of natural science and modern philosophy. René Descartes, Pierre-Simon de Laplace, and Isaac Newton, though with differences of emphasis, elaborated a new science of mechanism which deepened the spiritual and moral emptying of space and nature that coal mining and other early industrial procedures advanced in material culture. After Descartes and Newton, space, soil, and air were no longer said to be vitally *alive* as they had been from the Classical era until the late Middle Ages. Henceforth nature was available for human control, dominion, and reordering. But in the twentieth century, under the influence of the science of electromechanics, thermodynamics, and particle physics, Newtonian mechanistic cosmology was complicated, if not displaced, by a relational cosmology more reminiscent in some ways of pre-Copernican cosmology. In the new physics of James Clerk Maxwell, Albert Einstein, and Werner Heisenberg, the human observer, and other bodies and particles in the cosmos, were again understood as being held in relationship by unseen or hidden forces, mostly operative at the subatomic or quantum level. Similar recovery of a relational cosmology occurred in the late-nineteenth-century emergence of the science of ecology, which sustained a more communitarian and relational account of interactions between species in ecological communities or 'ecosystems'.

3. Lewis Mumford, *Technics and Civilisation* (New York: Harcourt, 1934), 67-75; see also Langdon Winner, 'Foreword', xi, in the Chicago University Press 2010 edition.

Ecological science, geology, and chemistry reached a new synthesis in the Gaian science of James Lovelock, which has significantly shaped the contemporary scientific account of climate change. The philosopher who more than any other anticipated Lovelock's account of an earth which, from the chemistry of its rocks to the oxygenating capacities of plants and the oxygen-breathing lives of animals to the constitution of the atmosphere, is *alive* was Alfred North Whitehead. Whitehead saw the need for a physics and metaphysics that overcame the dualisms of Descartes, and for a more radical empiricism than that of Newton. Whitehead challenged Newton's account of the fixity of the laws of nature, and he critiqued the descriptions of reason, of *a priori* mental categories and self-consciousness, of Immanuel Kant and his heirs. In this chapter, after reviewing the history of coal and of the rise of Newtonianism, I explore the post-Newtonian repair of Whitehead, who, while acknowledging the Copernican turn, argues that humanity nonetheless needs to recover a science-informed metaphysics of organism.

Coal and the Rise of Commerce in a Cold Climate

Materially, coal is the substance that made industrial society possible. It remains the fuel of choice for almost every country intent on making the transition from the agrarian to the industrial that England first made in the sixteenth and seventeenth centuries. But coal is the most planet-damaging fuel of all, though it is closely followed by bitumen (sand oil) and, arguably, shale gas. Coal is very dense carbon, and burning it releases not only greater quantities of CO_2 per unit of heat than other fossil fuels but also a range of other noxious gases and heavy metals. The ecological and human health effects of coal burning are extensive on those who breathe air polluted by it. Mercury and arsenic from coal smoke harm human foetal development, and sulphur dioxide, black soot, and carbon monoxide from those same chimneys are toxic to humans and other animals.[4] Ecological and human health effects of coal mining are also extensive. Deforestation and pollution of rivers and streams are near universal accompaniments to mining, regardless of the method of extraction used. Trees are used to support mine shafts and to fire the furnaces of the smiths who make and repair the min-

4. Lucijan Mohorovic, 'First Two Months of Pregnancy — Critical Time for Preterm Delivery and Low Birthweight Caused by Adverse Effects of Coal Combustion Toxics', *Early Human Development* 80 (2004): 115-23.

ing machinery and tools. Water is polluted because coal seams are located underground in sedimentary rock through which water passes into aquifers that are sources of potable water. If the health and ecological costs of coal — leaving aside climate change — were included in the upfront costs of burning coal in power stations, then solar, wave, and wind power would be much more cost competitive than they currently are. If the costs from more extreme weather events now and for future generations from burning coal are factored in, the case against coal is overwhelming.

According to James Hansen, coal-fired power stations are 'the single greatest threat to civilisation and all life on our planet.'[5] Coal is also the largest remaining fossil fuel reserve of carbon dioxide; there is sufficient coal underground to power the planet for more than two hundred years at present rates of energy use.[6] If it is all burned, the climatic effects would be so catastrophic as to extinguish most species, and most humans. But far from a moratorium on coal, the world is in the midst of a 'coal renaissance', with new coal-fired electric power plants being built in Brazil, China, Germany, the Netherlands, Poland, South Africa, and Turkey, and coal extraction growing exponentially in Indonesia and Australia.[7]

Coal is a beguiling and problematic substance. It is abundant in the earth's crust and burns hotter than timber or bush, and it has been known for two thousand years as a substance that was better than wood or charcoal for fuelling smiths' furnaces because of its great heat. Despite its unpleasant appearance and smell when burned, it is unsurprising that it was known as black gold. Every country that has it burns it, and some of the world's largest industrial and manufacturing economies — including Australia, China, Germany, Poland, Russia, and the United States — continue to depend on domestic or imported coal for the majority of their base load power.

Britain's love affair with coal began when the Romans in Northumbria first burned coal to warm their centrally heated villas in the vicinity of Hadrian's Wall. In the Middle Ages the coal-rich Tyne Valley was largely owned by abbeys and monasteries, as farmers and landowners had made

5. James Hansen, 'Coal-fired Power Stations Are Death Factories', *The Observer,* 15 February 2009.

6. Mikael Hööka, Werner Zittelb, et al., 'Global Coal Production Outlooks Based on a Logistic Model', *Fuel* 89 (2010): 3546-58.

7. 'The Unwelcome Renaissance: Europe's Energy Policy Delivers the Worst of All Possible Worlds', *The Economist,* 5 January 2013, at http://www.economist.com/news/briefing/21569039-europes-energy-policy-delivers-worst-all-possible-worlds-unwelcome-renaissance (accessed 23 February 2013).

successive gifts of land to the monks over many centuries. While coal was so abundant in the Tyne Valley that it would tumble out of the banks of rivers, the monks were not enthusiastic about building mine workings to haul it from underground. As Carolyn Merchant suggests, in the Middle Ages theologians — and most theologians were monks — conceived of the earth as a feminine partner of the divine Creator.[8] The ascription of femininity to the earth goes back at least to Plato in Western philosophy, who described the earth as 'nurse' in the *Timaeus.* Most Classical writers referred to the earth as Mother and believed her to be a living organic body of which the rocks were living extrusions. They consequently regarded mining with deep suspicion. For Pliny mining represented an invasive assault on the body of the earth, and its violence corrupted miners and those who employed them, just as the quest for gold led to avarice, and iron smelting led to war.[9] Noting the richness of earth, including the buried wealth 'which a kind lord had hidden', Ovid similarly deplored how miners 'dug into her vitals', and he considered precious metals they mined in 'stygian shadow' 'the root of evil'.[10] Seneca was particularly critical of miners' pollution of water and the creation of 'huge rivers and vast reservoirs of sluggish waters'. Seneca's words are strangely apposite as descriptions of the vast lakes of toxic water associated with the extraction process of Athabascan tar sands oil in Canada and the pollution of rivers and groundwater caused by coal extraction by mountaintop removal in Virginia and Kentucky.[11]

Negative attitudes to mining carried over from the Classical to the Christian era, and particularly among those influenced by Neoplatonism. Alain of Lille of the School of Chartres portrayed earth in his *De Planctu Naturae* as the woman servant of God grieving at the assaults of humans on her body which have torn her clothes and shamefully exposed her: 'by the unlawful assaults of man alone the garments of my modesty suffer disgrace and division'.[12]

> Many men have taken arms against their mother in evil and violence, they thereupon, in fixing between them and her a vast gulf of dissension,

8. Carolyn Merchant, *The Death of Nature: Women, Ecology and the Scientific Revolution* (San Francisco: Harper Collins, 1989), 29-41.

9. Pliny, *Natural History* 6.33, cited in Merchant, *Death of Nature,* 30-31.

10. Ovid, *The Metamorphoses* 1.6, cited in Merchant, *Death of Nature,* 31-32.

11. Seneca, *Natural Questions* 5.15.207-8, cited in Merchant, *Death of Nature,* 31-32.

12. Alain of Lille, *The Complaint of Nature,* trans. Douglas Moffat (New York: Henry Holt, 1908), cited in Merchant, *Death of Nature,* 10.

> lay on me the hands of outrage, and themselves tear apart my garments piece by piece, and, so far as in them lies, force me, stripped of dress, whom they ought to clothe with reverential honor, to come to shame like a harlot.[13]

Mining was immoral, and those who commissioned it sinned not only against Mother Earth but also against divine will. But as William of Ockham advanced the nominalist dissociation between the will of God and the physical 'accidents' of life on earth, the monasteries began to lose their theological objections toward mining. If, as Ockham argued, there is only an arbitrary relationship between physical created order and divine moral will, there are no theological grounds for restraining the reordering of the earth by miners and in the forges and metal works that coal fed.

In the early medieval period mines were run on the same manorial system as agriculture. Each family was assigned a seam of ore or coal, with the emphasis in production on equality rather than efficiency, as also in the medieval craft guilds.[14] Shared ownership and use rights in mines grew into a more formally coded system in the sixteenth century in Germany. Certificates were issued to participants in syndicates which developed large collectively owned mines, and the share accounts were kept by a shares clerk who recorded sales of shares. As William Parsons comments:

> in this procedure is seen the basis of the modern joint stock corporation with shares fully or partially paid, certificates for the shares, provisions for calls or dividends, an open list of stockholders, a responsible stock registrar and some public supervision of capital issues.[15]

This mutualised approach to mining gradually dissipated as the mines went deeper and required more capital investment. The demise of mutualism in mining led to bitter conflicts between miners and mine-owners, which culminated in the Peasants' War of 1525, after which the influence of the guilds on the organisation of mining gave way to the characteristic capitalist form of corporate share ownership and waged work. Hence mining became the material base of the emergent capitalism of Germany and saw the rise of

13. Alain of Lille, *Complaint of Nature,* 41, cited in Merchant, *Death of Nature,* 297.

14. Robert S. Lopez, *The Commercial Revolution of the Middle Ages 950-1350* (Cambridge: Cambridge University Press, 1976), 143.

15. William Barclay Parsons, *Engineers and Engineering in the Renaissance* (Baltimore: William and Wilkins Co., 1939), 185.

dynastic fortunes and corporations analogous in the extent of their power, wealth, and public influence to the twentieth-century rise of petroleum dynasties.[16]

In sixteenth-century Britain, commercial practices associated with the rise of mining were different from those in Germany. Free miners did exist in Britain, and very small-scale free mining continues today in the Forest of Dean in Gloucestershire. But most English mine workers were peasants who had lost their share in agricultural use rights to land as a result of the Tudor dissolution of the monasteries and the enclosures of the post-Renaissance English parliaments.[17] Coal mining set the pattern for the new form of wage labour which grew outside of the moral constraints of guild or manor after the failure of the English Revolution and the Restoration of the monarchy. Hence coal not only made industrial civilisation materially possible. It also shaped the relations of production that predominate in both capitalist and communist forms of mercantilism until the present.

More than half of all the coal burned in England from the Middle Ages to the present originated in the vast coal seam of northeast England. In an area more than fifty miles long and twenty wide, thick seams of coal lay mostly close to the surface and stretched from under the North Sea to the Pennines. The relative ease of access to the coal from drift mines, dug horizontally into hillsides and riversides, meant that when in the thirteenth century the area was experiencing a diminishment in the supply of wood, coal was increasingly used as a fuel for heating, as well as by smiths and smelters. The monasteries were large stone buildings much in need of heat, and with the rising cost of timber it was commercially a logical move to begin storing and burning coal. The monks had also been gifted much land in the area, and this land included existing guild-organised collieries. Hence by the thirteenth century the Abbeys of Durham, Finchal, Hexham, and Jarrow were already burning significant quantities of coal.[18]

London was the principal destination for the great majority of the coal dug in northeast England, and it went by sea. 'Sea coal', as coal became known, was so widely used by thirteenth-century Londoners that a street was named Sea Coal in 1228.[19] The use of coal was initially confined to

16. Mumford, *Technics and Civilisation*, 74-75.

17. John Hatcher, *The History of the British Coal Industry*, vol. 1: *Before 1700: Towards the Age of Coal* (Oxford: Clarendon Press, 1993).

18. Hatcher, *The History of the British Coal Industry*, 1:27.

19. William H. Te Brake, 'Air Pollution and Fuel Crises in Preindustrial London, 1250-1650', *Technology and Culture* 16 (1975): 337-59, 339.

smiths' shops and lime and brick kilns. But by the sixteenth century the rising price of firewood meant that coal rapidly became the fuel of choice for households and even palaces, although it was not until the seventeenth century that the use of coal to fuel workshops and homes became widespread throughout Britain. This was provoked by a near doubling of the English population between 1530 and 1650, which led to great pressure on woodlands and a growing scarcity of timber in a development sometimes called 'peak timber'.[20]

The first complaints about coal pollution in London are recorded in the thirteenth century in the report of a royal commission appointed in 1285 to investigate smoke from lime kilns and blacksmiths' workshops.[21] Edward I unsuccessfully banned the use of coal after complaints from the bishops and lords who had gathered for the parliament of 1306 and noted a foul smell and a dark aspect to London's air.[22] Having failed to ban it, the royal court regularly abandoned the Palace of Westminster for other royal residences at Hampton Court, and later Windsor, in part because of the smoke. But it was not for seven centuries that coal smoke in London was legally regulated. In 1952 a thick coal smog had descended on a wintry London, stopping trains, cars, and public events and eventually killing thousands of Londoners. Four years later the British Parliament passed the first Clean Air Act, banning coal smoke from London's chimneys.[23] Subsequently, coal in London would be burned as smokeless fuel created from raw coal in coke plants in south Wales before being transported to London to be burned in open fires and coke-fired cooking stoves like the ones my mother maintained in a 1960s London suburb.

Britain's history as the first major coal-burning nation means that *per capita* of those presently living, Britain has emitted more fossil fuel–related CO_2 into the atmosphere than any other nation. Britain has few deep mines left, but it still emits significant quantities of CO_2, as well as mercury, soot, and sulphur particulates, from coal-fired electricity-generating plants which rely largely on imported coal. Just one of Britain's coal-fired power stations, Drax B in Selby, Yorkshire, emits more CO_2 than a number of African nation-states, and one-third of British electricity presently is derived from such plants. But the chimneys of modern British coal plants are so

20. Hatcher, *British Coal Industry*, 37-51.

21. Te Brake, 'Air Pollution and Fuel Crises in Preindustrial London', 339, note 5.

22. Barbara Freese, *Coal: A Human History* (Cambridge, MA: Perseus Books, 2003), 1-2.

23. Eric Ashby and Mary Anderson, *The Politics of Clean Air* (Oxford: Clarendon Press, 1981).

high that it is the inhabitants of Scandinavia, not Britain, that endure the worst of the pollution.

It is a different story in the United States and China. The whole of the eastern seaboard of the United States has a permanent atmospheric aerosol which in summer turns to a heat haze in the principal cities that causes millions of cases of asthma and bronchial problems. Most of the haze emanates from the many coal-fired power stations situated close to coal seams in the Appalachians — many of them belonging to the Tennessee Valley Authority — from whence the coal smoke blows east.[24] Chinese cities more often site large coal-burning plants within city boundaries, and the pollution is therefore even more noxious, and much more visible, than in the United States.[25] Despite the health effects on their own populations of coal smoke, and the role of coal as the most significant source of greenhouse gas emissions, neither the United States nor China has regulatory plans to reduce the use of coal in power generation. Indeed, both China and the U.S. have commissioned a significant number of new coal-fired power stations during the period of the Kyoto Protocol, as have India, South Africa, and Turkey, among others.[26] Worldwide, more than 1,200 new coal-fired power plants are planned in the next twenty years.[27] These will be fuelled by vast new coal fields being opened up in Kalimantan and Australia, and major increases in exports from existing fields in Australia and the United States.[28]

The atmosphere is unlike land in every material respect, and people are not accustomed to thinking of the air they breathe — and into which their domestic animals, cars, and chimneys emit waste gases — as a finite resource needing customary governance arrangements such as those that historically developed around other commons, including fisheries, forests,

24. L. W. Anthony Chen, Bruce G. Doddridge, et al., 'Origins of Fine Aerosol Mass in the Baltimore-Washington Corridor', *Atmospheric Environment* 36 (2002): 4541-44.

25. China's coal output stood at 2549 million tons in 2007, which exceeded by a billion tons U.S. output in the same year. However, much of the coal China burns reflects displaced pollution from Europe and North America as corporations avoid environmental regulation, as well as labour laws, by relocating their factories and hence emissions to China; see further *Cleaner Coal in China* (Paris: International Energy Agency, 2009), 38-43.

26. Dieter Helm, 'Climate-change Policy: Why Has So Little Been Achieved?' *Oxford Review of Economic Policy* 24 (2008): 211-38.

27. Ailun Yang and Yiyun Cui, *Global Coal Risk Assessment: Data Analysis and Market Research* (Washington, DC: World Resources Institute, 2012).

28. Ria Voorhar and Lauri Myllyvirta, *Point of No Return: The Massive Climate Threats We Must Avoid* (London: Greenpeace, 2013).

grasslands, rivers, and water tables.[29] It took Londoners 650 years, after hundreds of thousands of premature deaths, to regulate coal smoke. This is despite medical testimony from the thirteenth century onwards that mortality was higher in London than outside its smoky environs. The reason in part was necessity. England is a temperate and densely populated country, and its inhabitants were the first to cut down and burn most of their ancient forests and replace them with grazing animals. Hence, until the discovery of oil, coal was the only alternate source of heat available.

The story of coal smoke in London indicates the magnitude of the cultural problem, and moral dilemma, presented by the scientific evidence that the earth is warming because the atmosphere is limited in its capacity to absorb industrial quantities of CO_2 and methane emissions. As we have seen, the morality of the problem is accentuated by differential burdens from, and responsibilities for, climate change. While greenhouse gas emissions resident in the atmosphere emanate in large part from the activities of relatively affluent city dwellers in developed, and increasingly in developing, countries, it is very low-income households and subsistence farmers and fishers — primarily in Africa and south Asia — who are already experiencing, and will experience, more life-threatening extreme weather events from climate change.[30] Here again there is an analogy with coal smoke in London. Rich Londoners, like the royal court, owned country houses away from town where their children mainly lived, educated and fed by servants, and to which their parents would repair between periods of residence and work in the smoke of the city. The poor, on the other hand, had no choice but to live, and die younger, in the smoke.

The coal industry that grew rapidly to become the dominant source of wealth in seventeenth-century England and Germany spread to the colonies over succeeding centuries. The United States in particular was powerfully shaped by the 'kingdom of coal' in the period of rapid industrialisation after the Revolution.[31] Child labour, excessive hours of work, and death

29. For a fuller discussion of commons governance arrangements see Elinor Ostrom, *Governing the Commons: The Evolution of Institutions for Collective Action* (Cambridge: Cambridge University Press, 1990).

30. See further WHO, *Protecting Health from Climate Change: Connecting Science, Policy and People* (Geneva: World Health Organization, 2009).

31. Daniel L. Miller and Richard E. Sharpless, *The Kingdom of Coal: Work, Enterprise, and Ethnic Communities in the Mine Fields* (Philadelphia: University of Pennsylvania Press, 1985); and Harold Platt, *Shock Cities: Environmental Transformation and Reform in Manchester and Chicago* (Chicago: University of Chicago Press, 2005).

from poorly regulated mines were all common in the mining communities which spread across the rich seams of anthracite that are found throughout the Appalachian Mountains. Canals, and then railroads, to bring the coal to the cities shaped the communications and geography of southern and northern states. Mining unions, and thence labour laws, shaped the pattern of industrial working relations in the United States until the great corporate outsourcing of industrial work that began in the Reagan era.

The Reagan era also saw the growth of a new kind of mining, designed to get around labour laws and controls over mines. Opencast or strip mining in places such as the American Appalachians, Kalimantan, and New South Wales creates cratered, lunar-like landscapes that far exceed the landscape scarring from waste heaps, deforestation, and collapsed mine workings associated with traditional drift and deep mine techniques. The technique for disposing of mountain tops in West Virginia is called 'valley fill'. This involves permanently flattening mountains to access the coal seams and putting the waste rock and earth into river valleys down the mountain and beyond, which are then aerially sprayed with grass seed. This results not only in pollution of river water and ground water, thus affecting the health of local residents and the viability of small farms in such areas. It also involves a complete restructuring of the topography of the area. The land is gradually flattened, and the remaking of the area is so extensive that maps have to be redrawn every couple of years. This kind of mining represents the 'total domination of nature' in this area by the mining companies responsible.[32] The desecration of the Appalachians in West Virginia and Georgia by mountaintop removal is a particularly brutal illustration of the mechanistic cosmology that underlies, and has been sustained by, the development of a coal-fuelled industrial culture, a cosmology in which land is conceived as empty space, unowned and disordered, until it is developed and reordered by humanity for the creation of human wealth.[33]

The Optics of Numbers and the Dehumanising of Space

The idea of empty space is a modern one. It originates in the atomistic split between being and matter advanced by Cartesian philosophy, and then by

32. Rebecca R. Scott, *Removing Mountains: Extracting Nature and Identity in the Appalachian Coalfields* (Minneapolis: University of Minneapolis Press, 2010), 95.

33. Scott, *Removing Mountains,* 210.

Copernican astronomy and Newtonian mathematics and mechanics. The early modern revival of classical atomistic thought originated in the attack on the scholastic philosophy of the late Middle Ages mounted at the Renaissance, and in particular in association with the observations of Copernicus, Galileo, and Johannes Kepler through the new high-powered telescopes that were invented in the early modern period. These optics changed the way of seeing the world, and humanity's place within it. Their creation was dependent on the coal-fuelled furnaces and workshops of the emergent craft industries of late medieval Europe, where coal gave superior and more stable heat to smiths and glass makers with which they created the precision materials that made possible the new timekeeping and optical devices — the clock, the telescope, and the microscope — which gave Renaissance humanity a whole new perspective on the heavens, and on the earth.

Lewis Mumford argues that the clock — and not the steam engine — is the crucial mechanical invention that advanced the modern industrial age.[34] The clock made possible the standardisation of production and of the measurement of space and time, and hence the reliable navigation of the whole earth, in ways that had not been possible before and are characteristically modern. Every new phase of industrial work is testament to the clock, from the invention of longitude to Taylorist manufacturing and the digital age. The post-industrial information worker does not 'clock on' at a physical time station like her forebears, but instead she clocks on when she enters the corporate software that provides her working environment, and any move out of the software — for bodily and social needs — is recorded as non-working time.

If the clock is the paradigmatic machine shaping *social* relations in the modern era, optical lenses and their capacity to reveal and represent the world on mechanical surfaces, and as recorded mechanical images, are the paradigmatic instruments which reshape human-nature relations in the modern era. They may even be said to be responsible for the modern split between human and natural histories, and hence the division of culture from nature. Before the telescope, human beings had thought they inhabited a world in which their planet was the centre, and the history and trajectory of the earth and its most powerful inhabitants were central to the movement of the sun and the planets. Similarly, before the microscope, humans envisaged themselves and their 'humours', ideas, and spirits as the prime determinants of the health and relations of bodies. Telescopes re-

34. Mumford, *Technics and Civilisation,* 14-15.

vealed that the earth is not at the centre of the cosmos nor its most powerful orb. Instead, the earth moves around the sun and not the other way around. Similarly, the microscope revealed a world of atoms, bacteria, cells, genes, and microbes which turned out to have a history far older than that of *Homo sapiens,* and *Homo sapiens* is merely one in a long line of previous species including earlier mammals and amphibian and marine organisms. As H. J. Schellnhuber argues, it was 'optical amplification techniques' which brought about the Copernican revolution.[35] This is because the Copernican revolution was underwritten by a new precision in the mathematical description of the universe, a development that was as momentous for the humanities as it was for the sciences, since it radically decentred humanity from natural history and the earth from the universe. After Copernicus, scientists ascribed absolute formal power to the motion of objects in space, independent of their relations to other objects or to consciousness, whether divine or human.[36] After Copernicus, scientific humanity set out on a five-hundred-year journey to empty the cosmos of agency, while increasingly imposing humanly intended technical interventions on this emptied cosmos. At the same time, in what increasingly came to be called the humanities — to use the word 'metaphysics' would still imply an element of *physis* in philosophy and the arts[37] — philosophers and artists from Descartes and Kant to Friedrich Schelling and G. W. F. Hegel were forced to reimagine the human world of the interior, the world of *spirit,* as the principal loci of rational purposiveness and spiritual meaning in history.

The material link between coal, optics, and the cosmology of mechanism underlines Mumford's claim that mining exercises a spiritual as well as material and social influences on the societies and peoples that engage in it, something that is also hinted at in the accounts of Solomon's mines in the Hebrew Bible.[38] The material culture advanced by coal, in which the earth is dominated and reengineered from its depths to its outer atmospheric limits,

35. H. J. Schellnhuber, '"Earth System" Analysis and the Second Copernican Revolution', *Nature* 402 Supplement (1999): C19.

36. Medieval innovations in optics transformed not only science and philosophy but art and literature as well; see further David Lindberg, *Theories of Vision from al-Kindi to Kepler* (Chicago: University of Chicago Press, 1976); and Suzanne C. Akbari, *Seeing through the Veil: Optical Theory and Medieval Allegory* (Toronto: University of Toronto Press, 2004).

37. Jacob Taubes, 'Nachman Krochmal and Modern Historicism', in Jacob Taubes, *From Cult to Culture: Fragments toward a Critique of Historical Reason,* ed. C. E. Fonrobert and A. Engel (Stanford, CA: Stanford University Press, 2010), 28-44.

38. Mumford, *Technics and Civilisation,* 74-77.

also sustains the cultural objects and values that until today make it difficult for industrial societies to respond to or resolve the ecological and climate crisis. At the heart of this lack of response is a new attitude to human making as a form of conscious rational dominion over a material earth that is empty of moral or spiritual significance apart from humanity's rational reordering. It reflects the emergent view in the late Middle Ages, which the Copernican revolution underwrote, that moral and spiritual worth do not intrinsically belong to matter or organisms but only to those rational and spiritual beings who by their uses of matter impose worth on it.[39] Hence, as Francis Bacon argued, only those who engage in controlling nature are fully human; reengineering the earth, and even the atmosphere, become the markers of *being* human, or the way in which humans become like gods and fill the earth and their own history with meaning, after the scientific revolution.[40]

The irony is that, while Copernican astronomy *decentred* humans from earth history and was therefore resisted by the Catholic Church, the scientific revolution recast humanity as the technical controller of the habitat of the earth. This produces the reverse of the situation in the Ptolemaic era, in which humans were the microcosm to the macrocosm but human beings lacked the technical capacities to enforce their will on the other parties. In the Ptolemaic era the technologies and networks which connected the two involved the multiple agencies of divinities, humans, and other than human beings. But in the Copernican era there may be multiple agencies, but they are engaged in a dialogue of the deaf: only humans have rational and purposive power, and they are faced over against this with brute matter, beasts, and organisms that cannot speak and whose experiences — if they even have them — do not morally or spiritually count. Hence, as Jacob Taubes argues, the Catholic Church was right to resist the Copernican revolution, because it 'not only overthrew an old astronomical theory but also destroyed man's dwelling in the cosmos'.[41]

René Descartes was the philosopher who mediated the implications of the Copernican turn to the human subject and the newly denatured realm of 'culture'. His description of being as unconscious apart from human subjectivity was a powerful response to the new science of motion and mechanics. It also helped to resolve contradictions between the new science advanced

39. Francisco J. Benzoni, 'The Moral Worth of Creatures: Neo-classical Metaphysics and the Value Theories of Rolston and Callicott', *Environmental Values* 18 (2009): 5-32.

40. Francis Bacon, *Novum Organum* 249.

41. Taubes, 'Dialectic and Analogy', in Taubes, *From Cult to Culture,* 165-76, 169.

by observation and the biology and physics of Aristotle which had been retrieved by the medieval schoolmen. The schoolmen had advanced a distinction between matter or substance, which every body shares, and form, which differentiates types of bodies and univocally determines their behaviours. In this perspective, gravity is a force of 'heaviness' that acts on bodies differently according to properties that reside in bodies, rather than being an external force.[42] It was this scholastic distinction between form and substance which also gave rise to the novel medieval account of the transubstantiation of the elements of the bread and wine in the eucharist, and it was therefore a major cause of the Reformation. With transubstantiation in his sights, Descartes countered scholastic accounts of the univocity of types of being with the claim that bodies cannot respond univocally to the attractive force of the centre of the earth because that would require that bodies had knowledge of the centre, 'and there can be no knowledge except in a mind'.[43]

This elegant fix to scholastic physics required an ontological break between physical bodies and their motions, and an epistemological break between consciousness and sensory knowledge of bodies and movement. That it is a repair, and not a complete rejection of the scholastic worldview, is indicated in Descartes' adoption of the Aristotelian and scholastic understanding of 'substance' as the principal category of being. Descartes argued that there are two kinds of substance: the first *res cogitans* exist only in minds, and only humans have minds; the second *res extensa* are bodies and extend into space. As Howard Stein argues, Descartes' account of extension is not entirely clear, and it leads him down a road where in order to speak of the motion of bodies it becomes necessary to speak of bodies existing serially in different spaces. Consequently, for Descartes, space does not exist apart from bodies. Space is the extension of bodies, or a 'plenum', and consequently Descartes has no account of place.[44]

For Jean-Luc Marion the clue to Descartes' epistemological innovation is his account of *natura simplicissima* or 'simple nature'.[45] Minds appre-

42. Daniel Garber, 'Descartes' Physics', in John Cottingham, ed., *The Cambridge Companion to Descartes* (Cambridge: Cambridge University Press, 1992), 286-334.

43. Descartes, *The Philosophical Writings of Descartes* II, cited in Garber, 'Descartes' Physics', 287.

44. Howard Stein, 'Newton's Metaphysics', in Bernard Cohen and George Edwin Smith, eds., *The Cambridge Companion to Newton* (Cambridge: Cambridge University Press, 2002), 256-307.

45. Jean-Luc Marion, 'Cartesian Metaphysics and the Role of Simple Natures', trans. John Cottingham, in Cottingham, ed., *Cambridge Companion to Descartes,* 115-35.

hend things not as they are in their physical natures but in the simple effects that objects have on consciousness; they do not therefore know the essence or *ousia* of the things themselves. They do not even so know themselves. Hence persons know themselves as selves because they think, rather than because they have biological parents or eat and drink or breathe air. But they do not know reliably nearly as much as they think they know about anything that is not a rational thought. Other bodies are mediated to consciousness by embodied senses, and these senses impose categories and appearances on other bodies that are not intrinsic to these bodies but are rather products of the mind itself. So then things known in their 'simple natures' are not natural at all: the simple nature is not organic or material but rather is the idea, figure, or shape that material things form in the conscious mind. Hence,

> We are concerned with things only in so far as they are perceived by the intellect, and so we term 'simple' only those things which we know so clearly and distinctly that they cannot be divided by the mind into others which are more distinctly known.[46]

The revolutionary implications of this idea continue to be played out in cognitive science and physics as well as analytical philosophy to this day. Since Descartes, it has seemed increasingly obvious both to scientists and to philosophers that 'the universe is fundamentally physical and existed and evolved for billions of years without any consciousness in it at all'.[47]

The evacuation of agency and subjectivity from the cosmos promotes the alienation of mind from matter and a new instrumentalist and domineering attitude to nature. In the medieval mind humanity was the microcosm of the macrocosm: there was an analogical relation between earth and heaven, human life and other beings, which extended from the soil and the ocean surface to the atmosphere and beyond to the movements of the sun, moon, and stars. This cosmology of multiple agencies was symbolised by the first clocks, which depicted the movement of celestial bodies as well as the hours of the day and night; some of the more complex designs included the work of farmers on the soil.[48] But Cartesianism, as the philosophical

46. Descartes, *Règles utiles et claires pour la direction de l'esprit'* XII, cited in Marion, 'Cartesian Metaphysics', 116.

47. Jonathan Shear, 'Introduction', in Jonathan Shear, ed., *Explaining Consciousness: The Hard Problem* (Cambridge, MA: MIT Press, 1995), 1.

48. The astronomical clock at Wells Cathedral, Somerset, is a superb and still working

outworking of Copernicanism, established a seemingly unbridgeable divide between nature and culture, physical objects and human ideas. Consequently, animals are no longer *other* animals as the medievals knew them; they are *wholly* other and become 'beasts'.[49] If animals lack consciousness, they lack the ability to form concepts or experience pain and pleasure as humans do. They may, as Descartes admitted, *feel* pain, but such feeling has no moral significance since they cannot have the thought that they would rather it ceased.

For Cartesians the medieval idea that there may be analogies of being and causation — such as 'a great chain of being' — between humans and other creatures is metaphysical speculation. Such metaphysical claims may be internally logical, but they are inconsistent with scientific descriptions of the material world of atoms and motion. Hence, as Jean-Luc Marion argues, Descartes inaugurated the split which Newton took up between physics and metaphysics, and later Kant between science and ethics, as well as the split between culture and nature.[50] Descartes also inaugurated the dualism in modern culture and philosophy between descriptions of conscious thought and embodiment, which, as many have argued, is deeply implicated in the origination of the modern ecological crisis.

This dualism also has implications for the relation of human minds to God. God for Descartes is uncaused and so is not knowable in the same way as simple natures are knowable. But nonetheless the *idea* of God resides in minds and precedes knowledge of other simple natures since it participates in a dualism between ideas and *ousia*. This dualism is characteristic of the way in which human beings — and all bodies — encounter God. And since only bodies can be extended, mind strictly speaking lacks material location. But Descartes also wanted to affirm the traditional theological belief in divine omnipresence. The only way through this conundrum was to argue — as he does in correspondence with the Cambridge Platonist Henry More

example of the horological depiction of a cosmology of layered agency from stars and the sun to the tillers of the soil and the fish in the ocean.

49. Keith Thomas charts the rise of a growing insensibility to animal suffering in sermons and other literature from the sixteenth century, and it is at this time also that animals are no longer tried for crimes such as manslaughter when involved in a human death, since they are no longer seen as agents; Thomas, *Man and the Natural World: Changing Attitudes in England, 1500-1800* (London: Allen Lane, 1983).

50. Marion, 'Cartesian Metaphysics', 117. For a contrary view to Marion's see John Carriero, *Between Two Worlds: A Reading of Descartes'* Meditations (Princeton, NJ: Princeton University Press, 2009).

— that while God in his actions is present throughout created order, God in his divine power is not.[51] Here the split between *res cogitans* and *res extensa* builds on the scholastic and nominalist duality between the power of God and human apprehension of that power in creation.

Isaac Newton was no monk but a lay philosopher, and proud of being so. He thought that Descartes' attempt to repair the scholastics was wrong. Instead, he proposed a more radical break with the cosmology and metaphysics of the late Middle Ages: he set out to explain motion and gravity and the relations of bodies mathematically and in ways that could be demonstrated experimentally. Newton also challenged Descartes' argument that mind and matter, bodies and knowledge are completely distinct categories of being, a challenge which had implications both for Newton's God and for Newton's account of substance.

For Newton, as for Descartes, there are material, embodied, and perceptible substances which are located in space. But their location in space is not simply a function of the extension of their material and embodied substance. He rejected Descartes' account that extension is the same as embodiment. Instead, he argued that bodies move through space and their movement is in part a function of their relations to other material substances. Here is the first crucial, and most famous, difference with Descartes. For Newton gravity is an attractive force, related to motion; heavier moving objects act upon lighter ones, as is the case in the relation of the earth to the sun, or the moon to the earth.[52] Therefore motion is not a function of substantial bodies and their extension into space but of *relations* between substantial beings in space; and space is not a plenum — it is not embodied but rather is the space between bodies.

The second radical departure from Descartes is that, while for Newton there are immaterial substances, like God and minds, that cannot be visually perceived and that do not refract light, nonetheless such immaterial substances are located in space. This move required another and more dramatic innovation, and one that gets to the heart of the difficulty with Newton's attempts to repair Descartes' mind-body dualism. If God, like minds, is located in space, and God is omnipresent as well as omnipotent, then space is a prerequisite for the being of God. Hence to preserve the

51. Andrew Janiak, 'Substance and Action in Descartes and Newton', *The Monist* 93 (2010): 657-77.

52. Howard Stein, 'Newton's Metaphysics', in I. Bernard Cohen and George E. Smith, eds., *The Cambridge Companion to Newton* (Cambridge: Cambridge University Press, 2002), 256-307.

scholastic understanding of the Aristotelian category of substance, while also holding that the world is in motion and that minds are located in space, Newton argued that God, as a mindful substance that is also capable of extension, did not create *ex nihilo* since God always required space in which to be.[53]

As Simon Oliver shows, Newton's accounts of bodies and motion reflect not only experimental knowledge but his theological beliefs, which included an Arian view of Christ and medieval voluntarism.[54] For Newton the Christian belief in the Trinity was an idolatrous accretion to primitive Christianity in which Christ was subordinate to God the Father, and the Council of Nicaea's Trinitarianism was a corruption of Christ's efforts to return humanity to an original pristine religion which had existed at the time of Noah. Newton saw himself as a prophet, doing work analogous to that of Christ, and cleansing theology and philosophy (in the seventeenth century science was included in the word 'philosophy') of superstition and idolatrous accretions. But Newton adopted a far from ancient account of the nature of God when he claimed that medieval theologians were right to argue that the will of God was the supreme attribute of the divine. The later patristic description of Christ as *pantokrator* ascribes to God an omnipotent dominion as ruler of the universe. For Newton, God does not rule the world as the soul rules the body, but rather by a perfect will which 'constantly cooperates with all things according to accurate laws, as being the foundation and cause of the whole of nature, except where it is good to act otherwise.'[55] On this account physical laws are not immanent within but imposed from outside upon nature by the all-powerful will of God. At the same time God can arbitrarily choose to change the laws. Hence it is incumbent on the physical scientist constantly to check experimentally that the laws as so far articulated by science remain in force. This creates a theological as well as a scientific motive for experimental knowledge, whose purpose is repeatedly to examine the 'constancy of the activities and laws of nature' as a way of judging the 'currents of the divine will as it replenishes a decaying creation.'[56]

Newton's theology influenced his cosmology in another profound way. Stephen Snobelen argues that Newton was a voluntarist and a Unitarian, al-

53. Janiak, 'Substance and Action in Descartes and Newton,' 668.

54. Simon Oliver, *Philosophy, God and Motion* (London: Routledge, 2005), 157-62.

55. Newton, *Memorandum*, David Gregory ms., 245, f. 14a, cited in Oliver, *Philosophy, God and Motion*, 159.

56. Oliver, *Philosophy, God and Motion*, 160.

though he could not publish such views beyond his disciples since he was a fellow of Trinity College and required to uphold Christian doctrine.[57] These theological positions leant him toward the view that behind the God that humans experience in space and time, and the Trinitarian God of the scriptures and creeds, there is a true God, or essence of God, who is ultimately unknowable apart from the actions of divine will in the natural order.[58] Revelation is therefore not a reliable glimpse into the divine life but a set of approximate descriptions of the absolute power of God manifest in the divine will.

Since for Newton God is not Trinitarian, then relationality does not exist in God but only in creatures. Consequently all theology becomes natural theology, since it is merely a weaker or 'vulgar' description of the mechanics through which the divine will determines the relations of beings and bodies in time and space, which are more precisely known through mathematics and experimental knowledge.[59] Scientific descriptions are certain and secure since they are *non-relational;* they do not depend upon the relation of the observer to what is observed.

Newton's mathematical descriptions of matter and motion, the force of gravity, and the movements of the planets reinvented physics and nurtured a new generation of mechanical engineers who invented the promethean machines that drove the age of steam. Steam engines turned coal from a dirty black rock that burned hot into motive power. Replacing wind, water, horse, and human muscle power, steam engines drove the industrial revolution. Steam also created mechanical motion of a kind that had not been available to any previous civilisation. With the advent of electricity, whose discovery and harnessing depended deeply on Newtonian physics, steam created a whole new kind of power. The availability of electricity is the defining infrastructural feature — along with drains and piped water — of developed societies, and its generation by steam-driven turbines remains the material backbone of industrial civilisation and the single largest source of climate-changing greenhouse gas emissions.

57. Snobelen notes that Newton's library contained many Socianian works which indicated that he was under the influence of the Polish unitarian movement. Further circumstantial evidence of his views is indicated by the fact that while in Newton's time all fellows of Trinity College were supposed to be ordained, Newton obtained royal dispensation not to be; Stephen D. Snobelen, '"God of Gods, and Lord of Lords": The Theology of Isaac Newton's *General Scholium* to the *Principia*', *Osiris* 16 (2001): 169-208.

58. Snobelen, '"God of Gods, and Lord of Lords"', 178-85.

59. Oliver, *Philosophy, God and Motion,* 160-61.

Newtonian physics, and the engineering and scientific innovations it inaugurated, not only gave to modern humanity tremendous material power over the earth. It also underwrote the Copernican decentring of humanity from nature, and nature and the earth from the universe. This decentring is deeply implicated in the cultural inability to respond appropriately to the natural signs which indicate that present human behaviours in burning coal, oil, and gas are changing the climate. Classical and Christian metaphysics set humanity at the centre of the universe, both spiritually and physically. From Aristotle to William Ockham, the human being is the microcosm of the macrocosm, and what occurs in the physical universe beyond the human mind and body maps onto the relations of minds and bodies in human society. It is therefore reasonable for premodern people to envisage a connection between order, or disorder, among earthly beings and order, or disorder, in the heavens. But under the influence of Newtonian cosmology and modern physics, moderns now dismiss such connections as 'magical' or superstitious. Precisely as these scientific theories have given to humanity powers to reorder the heat exchange of the earth and the sun, they have underwritten a deep dualism between culture and nature. Newton's mechanistic cosmology, combined with the Cartesian account of a disembodied consciousness, shapes new cultural perceptions of human-nature relationships. These perceptions make the claim that the earth system might be influenced by human behaviour counter-modern, even anti-scientific, for Copernican science.

Whitehead's Repair of the Nature-Culture Divide

The modern philosopher who sought more than any other to overcome the Cartesian and Newtonian legacy of the nature-culture divide is Alfred North Whitehead. Across major works in mathematics, philosophy, and religion, Whitehead argues for the remapping of consciousness and cosmos, matter and mindfulness, organism and ontology, psyche and physics, back onto each other. The driving idea of Whitehead's *ouvre* is, as Murray Code pithily puts it, to 'put Life back into Nature.'[60]

In *The Concept of Nature,* Whitehead asks, 'what is nature?' and notes the dualistic nature-culture riposte 'what is in our perception':

60. Murray Code, *Process, Reality and the Power of Symbols: Thinking with A. N. Whitehead* (New York: Palgrave Macmillan, 2007), 114.

> The fire is burning and we see a red coal. This is explained in science by radiant energy from the coal entering our eyes. But in seeking for such an explanation we are not asking what are the sort of occurrences which are fitted to cause a mind to see red. The chain of causation is entirely different. The mind is cut out altogether.[61]

In the conventional split between natural substance and mental apprehension, scientific materialists conceive of matter and substance only mechanically, leaving philosophers or poets to describe them mindfully, or soulfully. Scientists explain a red sunset as molecules in the atmosphere refracting the sun's electrical rays. Philosophers then explain that what the mind perceives as red — the colour red, the quality redness — is not what science describes but the secondary quality or concept an observer uses to interpret sense experiences, albeit when the concept red is to a large extent shared with other minds. Red is not *nature*'s quality or idea, and neither is the beauty of the sunset.[62] What poets describe and Romantics valorise — babbling brooks, beautiful sunsets, cacophonous rainforests, colourful sunsets, the drumming of rain on roofs, the silence of snow — are secondary qualities, present in the mind, not in the material world.

The gap between primary and secondary qualities is both manifestation and root of the nature-culture divide. Ethics, or what humans ought to do, no longer arises from descriptions of the primary qualities of the material world. When scientists describe the primary manifestations of anthropogenic climate change — thinning Arctic summer ice, increased atmospheric heat and water vapour, ocean acidification — these acquire cultural power only when they are also described as secondary qualities — floods and storm damage to human homes, reduced outputs on farmers' fields, declining catches of shellfish. Whence arises the authority to link particular primary qualities to particular secondary qualities? What in some minds is clear evidence of redness, or anthropogenic climate change, in others is clear evidence of crimson, or that grapes grew in Yorkshire in the Middle Ages.

For Whitehead the post-Newtonian bifurcation of nature from mind sets up *two* natures, one apprehended and the other causing the apprehension:

61. Alfred North Whitehead, *The Concept of Nature* (Cambridge: Cambridge University Press, 1920), 41.

62. Isabelle Stengers, *Thinking with Whitehead: A Free and Wild Creation of Concepts*, trans. Michael Chase (Cambridge, MA: Harvard University Press, 2011), 33, 37; I am much indebted to Stengers's luminous new reading of Whitehead in the account which follows.

> The nature which is in fact apprehended in awareness holds within it the greenness of the trees, the song of the birds, the warmth of the sun, the hardness of chairs, and the feel of the velvet. The nature which is the cause of awareness is the conjectured system of molecules and electrons which so affects the mind as to produce the awareness of the apprehension of nature.[63]

Against the division of mind and matter of the Newtonian schema, scientists and philosophers should attend more closely and simultaneously to 'sense' and to 'perceptual' and 'scientific' objects.[64] When the ethologist recognises that birdsong has a communicative value to the bird, she is so attending; she refuses the claim that only human minds think 'song' whereas birds merely follow genetic instincts.[65]

For Newton and his heirs, time and motion, atoms and bodies endure in the 'container' which is empty space, and the categories time, space, matter, and motion refer to material realities that exist independently of the apprehension of them by human or divine minds. Against this, Whitehead argues that time is more than the endurance of matter, more than the connection between different states of material relations. This more than material quality of time is indicated in the human capacity to experience time not only as the endurance of things past but also as the coming into being of the future in the present: the sense that a new day is new, while the material relations in which it is embedded are nonetheless approximately as they were the day before.[66] The apprehension of the simultaneity of endurance and novelty, of relations in time between past and present, and present and future, is not confined to the human mind. It is also a feature of the reality of time, and hence time cannot be reduced to nature: 'In passage we reach a connexion of nature with the ultimate metaphysical reality. The quality of passage in duration is a particular exhibition in nature of a quality which extends beyond nature.'[67]

In *Process and Reality,* Whitehead argues that his account of time corrects both Descartes and Hume on time. Descartes emphasised change — that things now are not the same as things as they were — as the peculiar

63. Whitehead, *Concept of Nature,* 31.

64. Stengers, *Thinking with Whitehead,* 83.

65. Stengers, *Thinking with Whitehead,* 83

66. Stengers, *Thinking with Whitehead,* 53-55.

67. Whitehead, *Concept of Nature,* 55; Whitehead acknowledges a debt to Bergson for this conception of time.

feature that the idea of time involves in suggesting to the mind that it is the same as it was in previous moments, even though the parts the mind observes are constantly changing. Hume emphasised endurance as the feature to which the idea of time refers, since 'time is composed of parts that are not coexistent' and therefore time itself cannot give human beings the idea of time. For Whitehead the key is to recognise that '*experience* involves *becoming,* that *becoming* means that *something becomes,* and that *what becomes* involves *repetition* transformed into *novel immediacy*'; only when being and becoming, novelty and repetition are held together — as they are in nature as well as in minds — can the shared assumption of Descartes and Hume of 'the individual independence of successive temporal occasions' be challenged.[68]

The relevance of this to climate change may be underlined. Most contemporary scientists claim something that their Newtonian forebears would have found it hard to accept, to wit that weather events now, and in fifty years' time, are related not only to *natural* temporal variability in the enduring climate system but also to variability in humanly generated greenhouse gas emissions. The climate, in other words, is becoming new, even while particular states of climate — weather — at particular moments may repeat particular states of weather in particular places in the past. The idea of anthropogenic climate change challenges the relations of nature and culture in space and time that are fundamental to the Newtonian schema. It suggests that there exists a relational continuum between weather states within the climate system, which includes bodily behaviours and human perceptions of the impacts of those behaviours, and not just contiguous atomic states of atmospheric molecules; to recognise that the climate might be novel calls forth a different perception of time and space where 'bodily life' is 'inside present duration' rather than bodies, time, and space being atomically distinct categories of existence.[69] Bodily life is constituted by relations and also *constitutes* them: the mind is and endures within bodies and participates in these relations.

If the category time is reducible neither to an idea in the mind nor to an immanent category in nature, but is one that transcends them both while also participating in the endurance of minds and matter and the sense of endurance experienced by minds, then the same is true for the category space. In the preface to *Process and Reality* Whitehead lists nine modern

68. Alfred North Whitehead, *Process and Reality* (New York: Free Press, 1978), 136-37.

69. Stengers, *Thinking with Whitehead,* 68.

habits of thought that he intends to repudiate, including the 'Kantian doctrine of the objective world as a theoretical construct from purely subjective experience.'[70] Against this he intends to find a way to bring the Platonist cosmology of the *Timaeus* into dialogue with the seventeenth-century cosmology of Galileo, Descartes, Newton, and Locke. He subsequently adds a third overall essential aim, which is to create the possibility to think past memories about past events in a way that admits that the *experiences* of these events endure in the present constitution of minds and bodies and of the cosmos.[71] This is particularly relevant to the problem of anthropogenic climate change, which concerns past emissions and their present effects and the temporal lag between emissions and effects.

If sense objects are not attributes of objects in space and time, then how are they related to 'real' material objects? The traditional answer is that through their sense objects human beings *participate* in the material objects which occasion them.[72] But this is not language Whitehead adopts, and this would shortcut his own answer to the question. For Whitehead there is no going back before Descartes, Newton, and Kant. A simple correspondence theory of truth will no longer do. Things as we see them and things as they are *are* different: the sun may appear to human sense perception to move in the sky, but Copernican astronomy reveals that the earth moves continuously around the sun and on its own axis and this is why the sun appears to move across the sky. Things are differently experienced when heard or touched or smelt and when they are examined through a telescope or displayed as data or visuals on a computer screen. Instead of going back, the scientific key to overcoming the bifurcation of sensory and scientific knowledge, nature and culture, is to examine the biological *and* metaphysical implications of post-Newtonian physics. For Whitehead there are two crucial ways of thinking — one ancient and one post-Newtonian — that need to be combined in order to do this. The first is the metaphor, and reality of, organism, which is the central concept in *Science and the Modern World.* The second is Einstein's theory of relativity, which Whitehead treats in *Process and Reality.* He gives to both a new twist.

First, organism. In *Science and the Modern World,* Whitehead sets out to correct the wrong turn taken by the human sciences when philosophers

70. Whitehead, *Process and Reality,* xiii.

71. Code, *Process, Reality and the Power of Symbols,* 87.

72. Owen Barfield, *Saving the Appearances: A Study in Idolatry,* 2nd ed. (Middletown, CT: Wesleyan University Press, 1988).

describe mental concepts, and social scientists describe 'social facts', by analogy with the Newtonian descriptions of matter, time, and space and the 'laws' which govern them. This wrong turn produces what Whitehead calls the 'fallacy of misplaced concreteness'. Economists' and politicians' faith in the effectiveness of 'markets' as instruments of human governance illustrates this fallacy. Economists model human exchange relations on the putatively 'natural law' of price and the behaviour of supply-and-demand curves. Exchange and productions modelled on competitive behaviours in markets which conform more closely to the idealised supply-and-demand curves of economistic models are said to be more 'efficient' than other kinds of exchange and production relationships. But when the markets behave in unpredictable ways, such as in the global financial crisis of 2008, which very few economists anticipated, this does not lead economists to suspect their 'facts' because these are modelled on physical facts and so are beyond question.[73]

Whitehead's critique of misplaced concreteness as an outcome of Newtonianism in the human sciences is central to the climate change dilemma. Fossil fuelled economic growth is said to be the origin of increased human flourishing in industrially developed societies. Climate change scientists argue that continued growth in the use of fossil fuels to sustain growing production of industrial goods and services is inconsistent with a stable climate. Social scientists respond by *underwriting* the market-led strategy of growthism when they identify market-based solutions to the collective action problem of climate change in the form of 'carbon emissions trading', on which more below.

Economistic, market-led approaches to the politics of climate change exemplify the 'enfeebled thought' which Whitehead identifies as the malign impact of Cartesianism and Newtonianism on the human sciences. The new scientific rationality which promised liberation from the narrow scholasticism of the late Middle Ages produces a new narrowness of academic disciplines and professions that do not interact, so that engineers build bridges without reference to aesthetics and social scientists build emissions markets that do not reduce climate-damaging fossil fuel extraction.[74] These disciplinary straitjackets are the poisoned fruit of an educational re-

73. See also Herman Daly and John B. Cobb, *For the Common Good: Redirecting the Economy toward Community, the Environment and a Sustainable Future,* 2nd ed. (Boston: Beacon Press, 1994).

74. Alfred North Whitehead, *Science and the Modern World* (New York: Free Press, 1967), 196.

gime that imparts information but kills wonder, that promotes misplaced concepts but kills life.[75] The problem originates in scientific reductionism, which forces a split between scientific facticity and social and moral ends. Thinking physical reality without ends promotes the modelling of human behaviour without ends, and underwrites the nature-culture divide.

Whitehead attempts a repair to scientific reductionism in the natural and human sciences through an empirical method articulated around the idea of 'prehensile occasions', which are 'the most concrete finite entities' and represent the coincidence in the myriad events that constitute the world from its beginning until now, 'the operation and production of reality.'[76] Human beings have the capacity to take into account at the same time the possibility of action, thoughts about actions, and the relations of actions to material objects and context, a capacity which is named mindfulness. Whitehead suggests that every really existent part of reality possesses this capacity in some degree, and that the unification of these occasions represents the process by which myriad individual events move reality from 'nature' to the 'order of nature'. The starting point for this account of process is

> the self-knowledge of our bodily event. I mean the total event, and not the inspection of the details of the body. This self-knowledge discloses a prehensive unification of modal presences or entities beyond itself. I generalise by the use of the principle that this total bodily event is on the same level as all other events, except for an unusual complexity and stability of inherent pattern.[77]

The value of nature which the poet perceives but science misses is grounded in the unity of all prehensive occasions which together constitute the organic processual character of all that is. As Stengers puts it, for Whitehead,

75. Rachel Carson gave a similar critique of the role of disciplined and reductionist knowledges in killing wonder; see further Lisa Sideris, 'The Secular and Religious Sources of Rachel Carson's Sense of Wonder', in Lisa Sideris and Kathleen Dean Moore, eds., *Rachel Carson: Legacy and Challenge* (Albany: State University of New York Press, 2010), 232-50.

76. Whitehead, *Science and the Modern World,* 196.

77. Whitehead, *Science and the Modern World,* 73. By psychological field Whitehead refers to the philosophy of William James, whose account of psychological experience Whitehead believed resolved in a decisive way problems opened up by Cartesianism; the essay of James to which Whitehead refers is 'Does "Consciousness" Exist?'; see further Stengers, *Thinking with Whitehead,* 154-55.

> the endurance of things has its significance in the self-retention of that which imposes itself as a definite attainment for its own sake. That which endures is limited, obstructive, intolerant, infecting its environment with its own aspects. But it is not self-sufficient. The aspects of all things enter into its very nature. It is only itself as drawing together into its own limitations the larger whole in which it finds itself.[78]

For Whitehead, experience is something that is shared by all entities, from atoms and cells to planets and stars. Subjectivity is present in entities in that they respond locally and relationally to the state of things around them at each moment. Hence for Whitehead, as Francisco Benzoni argues,

> There can be no ontological divide between human beings and other creatures. All metaphysically fundamental entities . . . are 'subjects' and as such have intrinsic value. Subjectivity characterizes all levels of reality, and so the world cannot be divided into . . . principals and instruments.[79]

Whitehead makes a distinction not between subjects and objects, as Descartes and his heirs do, but between subjectivity and consciousness. While consciousness is reserved for a narrow class of beings — for Descartes humans, for some others humans and other animals — subjects are centres of experience, and they adjust their behaviours in the light of their experiential encounters with other entities in their locality. But this account of experience makes mind-body dualism incoherent, since all true individuals that together construct reality are capable of responding to their experiences of previous states in the way they behave in successive states.[80] The process of response to previous states occurs in every organism, and by extension in every entity that endures in the world, because each is relationally situated toward other organisms:

> The concrete enduring entities are organisms, so that the plan of the whole influences the very characters of the various subordinate organisms which enter it. In the case of an animal, the mental states enter into the plan of the total organism and thus modify the plans of the succes-

78. Stengers, *Thinking with Whitehead*, 157.

79. Francisco Benzoni, *Ecological Ethics and the Human Soul: Aquinas, Whitehead, and the Metaphysics of Value* (Notre Dame, IN: University of Notre Dame Press, 2007), 129.

80. Benzoni, 'The Moral Worth of Creatures', 8.

sive subordinate organisms until the ultimate smallest organisms, such as electrons, are reached.[81]

Whitehead's project involves the recovery and repair, in the light of modern scientific discoveries including relativity theory and ecosystem theory, of the ancient cosmology of *organism* which Newton, Descartes, and their heirs displaced with mechanism. For Whitehead values arise in the 'plans' of all individual entities, and this is possible because each entity is not contained, as Newton argued, in a space-time continuum where space and time are non-relational. Instead, every individual entity is in a process of continuous learning from and response to other entities, such that 'every body defines its own spatio-temporal stratifications, its own discrimination of what space and time are.'[82] Hence subjective experience is implicated in the temporal endurance of atoms, biota, and cells as a *factor* in that endurance. And hence the composite of beings in the universe, or in a body, can be properly described only as being in the process of becoming and not as a collation of fixed or stable entities.

The power of Whitehead's resistance to Cartesianism, and to the Newtonian mathematisation of physical reality, is revealed in the way he uses the word 'society' to describe microscopic and macroscopic levels of reality. For Whitehead ecosystems are *societies,* which persist just to the extent that micro- and macro-organisms within them conform to a certain set of behaviours, a 'social order', and hence make possible endurance within the system, for 'every society must be considered with its background of a wider environment of actual entities, which also contribute their objectifications to which the members of the society must conform.'[83]

For Whitehead organism rather than mechanism is the driving metaphor of metaphysics. To describe the earth as an organism, electrons as organisms, and human beings as organisms denies the non-relational categories of time and space, and the quest for scientific certainty of the laws that determine the behaviours of particles and other moving objects within them. This is not to deny that within their specific spheres of operation such laws are identifiable, and hence that the engineering of a concrete and steel structure, or a modified gene, depends upon this sphere-specific knowledge. But it is to deny that this knowledge is sufficient to explain

81. Whitehead, *Science and the Modern World,* 79.

82. Stenger, *Thinking with Whitehead,* 168.

83. Whitehead, *Process and Reality,* 8; cited in Benzoni, 'Moral Worth of Creatures', 8.

the behaviour of genes or electrons in relation to other fields, such as the ecosystems in which they are situated, or the data sets and the observers who interrogate them through microscopes or computer models. Hence an ugly bridge for Whitehead is an offence both psychological and ecological, even if the physical mechanics used in its creation are correct in a sphere-specific way.[84]

Is this simply a 'parts and wholes' problem? Well, yes and no. Yes, in that sphere-specific knowledge has its place — or better, its spaces — in the modern world. No, in that the reductive nature of mechanistic sphere-specific scientific knowledge involves the misconstruction of the meaning of life in a metaphysical sense, and misidentifies questions of meaning and value with mathematical calibrations and statistical probabilities. Whitehead is not saying that if only *scientists* would develop a unified field of knowledge then all will be well with the modern world. The modern world is misshaped by the form of rationality that mechanistic sphere-specific scientific knowledge has advanced in *all* spheres of knowledge and action, and not just the scientific. This misshaping is at the heart of the ecological misalignment of modern rationality to the biological and physical environment of the earth and the solar system, of which the inability to alter habitual behaviours in the light of anthropogenic climate change is an extreme example.

Dualism between mind and body is a central feature of the rationality advanced by mathematisation and the cosmology of mechanism, but so too is the dualism between minds in bodies from their environments. This dualism can only be overcome for Whitehead by the recognition that bodily endurance is event specific. Minds reside in bodies and mindful behaviours appear to constitute their flourishing. But bodies endure not because minds make them but because they are part of a larger body, 'nature', that they are never truly separate from, neither in life nor in death. Water and air constantly pass into and out of the human body. Without these the mind would not endure. That the mind-body of human beings relies on this natural substrate makes it hard to answer the question of where the body stops and the environment begins. That moderns seem to desire a secure answer to this question is the peculiar nature of modern rationality that Whitehead sets out to repair. For Whitehead the problem is the inability of modern rationalities — in the natural and human sciences — to recognise the given *and* processional character of the organic environment, and the constitutive

84. Stengers, *Thinking with Whitehead*, 169-84.

role in the creation and maintenance of that environment of *experience* in the connective field of myriad concrete events, including the endurance of electrons and microorganisms as well as rational agents.

The Fallacy of Misplaced Concreteness and the Enduring Kingdom of Coal

If coal — and fossil fuel extraction and use more broadly — has advanced an attitude to matter and the earth as dead, insensible, and lacking in consciousness or purposiveness, then optics made from coal and the subsequent Copernican and Cartesian turns undergirded this attitude during and after the scientific revolution. But there is no getting away from the fact that, in a science-informed culture, nature and human perceptions of natural objects and events no longer correspond and that this produces an alienation between nature and humanity which is genuinely novel in human history. To the primitive mind, such as present-day inhabitants of the rainforest in Pahang, Malaysia, when the climate warms and rivers flood because of human activities such as burning forests and fossil fuels, the forest is destabilised. And not only the forest, but human dwelling beyond the forest, is put at risk.[85] To post-Cartesian and post-Newtonian scientists, when the primitive mind ascribes animality to species, the wind, or the sun, which for the scientist lack purposiveness, this is animism. Modern scientific representations of reality as material substrate lacking agency or subjectivity have a factual purchase on reality which is stronger than primitive conceptions. Hence the 'primitive' fear that human disturbance of the given order of nature, such as the destruction of a tropical rainforest or the burning of a mountain of coal, will disturb human purposes is dismissed as irrational superstition.

After Copernicus and Descartes, there is no correspondence between the given order of nature and how humans are to live in it. Science-informed technological interventions from rocky substrate to the atmosphere in the earthly habitat might offend human sensibilities, but they do not disturb the Cartesian claim that matter and substance lack agency and only acquire meaning when re-created by human beings for human rational purposes. Hence a civilisation devoted to the amassing of energetic power over nature

85. See further Lye Tuck-Po, *Changing Pathways: Forest Degradation and the Batek of Pahang, Malaysia* (Oxford: Lexington Books, 2004).

in and through the mining of coal, and later oil and gas, does not desist from this path because it is damaging 'nature' since nature is no longer agential, experiential, or subjective; it is no longer neighbour in the human neighbourhood but instead is a stable material backdrop to human culture and creativity. And this is especially true of the air. Philosophers such as Tom Regan have developed sufficient empirical knowledge of other animals to reduce the circle of beastly disregard in which scientists reject the claim that animals are animated by a spirit analogous to that which animates humans. Philosophers such as Holmes Rolston extend subjective recognition of the intrinsic value of persons to species and ecosystems.[86] But no philosophical strategy of intrinsic value imputation that does not also reject the subject-object divide of modern science can recover the sensibility of an animist, or a Hebrew, that the same spirit — *nephesh* — that moves warm-blooded mammals moves the wind.

The Anthropocene reveals the impossibility of the dualisms of the Cartesian and Copernican turns in a way that philosophy could not. When in the context of a 50 percent increase of carbon dioxide levels over pre-industrial levels in the atmosphere we experience extreme weather, we can no longer dismiss the weather as non-agential. Even if angels do not direct the storms, human activity might be driving them. Similarly, when a carbon atom from new Arctic ice is studied under an electron microscope, the scientist may be looking at a carbon atom that has come from the fossil fuels burned in the airplanes that keep the Arctic laboratory supplied with diesel oil and food. Even if it was possible before the Anthropocene to deny that there is subjectivity in atoms and air, after the Anthropocene it is no longer possible to deny this. Climate science therefore requires a post-Copernican turn, and it is just such a turn that Whitehead's neoclassical metaphysics offers. And Whitehead also offers an insightful account of the fallacious misplaced concreteness of industrial development and economic growth without limits, which, even as the Anthropocene unfolds, continues to characterise the energetic reconstruction of the earth's atmosphere through the mining and burning of coal.

Coal remains front and centre in the causation, and potential mitigation, of global warming. There are currently approximately 2,300 coal-fired electric power stations operating around the world. The nations are planning to build *2,000* new coal-fired plants in the next ten years, the majority

86. Holmes Rolston, *Environmental Ethics: Duties to and Values in Nature* (Philadelphia: Temple University Press, 1988).

in China and India. Annual coal exports from Indonesia, Australia, Vietnam, Russia, Canada, and the United States, which together amounted to 2.1 million metric tons in 2000, increased to *176 million metric tons* in 2010, with 100 million metric tons from Indonesia and Australia alone.[87] This quantity of burned coal is sufficient, if it continues for just ten years from 2020 to 2030, to move the planet into an unstoppable warming event which will see temperatures rise in the next fifty years in excess of three, four, five, or even six degrees Celsius.

Instead of resisting the rapid rise of coal extraction, export, and burning in the countries responsible, climate change scientists, lobbyists, and activists focus their efforts on a United Nations negotiating process, the main outcome of which — the Kyoto Protocol — does not even refer to fossil fuel *extraction.* By agreeing only to limit their domestic greenhouse gas *emissions,* national signatories to the Protocol are still able to increase their extraction and export of fossil fuels while continuing to import goods which contain the embedded carbon they claim no longer to produce. The UK government claims to have more than met its Kyoto Protocol greenhouse gas reduction targets, with domestic carbon production falling 15 percent between 1990 and 2005. But UK consumption of embedded carbon emissions increased by 19 percent in the same period.[88] This is because the UK exported its dirtier industries, visiting depression and unemployment on domestic working-class communities. At the same time the UK permits the City of London and retail banks to sustain a debt-fuelled consumer boom to enable its citizens to continue to consume the manufactured goods and construction materials they no longer have a hand in making from China. The latest stage of the United Nations Framework Convention on Climate Change (UNFCCC) treaty process, at time of writing, is that the UNFCCC signatories agreed in Durban in 2011 merely to the intention to agree by 2015 on country-by-country greenhouse gas reduction targets to commence in 2020 for all signatory nations. But if coal extraction and burning doubles by 2020, the core aim of the UNFCCC — to prevent dangerous destabilisation of the climate — will have been lost. The UNFCCC treaty process is therefore a supreme example of the fallacy of misplaced concreteness.

As Dieter Helm argues, European politicians and economists have attempted to fool themselves, and the public, that climate change mitigation

87. Dieter Helm, *The Carbon Crunch: How We're Getting Climate Change Wrong — and How to Fix It* (New Haven: Yale University Press, 2012), 41.

88. Helm, *The Carbon Crunch,* 7.

can be achieved without significant *sacrifice* in consumption patterns or reductions in economic growth. Thus the UK government–commissioned Stern Report argued that the cost of decarbonisation would be as little as 2 percent of gross domestic product if begun before 2020.[89] But in reality to reduce emissions means to reduce consumption. There is no currently feasible way to decouple the smelting of aluminium and steel, the making of cement, the manufacture and fuelling of airplanes, cars, computers, domestic appliances, entertainment devices, and mobile phones from coal and oil since all these industries rely on the intense carbon embedded in fossil fuels. If such fuels are no longer burned in Europe, while *consumption* remains unchallenged, they will be burned elsewhere.[90] Genuine climate change mitigation requires reduced consumption. There is no way around this.

The unavoidable means to climate change mitigation in the short term is ending the burning and extraction of coal. Just ten years of ongoing consumption of coal at current rates of expansion will likely tip the planet beyond the point of no return. Far from promoting an end to coal, Kyoto Protocol signatory countries *promote* coal plants not only by commissioning new domestic coal power and licensing increased coal extraction, but by providing World Bank development funds to South Africa, India, and other developing countries to build new coal plants as aids to industrial development. They also permit fossil fuel companies to include quantities of coal, which cannot be burned without taking the planet beyond four degrees of warming, in their asset portfolios and stock exchange valuations.

The banning of the export, extraction, and use of coal, and the imposition of punitive taxes on coal and coal-derived products to facilitate this, has to proceed country by country in the *present,* not the future.[91] The only conceivable political and moral authority for such action is that of national governments. But there can be no free-riders. If by 2020 Indonesia, Australia, Mongolia, Vietnam, Russia, Canada, and the United States — the principal coal-exporting countries — have not ended coal exports, and if developing as well as developed countries have not mothballed most coal-based electric power plants, there is little chance of avoiding the more extreme climate change events outlined in chapter 1 above.

89. Nicholas Stern, *The Economics of Climate Change: The Stern Review* (Cambridge: Cambridge University Press, 2007).

90. Helm, *The Carbon Crunch,* 32-56.

91. I argue this in more detail in Michael S. Northcott, *A Moral Climate: The Ethics of Global Warming* (London: Darton, Longman and Todd, 2007).

To argue that the nations severally and then cooperatively, rather than individual consumers or corporations, have the prime responsibility for acting *now* on carbon mitigation requires a narrative of the ecological boundaries of the nation-state and ecological limits on its reengineering of the environment of the kind I develop below. But at the origin of the scientific revolution, Francis Bacon envisaged expansion in the nation's powers *over* nature as the root of the salvific potential of the nascent nation-state of Elizabethan England and its emergent empire. For Bacon knowledge as power and control over nature has the potential to redeem the human condition from suffering. Many now argue in Baconian mode that, since politics cannot repair climate change, scientific interventions in the atmosphere can. Against Baconianism, the Italian Giambattista Vico argued that the dangers as well as potential of the new science could be properly understood only through cultural knowledge of the providential ordering of peoples from the first civilisations to the scientific era. Through a conversation between Bacon and Vico, I argue in the next chapter that the objectifying nature of scientific consciousness, apart from 'prehensive' participation, and the related turn to misplaced concreteness in the human sciences can be repaired only through a political theology of the ecological and spiritual roots of cultures and nations.

3. Engineering the Air

After coal mining and coal-based stationary power, the next greatest humanly generated source of greenhouse gas emissions is the oil industry and the ubiquitous networks of pipelines and grids, cars, trucks, ships, and planes that oil enables and sustains. Although down from the peak, oil still constitutes 35 percent of global energy use and 95 percent of vehicular mobility energy.[1] While oil is substitutable by coal, natural gas, nuclear power, and renewables in electricity generation and heating and cooling systems, it remains an intrinsic component of modern mobility and speed. In many countries oil-based mobility has become essential for travel to work, college, the food store, and even religious worship. In many more, oil-fuelled shipping brings food and other essential commodities to the store and provides an essential link to foreign markets for farmers and businesses. Oil is the principal source of nitrogen for agricultural fertiliser and herbicides and pesticides, and oil-fuelled ships and vehicles mine and transport potassium and phosphates, the other essential artificial fertilising elements. Oil is the feedstock for plastics which are ubiquitous in most societies, constituting everything from computers and drinks containers to drainage pipes and medical supplies. Oil also fuels increasing levels of professional air travel, air cargo, and travel for tourism, though most of this is nonessential and can be discounted as luxury emissions.

1. Some argue that peak oil and not climate change will drive the transition to a sustainable post–fossil fuel society, e.g., Colin Campbell, *The Golden Century of Oil, 1950-2050* (Dordrecht: Kluwer Academic, 1991), but there is sufficient unconventional oil in shale, in deep water and near the Poles to sustain growing oil consumption into the second half of the present century; Robert M. Mills, *The Myth of the Oil Crisis: Overcoming the Challenges of Depletion, Geopolitics, and Global Warming* (Westport, CT: Praeger, 2008).

The Geopolitics of Oil

Given the ubiquity of oil in industrial societies and the very great dependence of the United States, as well as much of Europe, on oil in fundamentals such as agriculture and infrastructure, and the growth of oil-related infrastructure in the rest of the world, oil figures very high in world geopolitics and in the geopolitics of climate change. Daniel Yergin argues that oil has constituted the material base of geopolitics, and in particular the imperial power of the United States, since the late nineteenth century.[2] This is in part because the United States was the first oil economy, much as Britain was the first coal economy. The heavy consumption of oil in the United States was such that its corporations were driven after the Second World War to look overseas to replace rapidly diminishing domestic supplies. The United States used 60 percent of the world oil supply until the 1960s and still uses more than a third of global oil output.

The search for oil became the key driver of colonial expansion in the twentieth century. Much of this was under the aegis of two very powerful companies — Standard Oil (now Exxon/Esso) and Royal Dutch Shell Oil — which not only had great influence on colonial policies but also great power over and influence with national governments.[3] Oil and war were also closely associated in the twentieth century and beyond. Access to oil in the great Baku oil fields, and then in the Middle East, was a major factor in the First and Second World Wars, and oil has continued to dominate the politics of the Middle East.[4] The Second World War also saw a new kind of oil-fuelled war with Hitler's *blitzkrieg* and the oil-fuelled indiscriminate aerial bombing, by both Germany and the Allies, of cities and civilian infrastructure.

In the developing world, there are few nation-states with major oil reserves which have not been marred by systemic government corruption, civil conflict, and in many cases all-out war.[5] Michael Watts argues that

2. Daniel Yergin, *The Prize: The Epic Quest for Oil, Money, and Power,* updated paperback edition (New York: Free Press, 2008).

3. Yergin charts many instances of corrupt influence by Standard Oil — and John D. Rockefeller — over American politics in *The Prize.*

4. Bill Richardson, U.S. Secretary of Energy, in 1999 claimed that 'oil has literally made foreign and security policy for decades'; as quoted in Mary Kaldor, Terry Lynn Karl, and Yahia Said, eds., *Oil Wars* (London: Pluto Press, 2007), 'Introduction', 1-40.

5. George Philip, *The Political Economy of International Oil* (Edinburgh: Edinburgh University Press, 1994).

oil, criminality, and violence have been associated since the origins of the industry, and that oil has largely determined the shape of imperial relations between developed and developing countries in the twentieth century.[6] Oil is also associated with the 'resource curse' afflicting nations over-dependent on the extraction of one natural resource or commodity: when oil takes up more than 20 percent of the gross domestic product of a nation, the chances of civil conflict or all-out war increase by approximately the same percentage.[7]

Oil, more than coal, creates and sustains a 'politics of speed' and extreme mobility which is the cause of a very high number of 'accidental' deaths in the developed and developing worlds.[8] Speed is morally problematic not just because of the number of deaths caused by speed — including nonhuman species as well as people — but also because of the ecological and social dislocation associated with it. Modern air and sea freight rapidly move vast quantities of goods, raw materials, and species around the planet. This mobility of goods underlies the continuing expropriation of the natural resource wealth of rich from poor, and developed from developing nations, and the accelerating rate of species extinction. It also provides the technical means for developed world economies to draw upon an 'ecological footprint' of goods and services many times their land area.[9]

The modern 'global economy' is essentially an oil-fuelled economy.[10] Without oil the present scale of movement of people and goods could not be sustained. Most economists since Adam Smith have argued that the wealth of nations can be advanced only through the global exchange of goods and the associated international division of labour. That this exchange relied in Smith's time on colonial expropriation did not disturb Smith, though it has troubled some of his heirs. To the spatial inequities involved in the expropriation of other people's environments is now added a new kind of spatial

6. Michael Watts, 'Petro-Violence: Community, Extraction and Political Ecology of a Mythic Commodity', in Nancy Lee Peluso and Michael Watts, eds., *Violent Environments* (Ithaca, NY: Cornell University Press, 2001), 189-212.

7. Paul Collier, *The Economic Causes of Civil Conflict and Their Implications for Policy* (Washington, DC: World Bank, 2000).

8. See further Paul Virilio, *Speed and Politics,* trans. Mark Polizzotti (Berkeley: University of California Press, 2007).

9. For a fuller discussion of Virilio and the politics of speed, and accidents, see Michael S. Northcott, 'The Desire for Speed and the Rhythm of the Earth', in Sigurd Bergmann and Tore Sager, eds., *The Ethics of Mobilities: Rethinking Place, Exclusion, Freedom and Environment* (Burlington, VT: Ashgate, 2008), 215-32.

10. Yergin, *The Prize,* 112-64.

expropriation in which the means for sustaining spatial expropriation also involves atmospheric expropriation. This new form of spatial expropriation will impact future generations, as well as presently existing people, and particularly in the tropical and subtropical regions that themselves have been the principal domains for colonial expropriation.

Added to the climatological costs of oil are the direct ecological costs of oil production and consumption. Spillage from oil production and transportation is of note politically only when it affects iconic natural areas or developed country lands or oceans. Thus significant pollution incidents in North America — including the Yellowstone Valley in Wyoming, Prince William Sound in Alaska, and the Gulf of Mexico — attracted much media coverage. However, to give just one example, more oil is spilled *annually* in the Niger Delta — which contains Sub-Saharan Africa's largest easily accessible deposit of oil — than was spilled in the well-publicised *Deepwater Horizon* blowout in the Mexican Gulf in 2010. Oil from leaking pipes and wells has polluted large areas of the Niger Delta, spreading from surface waters and lands deep into the water table; this sixty-year environmental catastrophe has greatly reduced life expectancy in the Delta and has led to growing levels of violence and social collapse.[11] Despite fines and extensive local protests, the oil companies have made no efforts to clean up the pollution.[12]

Pollution and ecological destruction associated with conventional oil production are magnified significantly in the new kinds of oil exploration and production that the industry is now pursuing as it seeks to utilise less easily accessible oil sources. Novel technologies associated with deep water oil extraction three miles below the surface of the ocean were implicated in the Gulf of Mexico oil blowout and ongoing pollution incident in 2010. In Indonesia, drilling in an earthquake zone occasioned the world's first humanly created volcanic flow of hot subterranean mud.[13] Shale oil and gas are far more polluting than conventional oil extraction. Bitumen sands

11. For a firsthand account of the tragedy in the Niger Delta, and a full range of references, see the thesis of my PhD student Nkem Osuigwe, 'Crude Oil, Conflict and Christian Witness in Nigeria: Baptist and Pentecostal Perspectives' (University of Edinburgh, 2010), at http://www.era.lib.ed.ac.uk/handle/1842/4016.

12. Ike Okonta and Oronto Douglas, *Where Vultures Feast: Shell, Human Rights, and Oil* (London: Verso, 2003).

13. Michael S. Northcott, 'Anthropogenic Climate Change and the Truthfulness of Trees', in Sigurd Bergmann and Dieter Gerten, eds., *Religion and Dangerous Environmental Change: Transdisciplinary Perspectives on the Ethics of Climate and Sustainability* (Berlin: LIT Verlag, 2010), 103-18.

— of which the largest deposits include those of Canada and Venezuela — contain as much oil as viscous oil deposits in sedimentary rock. But the extraction and refining of a barrel of viscous oil from sand requires three times the energy of conventional oil, as well as considerable quantities of water since the oil has to be steamed or boiled out of the sand either in situ in the ground or after extraction. The Athabasca River is the principal source of water used in the extraction of oil, by steam, from Albertan tar sand; vast lakes of polluted water line the river for hundreds of miles and are the largest body of stored toxic water on earth.[14]

The status of oil as the *über* commodity of the twentieth century has not changed in the twenty-first. Michael Watts suggests that there is an intrinsic relationship between the mythic character of oil and the ecological, political, and social destruction associated with its extraction.[15] Participation in the oil economy by nation-states who possess and extract it involves the construction of a social structure of land rent and engagement with transnational forces which issues in a new kind of social contract. Oil is denominated in dollars, and its extraction creates a viscous flow of both liquid and monied wealth, which is largely one way — from extraction zones to local and global metropolitan centres. The rents associated with this extraction give legitimacy to a newly empowered and increasingly centralising state where wealth magically increases among the elites who control oil rents. The accumulation of wealth without effort by the renters corrupts the body politic by producing a concentration of power and wealth which extracts social power from local communities and in effect turns the national economy into the national equivalent of a petroleum company town. The local communities which lose most power in this process are those which are subjected to the negative costs of oil extraction. Since oil rents in many domains are claimed by the state as a subterranean mineral, regardless of who owns the land, the state becomes the object of conflict, often provoking civil war:

> Oil, then, simultaneously elevates and expands the centrality of the nation-state as a vehicle for modernity, progress, civilisation, and at the same time produces conditions that directly challenge and question those very same, and hallowed, tenets of nationalism and development.[16]

14. Kevin P. Timoney and Peter Lee, 'Does the Alberta Tar Sands Industry Pollute? The Scientific Evidence', *Open Conservation Biology Journal* 3 (2009): 65-81.

15. Watts, 'Petro-Violence', 189 and 204-7.

16. Watts, 'Petro-Violence', 208.

The *Deepwater* Gulf oil spill in 2010 occasioned the observation that the oil industry was recklessly drilling into 'mother earth' and 'spewing her guts' on the surface of the ocean.[17] Watts deploys the analogous metaphor of a vampire, sucking the blood out of the veins of the earth and so sucking the lifeblood out of the people who live on it.[18] The oil industry is a vampire in another sense, for the corporations established by the pioneering architects of the industry endure beyond the lifetimes of their founders because they possess a form of legal existence known as fictive personality. In this capacity they exercise the rights of persons but they lack the moral capacities that persons normally possess.[19] Consequently, oil companies 'see' in ways that are immune to human suffering and violence and are incapable of experiencing shame or compassion. In Africa and the Middle East they act collusively with corrupt governments and undermine civil peace and lawful order. Oil in these countries is extracted by enclaves of international oil workers who are detached from the local populace and often defended by private armies of security guards and military personnel. Thirteen thousand security personnel were employed by one oil company in Iraq to ensure that oil flowed even in the midst of national-scale conflict.[20]

The corporate mind exercises a hold on those who work in corporations analogous to spiritual possession. Company loyalty competes with other human loyalties and, where necessary, corrupts and subverts such conventional attachments. Oil-fuelled mobile technologies of satellite communications, helicopters, and jet planes facilitate this subversion, creating and sustaining company enclaves that are politically, socially, and even ecologically independent of the countries they are located in, with their own energy and filtered water supplies and sewage treatment facilities. The corporate mind can therefore contemplate sucking dry the veins of the earth without reference to the ecological or social conditions of the lands and oceans in which its agents operate. The same corporate mind is at work when oil corporations fund and promote confusion and controversy over

17. See further Michael S. Northcott, 'BP, the Blowout and the Bible Belt: Why Conservative Christianity Does Not Conserve Creation', *Expository Times* 122 (2010): 1-10.

18. Michael Watts, 'Economies of Violence: More Oil, More Blood', *Economic and Political Weekly* 38 (2003): 5089-99.

19. See further Michael Northcott, 'Artificial Persons against Nature: Environmental Governmentality, Economic Corporations, and Ecological Ethics', *Annals of the New York Academy of Sciences: The Year in Ecology and Conservation Biology* 1249 (2011): 104-17.

20. James Ferguson, 'Seeing Like an Oil Company: Space, Security, and Global Capital in Neoliberal Africa', *American Anthropologist* 3 (2005): 377-82.

climate change science in heavily oil-dependent countries such as Australia, Canada, Russia, Saudi Arabia, and the United States. The spirit of John D. Rockefeller — whose petrodollars corrupted state governments and funded an attempted coup in Washington, D.C. — lives on when Exxon funds an array of think tanks, scientists, media outlets, and lobby groups to debunk climate change science.

The United States since the days of John D. Rockefeller has been an oil state, as is contemporary Malaysia or Saudi Arabia. The interests of government and the oil industry are said to be seamless. Neither social justice nor ecological sustainability are of much interest to oil companies; they do not expect these to be treated as policy priorities above the continuing profitability of the oil industry by governments they influence. The future condition of the planet is of concern to oil companies only inasmuch as it presents opportunities for their continued profitability. This is why Shell Oil invests so extensively in geophysics and futures thinking, and employed at its Texas lab geophysicist M. King Hubbert, who invented the theory of 'peak oil'.[21] This also explains the contradiction Nicholas Stern identifies in market responses to climate change. The claimed fossil fuel reserves of oil companies are estimated in the trillions of dollars and appear on their balance sheets; if burned they would release approximately 319 billion metric tons of CO_2 and represent more than one-half a degree Celsius of global temperature change.[22] If claimed fossil fuel reserves cannot be used without destabilising the earth's climate, then it is doubtful that they should still be entered into estimates of corporate assets.[23]

While Hubbert's account of peak oil remains controversial, and especially because he neglected the potential of a rise in the oil price to generate new oil findings and oil extraction methods, conventional oil production peaked, according to the International Energy Authority, in or around 2008. After that date, oil from non-conventional sources including deep water, bitumen sands, and shale met growth in oil consumption, and by 2025 non-conventional oil will be the majority source.[24] Non-conventional oil has a climate footprint close to that of coal. As with coal, unrestrained

21. M. King Hubbert and Shell Development Corporation, 'Nuclear Energy and Fossil Fuel', *Drilling and Production Practice* 7 (1956): 1-19.

22. *Unburnable Carbon: Are the World's Financial Markets Carrying a Carbon Bubble?* (London: Carbon Tracker Initiative, 2012), 7.

23. Nicholas Stern, 'A Profound Contradiction at the Heart of Climate Change Policy', *Financial Times*, 8 December 2011.

24. International Energy Agency, *World Energy Outlook* (Paris: IEA, 2008).

deployment of this resource will produce greenhouse gas emissions so extensive as to commit the planet to unstoppable global warming.

Fossil fuelled climate change will be mitigated only when sufficient of the nations and corporations that extract and market fossil fuels decide to leave most of what remains in the ground. That there is a limit to energy production is counterintuitive to most modern economists, and hence to corporate and political leaders. Economists conventionally assume that any resource when it nears depletion will rise in price and that the price rise will stimulate alternative resources or technologies. But enough new energy supplies are still being identified to sustain present rising levels of energy demand, provided extreme weather does not disrupt long global supply chains, so rising prices in fossil fuels alone will not create a post–fossil fuel economy.

If oil has had a profound shaping influence on the history and political shape of industrial societies in the twentieth century, the materiality of oil is implicated in the character of this influence. Oil is not like coal. Its viscosity, and the kilojoules a quantity of oil represents, are unique compared to any other material substance. The concentrated energy of oil as a material substance is related to the capacity of this substance to generate concentrations of social and economic power that disable conventional human mores of respect for persons, for species, and for the condition of the soil, air, and water. The metaphor of oil as liquid gold indicates the close association between the liquidness of oil and the growth in monetary liquidity in the twentieth century. The rate of consumption of oil and gas, together with coal, correlates to the growth in the money supply and to the growth in GDP in almost every nation-state. If oil is money, and money buys increased prosperity and consumption, then it is for this reason that nation-states do not envisage reducing their use of energy in the face either of climate change or of peak oil. Instead, the dominant responses to the decline of conventional oil are the following: extract unconventional oil; turn gas and coal into liquids; substitute coal, oil, and gas with renewables such as solar, tidal, wave, and wind power; capture the carbon from fossil fuels and bury it in oil and gas fields. De-energising the economy is not part of the envisaged strategy. Nor is keeping fossil fuels in the ground.

Reengineering the Anthropocene

Given the failure of the nations to reduce their extraction and use of fossil fuels, some scientists, and many engineers from the oil and gas industries, propose that it will be necessary intentionally to engineer the earth system to reduce the heating potential of present and future greenhouse gas emissions. Geoengineering is defined as 'large-scale engineering of our environment to counteract the effects of changes in atmospheric chemistry.'[25] One of the first exponents of this approach, Brad Allenby, summarises the rationale for 'earth systems engineering' as follows:

> Anthropogenic perturbations of fundamental natural systems — for example, the nitrogen and carbon cycles and heavy metal stocks and flows — have been both unanticipated and problematic. Reducing such unintended consequences of human activity will require development of the ability to rationally engineer and manage coupled human-natural systems in a highly integrated fashion.[26]

Allenby notes that, at least since the 1980s, scientists have had the awareness that there is no part of the earth that is untouched by industrial engineering, and that, as a species now capable of transforming the planet, 'self-conscious intelligent management of the earth is one of the great challenges facing humanity as it approaches the 21st century.'[27] The 'emergence of an anthropogenic world' means that the human mark can no longer be ignored by politicians or citizens and requires a new human consciousness of 'responsibility for global systems.'

Allenby claims that humanity under the influence of scientific materialism has become aware of a responsibility for the state of the earth which humans formerly lacked, though as I have already argued it is only modern humans who fail to connect their actions to perturbations in the weather and other natural systems. Because of their 'new' influence on the planet, Allenby argues, humans have to engage consciously in active management and engineering of earth systems.[28] There are a range of geoengineering techniques under consideration which come into two categories: Solar Radiation Management techniques are intended to shade the earth from the heat of the sun and so reduce solar heating for a given quantity of atmo-

25. T. M. Lenton and N. E. Vaughan, 'The Radiative Forcing Potential of Different Climate Geoengineering Options', *Atmospheric Chemistry and Physics* 9 (2009): 5539-61.

26. Brad Allenby, 'Earth Systems Engineering: The Role of Industrial Ecology in an Engineered World', *Journal of Industrial Ecology* 2 (1999): 73-93.

27. W. C. Clark, 'Managing Planet Earth', *Scientific American* 261 (1989): 47-54; cited in Allenby, 'Earth Systems Engineering', 73.

28. Allenby, 'Earth Systems Engineering', 75-76.

spheric CO_2; Carbon Dioxide Removal techniques are intended to remove CO_2 from the atmosphere, or during the burning of fossil fuels, and either store them or convert them into other substances.[29]

Solar Radiation Management (SRM) involves efforts to reduce the warming potential of greenhouse gas emissions in the atmosphere by modifying the reflectivity of the atmosphere or the earth's surface. The most widely advocated form of SRM is the injection into the atmosphere of sulphates, which, Paul Crutzen argues, mimics volcanic activity and is therefore so close to nature as to have no long-run risks.[30] Against Crutzen, others point out that significant volcanic outputs of ash and sulphur into the atmosphere significantly reduce precipitation in agricultural areas in the Southern Hemisphere and especially those already at risk from drought-induced crop reduction, and will also reduce the productivity of solar panels, one of the most crucial renewable energy sources.[31] The National Centre for Atmospheric Research in 2007 showed that the Mount Pinatubo eruption in 1991, cited by Crutzen as a precedent for sulphates in the atmosphere, had dramatic hydrological effects, reducing precipitation, soil moisture, and river flow in many regions. Another study shows that China is already in effect masking the effects of global warming along the lines advocated by Crutzen with its dramatic increase in coal power plants and related emissions of soot and sulphur into the atmosphere between 1998 and 2008.[32] Atmospheric sulphate dissolves atmospheric ozone, which is a toxic pollutant to humans and other species at ground level. Atmospheric sulphates are washed out through clouds and deposited as acid rain on the earth's surface, causing further ecological harms. Increased sulphate levels may also seed cirrus clouds, which have a known warming effect. Sulphates also whiten skies, reducing blues and significantly enhancing the redness of sunsets, and these effects have both aesthetic and psychological implications.

Given the grave risks that sulphate injection carries for anyone depen-

29. Kelsi Bracmort, Richard K. Lattanzio, and Emily C. Barbour, *Geoengineering: Governance and Technology* (Washington, DC: Congressional Research Service, 2011).

30. Paul Crutzen, 'Albedo Enhancement by Stratospheric Sulfur Injections: A Contribution to a Policy Dilemma?' *Climatic Change* 77 (2006): 211-20.

31. J. Pongratz, D. B. Lobell, et al., 'Crop Yields in a Geoengineered Climate', *Nature: Climate Change* 2 (2012): 101-5.

32. Robert K. Kaumann, Heikki Kaui, Michael L. Mann, et al., 'Reconciling Anthropogenic Climate Change with Observed Temperature 1998-2008', *Proceedings of the National Academy of Sciences*, DOI: 10/1073/pnas.1102467108.

dent on marginal agricultural crop production for their livelihood — and survival — the consequential argument against this approach, despite its popularity and relative low cost compared to all other proposals, is extremely strong, as Stephen Gardiner argues.[33] Those who suggest that it is a 'lesser evil', and that, for example, it will 'save the ice' from melting if undertaken soon enough, neglect its impacts, and especially on the many people without access to stored wealth and grains who will likely go hungry if the project were undertaken at a scale sufficient to shade the earth by an amount sufficient to prevent the ice melting.

Less invasive SRM proposals include efforts to increase the solar reflectivity of the built environment through the grassing of roofs, tree planting to shade roads and city streets, and the use of white pigments in building, roofing, and road materials. Some of these techniques would have clear aesthetic and ecological benefits, in particular tree planting and grassing of roofs. Most of these techniques would help reduce the urban heat island effect, making cities cooler and reducing energy consumption for air conditioning, while grassing and greening would reduce water runoff. The use of white paint and light pigments on buildings is already widely practiced in Mediterranean villages. However, such albedo modification efforts would have only a modest effect in reducing global temperatures.[34] Painting roofs white is also relatively expensive for a very small quantity of reduced radiative forcing.[35]

Technological efforts to increase the solar reflectivity of glaciers and deserts are also proposed but are even more controversial and would be immensely costly. Covering glaciers or deserts with reflective materials, such as polyethylene-aluminium sheeting, would also represent a considerable aesthetic and ecological detriment to wild areas that humans and other species inhabit or visit and from which they derive benefit.[36] Such efforts would be more costly than reductions of black soot from existing pollution sources, especially coal-fired power stations and diesel engines, which are presently implicated in darkening Arctic ice and glaciers. Efforts to change

33. Stephen Gardiner, *A Perfect Moral Storm: The Ethical Tragedy of Climate Change* (Oxford: Oxford University Press, 2011), 347-52.

34. Lenton and Vaughan, 'The Radiative Forcing Potential of Different Climate Geoengineering Options', 5559-68.

35. The Royal Society estimates annual costs of $500 billion; *Geoengineering the Climate,* report by the Royal Society, September 2009, table 3.6, note (a), 35.

36. A. Gaskill, 'Desert Areas Coverage: Global Albedo Enhancement Project', cited in Royal Society, *Geoengineering the Climate,* 26.

the albedo of agricultural crops are less ecologically invasive than glacial albedo modification.[37] But they would require global direction and management on a large scale, and although this might be more effective than urban albedo management it is hard to envisage the political and economic conditions that would make it possible short of totalitarian control of farms and farmers.

Carbon Dioxide Reduction (CDR) represents the other main type of geoengineering. It names a class of actions proposed for removing CO_2 from the atmosphere and so reducing the retention of reflected solar heat in the atmosphere. There are two broad approaches: the first is to enhance the CO_2 absorption capacities of land and oceans, and the second is to attempt mechanically to remove CO_2 from the atmosphere and store it in underground reservoirs.

Ocean fertilisation — primarily using iron filings — has already been attempted experimentally on a small scale. A fifty-square-kilometre area of the Southern Ocean was fertilised with iron in 2000. Blooms of one particular phytoplankton were recorded in the days following the deposition, but there was no increase in krill or other species.[38] Eutrophication is a grave risk in ocean fertilisation.[39] So too is a runaway condition in which algal blooms spread to the point that ocean ecosystems experience mass die-off. The oceans are in any case already rising in acidity from having absorbed so much recent atmospheric CO_2, as indicated in declines in ocean fertility and shellfish quantities. Attempts to dissolve more CO_2 through ocean fertilisation would therefore seem very imprudent. Furthermore, a recent set of tests run on ocean water to the east of Iceland after the large eruption of a volcano in 2010 revealed no beneficial effect to phytoplankton from raised levels of iron particles in fissile matter.[40]

The burial of burned biomass as charcoal, or 'biochar', appears to offer more positive prospects for a method of sequestering CO_2 in the earth in a way that carries no obvious risks and that potentially enhances crop productivity. The method is most beneficial when it utilises agricultural,

37. Royal Society, *Geoengineering the Climate,* 24-25.

38. Philip W. Boyd, Andrew J. Watson, Cliff S. Law, et al., 'A Mesoscale Phytoplankton Bloom in the Polar Southern Ocean Stimulated by Iron Fertilization', *Nature* 407 (2000): 695-702.

39. David W. Keith and Hadi Dowlatabadi, 'A Serious Look at Geoengineering', *EOS Transactions American Geophysical Union* 73 (1992): 289 and 292-93.

40. Eric P. Achterberg and C. Mark Moore, 'Natural Iron Fertilization by the Eyjafjallajökull Volcanic Eruption', *Geophysical Research Letters* 40 (2013), DOI: 10.1002/grl.50221.

paper mill, and plantation waste that would otherwise have been burned and released as carbon into the atmosphere, and when such waste is turned into charcoal in an electricity generating plant or used to produce liquid fuel. Once created, charcoal is used as a soil enhancer, thus linking power production with a procedure that sequesters a proportion of carbon from power production in soils. Deployment of biochar is growing and is associated with improvements in soil structure that promote water retention and bacterial growth, and hence aid crop productivity.[41] One estimate indicates that if all existing agricultural crop residues were converted to biochar on a global scale this might have the potential to sequester around 1 gigaton of CO_2 annually, which is *one-seventh* of current annual CO_2 production from fossil fuels and cement making.[42]

The claim that biochar genuinely locks CO_2 in the soil would have to include full life-cycle analysis, including consideration of the land and soils whence the biomass had originated and subsequent use of treated soils.[43] Agricultural waste on these terms is a safe source for biochar since otherwise it is ploughed in or burned. There are similarly good prospects for plantation waste. However, it would add nothing to net terrestrial carbon capture if large-scale adoption of biochar as a geoengineering solution promoted conversion of forests to fast-growing biomass plantations and for use in electricity generation.

Destruction of old-growth forests and their replacement with oil palm plantations for the production of biofuels and cooking oil in Indonesia, Malaysia, Amazonia, and central Africa is already the source of around 20 percent annually of current anthropogenic global greenhouse gas emissions.[44] Indonesia alone is responsible for around 10 percent of current global greenhouse gas emissions owing to the extent of forest burning, clearing, and drying of tropical soils in Borneo and Sumatra. Efforts to reduce defor-

41. L. Van Zwieten, S. Kimber, S. Morris, et al., 'Effects of Biochar from Slow Pyrolysis of Papermill Waste on Agronomic Performance and Soil Fertility', *Plant Soil* 327 (2010): 235-46.

42. Simon Shackley, Saron Sohi, Stuart Haszeldine, et al., *Biochar, Reducing and Removing CO_2 While Improving Soils: A Significant and Sustainable Response to Climate Change,* UK Biochar Research Centre Working Paper 2 (Edinburgh: UK Biochar Research Centre, 2009).

43. D. S. Powlson, A. P. Whitmore and K. W. T. Goulding, 'Soil Carbon Sequestration to Mitigate Climate Change: A Critical Re-examination to Identify the True and the False', *European Journal of Soil Science* 62 (2011): 42-55.

44. J. M. Melillo, R. A. Houghton, D. W. Kicklighter, et al., 'Tropical Deforestation and the Global Carbon Budget', *Annual Review of Energy and Environment* 21 (1996): 293-310.

estation and to promote reforestation therefore merit the highest priority of all proposed projects to enhance terrestrial CO_2 absorption; they would reduce a significant terrestrial source of anthropogenic greenhouse gas emissions by reducing atmospheric pollution from burned forest biomass and soils, while also sequestering CO_2 into regrown biomass and soils in the future. However, it has proven very difficult in practice to end current government and corporate destruction of tropical forests. Furthermore, unwise promotion of 'biofuels' in Europe, under the European Renewables target, has prompted Malaysia and Indonesia to process palm oil into biodiesel and has provided a further commercial stimulus to the conversion of tropical forests into oil palm plantations. Analogously the United States federal government has prompted conversion of corn into ethanol with a government directive and subsidy regime designed to reduce dependence on imported oil. But the carbon footprint of plant-derived 'biofuels' in vehicular fuels is equivalent to, or in excess of, the carbon footprint of conventional oil. As well as promoting deforestation, the conversion of forests into oil palm plantations is also contributing to rising global food prices.[45] Were the land area used for these fuels to be kept as forest, or converted back to forest, it would sequester more carbon and provide other ecosystem services such as groundwater retention.[46]

Another CDR technique involves quarrying and crushing olivine rocks and spreading these on bare and unproductive land such as deserts and coastal sand dunes. Weathering of olivine rock fragments is a carbon-absorbing process which, on a large enough scale, has the potential to draw down CO_2 from the atmosphere at relatively low cost compared to other kinds of CO_2 sequestration.[47] Grinding olivine rocks to micron-sized particles and depositing them in the ocean is also proposed as a way to increase ocean absorption of CO_2 while also reducing its acidity, since olivine is alkaline.[48]

45. Timothy Searchinger, Ralph Heimlich, R. A. Houghton, et al., 'Use of U.S. Croplands for Biofuels Increases Greenhouse Gases through Emissions from Land-use Change', *Science* 319 (2008): 1238-40.

46. Renton Righelato and Dominick V. Spracklen, 'Carbon Mitigation by Biofuels or by Saving and Restoring Forests?' *Science* 317 (2007): 902.

47. R. D. Schuiling and P. Krijgsman, 'Enhanced Weathering: An Effective and Cheap Tool to Sequester CO_2', *Climatic Change* 74 (2006): 349-54.

48. Peter Kohler, Jesse F. Adams, et al., 'Geoengineering Impact of Open Ocean Dissolution of Olivine on Atmospheric CO_2, Surface Ocean pH and Marine Biology', *Environmental Research Letters* 8 (2013): 014009.

Discussions of the propriety of geoengineering tend to focus on two broad categories of concerns. The first is feasibility: Will the techniques work to mitigate climate change, and at an economic cost that is affordable, and without undue risk of untoward consequences? The second concerns questions of longer run consequences, duty, and agency: Will the techniques — or the prospect of such techniques if research is commissioned — encourage nations and corporations to defer other urgent actions to mitigate climate change, and in particular reduce the extraction of fossil fuels and the use of energy? Will the techniques promote inequity in their outcomes — will some nations suffer greater costs than others from the deployment of such techniques? Who will be responsible for authorising such techniques? Who will undertake them and pay for them? Will techniques impose differential burdens on different nations or groups of people? Will their use impose increased burdens on future generations? Will the techniques enhance the risk of sudden climate change if they suddenly become inoperative? Will the resultant planet be less habitable, and valuable, than the present one? Will geoengineering be used in a hostile and hence warlike fashion by some countries actively to expose other countries to greater climatic risks?

Jay Michaelson describes global geoengineering as a climatic 'Manhattan Project' after the code name for the atom bomb project in Los Alamos:

> The metaphor of the Manhattan Project indicates that intentional geoengineering of the climate would, like the splitting of the atom, represent an earth system intervention capable of endangering the planetary conditions which sustain all life.[49]

Gardiner uses the analogous metaphor of 'arming the future'.[50] Both metaphors highlight the intergenerational and long-run consequences involved in geoengineering. Those who argue that it is the lesser evil claim that it will give more time to present generations to reduce greenhouse gas emissions while reducing the burden of their procrastination on climate change effects experienced by future generations. But this will be even more the case if geoengineering commits future generations to an enduring responsibility for carbon and climate management which — together with the growing

49. Jay Michaelson, 'Geoengineering: A Climate Change Manhattan Project', *Stanford Environmental Law Journal* 17 (1998): 74-138.

50. Stephen M. Gardiner, 'Is "Arming the Future" with Geoengineering Really the Lesser Evil?' in Stephen M. Gardiner, Simon Caney, et al., eds., *Climate Ethics: Essential Readings* (Oxford: Oxford University Press, 2010), 284-315.

cost burden of adapting to climate change — will represent a much greater burden than national- and global-scale mitigation of climate change would represent for present generations.

Given the potentially destructive consequences of large-scale engineering interventions in the atmosphere and oceans, the 2010 United Nations Convention on Biodiversity includes a clause explicitly banning such interventions as a potentially grave threat to biodiversity. Neither UN bodies nor any national governments have plans to pursue any of the large-scale geoengineering options. However, the ongoing advocacy of geoengineering in scientific, NGO, and civil society contexts indicates the attraction of this approach to the intractable problem of climate change.

Bacon on Knowing as Power over Nature

One study of public attitudes found that when presented with geoengineering solutions to climate change, individuals are less prone to deny climate change science.[51] This finding points to the core conundrum of the scientific narrative of climate change: it runs directly counter to the modern story, which is that human control of nature is the means to advance human flourishing, a story first told by Francis Bacon. Bacon is widely acknowledged as the originator of the modern scientific revolution, and he was the first to describe the post-Aristotelian scientific method of empirical investigation. But Bacon's metaphorical language for the experimental and mechanical investigation of nature by the 'violence of impediments' arose from the frequent use of torture against heretics, rebels, and witches in the sixteenth century. In an address to James I, Bacon suggests that the interrogation of sorcerers, charms, and witches shed 'useful light' on the scientific quest for the 'further disclosing of the secrets of nature'. Just as men should not scruple to assault and torture the bodies of women accused of witchcraft or sorcery, 'neither ought a man to make scruple of entering and penetrating into these holes and corners, when the inquisition of truth is his whole object.'[52] For Carolyn Merchant, Bacon's language indicates the mixing of the sexual politics of the sixteenth century and the scientific method:

51. Nick Pidgeon, Adam Corner, et al., 'Exploring Early Public Responses to Geoengineering', *Philosophical Transactions of the Royal Society* 370 (2012): 4176-96.

52. Francis Bacon, 'De Dignitate et Augmentis Scientiarum', in Bacon, *Works*, vol. 4, 296, cited by Carolyn Merchant, *The Death of Nature: Women, Ecology, and the Scientific Revolution* (San Francisco: Harper Collins, 1989), 168.

> The interrogation of witches as symbol for interrogation of nature, the courtroom as model for its inquisition, and torture through mechanical devices as a tool for the subjugation of disorder were fundamental to the scientific method as power.[53]

Under the influence of the violent anti-Cathar movement in France, women accused of witchcraft in the infamous Pendle trials in Lancashire were also accused of fornication and other sexual sins. The sexualisation of heresy, and its investigation in the reign of James I, involved the adoption of patriarchy and an ontology of violence in which women and nature were both seen as essentially disordered realms requiring the controlling and ordering power of the male mind and of the mechanical arts to bring them into subjugation. This was the same monarch who as James VI of Scotland had promulgated a 'vermin act', which mandated the hunting and extermination of myriad wild creatures, from bats and kingfishers to eagles and osprey, as threats to agricultural production in Scotland, before he took up the English throne, and he and his heirs to that throne enacted similar laws in England.[54]

Nature had long been considered as feminine, and particularly in the Christian Platonism of the Middle Ages. However, what was new in the sixteenth and seventeenth centuries was a changed attitude to women, who, in Eve, were said by late medieval Catholic and most Protestant theologians to be the source of man's primeval fall and an enduring source of temptation and chaos, and hence in need of control and subjugation by men. In the context of this growing theological suspicion of women, the imaginary of nature as feminine turned from a metaphor which promoted a respectful attitude to the earth to one which promoted a violent, domineering, and controlling attitude to nature. In Girardian terms this new sexual politics reflected the gradual demise, under the influence of the rise of capitalism, of the penitential and sacrificial system of the Church at the Renaissance. As religion declined in influence, the ambivalent relationship between violence and the sacred found a new manifestation in the scapegoating of women and of various species,

53. Merchant, *The Death of Nature,* 172. Some argue that Merchant places too much emphasis on a metaphor Bacon in fact rarely uses; see for example Peter Pesic, 'Proteus Rebound: Reconsidering the "Torture of Nature"', *Isis* 99 (2008): 304-17.

54. On extinctions of species, many of them no threat to agricultural production, precipitated by the vermin acts see Roger Lovegrove, *Silent Fields: The Long Decline of a Nation's Wildlife* (Oxford: Oxford University Press, 2007).

whose joint persecution became the solidaristic sacrificial rituals of early modern England and Scotland.[55]

The identification of women with sacred violence reached a crescendo of trials and killings of women midwives and herbalists as witches all over Britain in the seventeenth and eighteenth centuries (including on the Mound in Edinburgh, metres from the front gate of New College where I work) and continued in the Puritan colonies.[56] For Bacon, nature as feminine, like women, requires control, for she

> exists in three states, and is subject, as it were, to three kinds of regimen. Either she is free and develops herself in her own ordinary course, or she is forced out of her proper state by the perverseness and insubordination of matter and the violence of impediments, or she is constrained and molded by art and human history. The first state refers to the 'species' of things; the second to 'monsters'; the third to 'things artificial'.[57]

Elsewhere Bacon holds that 'the earth, from its entire and unperfected cold, and the extreme contraction of matter, is most cold, dark, dense, and completely immovable.'[58] In similar vein, Shakespeare has a king announce 'she's dead as earth' in *King Lear,* act V, scene 3.

As Edwin Reed argues, both Shakespeare and Bacon, in response to the new science, abandoned the Christian Platonist account of the earth as mother of all living and instead revived the Classical view of nature according to which, of the elements that give life — earth, air, fire, and water — only the latter three are primal or energetic. Thus Bacon has it that the earth is 'quiet, torpid, inactive' and 'submits patiently to the heaven, fire, and other things.'[59] Matter for Bacon is dead, and this is why nature, when

55. See further William Leiss, *The Domination of Nature,* 2nd ed. (Montreal: McGill-Queen's University Press, 1994); Leiss's account of Bacon was the inspiration for Merchant's account of Bacon, as she acknowledges in *Death of Nature,* 318, n. 7.

56. Christina Larner, *Enemies of God: The Witch-Hunt in Scotland* (Baltimore: Johns Hopkins University Press, 1981).

57. Francis Bacon, *Preparative Toward Natural and Experimental History* (London, 1620), 'Aphorism I', 3.

58. Francis Bacon, *De Principiis et Originibus,* cited in Edwin Reed, *Bacon and Shakespeare Parallelisms* (London: Gay and Bird, 1902), 329.

59. Reed, *Bacon and Shakespeare Parallelisms,* 329. Reed intended to show by the extent of parallelisms that Bacon *wrote* Shakespeare's plays under a pseudonym; that he was wrong on this does not detract from the important parallels between them.

subject to the perverse power of matter, becomes monstrous, as did Caliban in *The Tempest,* who is described as 'a devil, and no monster'.[60]

If the crust of the earth, soil, and matter are essentially dead, then only the imposition of human artifice and ideas upon them can give them life. Here we see an analogy with Descartes, as well as novelty. For Bacon, mental concepts can be tested through the experimental method against reality, and thus sense perceptions of material reality can be corrected and the concepts arising from these affirmed or disproved. Bacon's account of the scientific method was fashioned in part from his observations of ore extraction, smelting, and craft techniques in the mines and metal workshops of early modern England. His work is replete with references to mining and metallurgy, and he envisaged a project he never completed of a 'history of trades', which was taken up by Robert Boyle in the first history of mining.[61] Bacon's experimental method was fashioned on the mechanical or practical arts, and he envisaged the generation and sharing of scientific information on the analogy with the workshop practice of recording and refining craft knowledge and techniques so they could be passed from one practitioner and workshop to the next.[62]

For Bacon, human beings truly know only what they make. This is why the mysteries of nature are mostly beyond human ken. Only those parts of nature that are brought under control and remade in the workshop and the laboratory can be said fully to be understood. This recognition has profound epistemological implications. Descartes responded to the Copernican critique of sense perceptions — which indicate that the sun and moon and stars move around a stationary earth, whereas in reality the earth moves around the sun — by setting up the modern split between nature as material reality governed by science and culture governed by idealist philosophy. By contrast, Bacon found a way back from the mind to the real world after Copernicus through a revised and more empirical approach to sense perception: the correspondence between human sense perceptions and physical reality could be reconstructed *through* the empirical method. In so doing Bacon also recovered the cosmic powers and

60. Shakespeare, *The Tempest,* act 2, scene 2 (London, 1623), cited in Reed, *Bacon and Shakespeare Parallelisms,* 329.

61. Pamela Long, 'The Openness of Knowledge: An Ideal and Its Context in 16th-Century Writings on Mining and Metallurgy', *Technology and Culture* 32 (1991): 318-55.

62. Jim Bennett, 'The Mechanical Arts in 1500', in Katherine Park and Lorraine Daston, eds., *The Cambridge History of Science,* vol. 3: *Early Modern Science* (Cambridge: Cambridge University Press, 2006), 677-95.

uniqueness of humanity against the decentring implications of Copernican cosmology.[63] Bacon also anticipated the Anthropocene: it is precisely the extent of modern scientific interventions into the depths of the earth, and in the earth's forests and oceans, which have given back to modern humanity seminal responsibility for the fate of the cosmos that Copernicus seemed to take away.

What is less often acknowledged in accounts of Bacon's scientific method is that Bacon was an idealist. His inductive method was more a rhetorical device than part of an empirical philosophy. He was so devoted to the prior truths of his ideas that on a number of occasions, as James Stephens observes, he compared his new science to a 'second scripture' and considered 'the obstacles to his success in terms of those which Christ faced'.[64] Bacon's idealism is manifest above all in the way his inductive method worked. This method involved the accumulation of knowledge of the workings and laws of nature through a constant process of refinement between mental concepts, ideas, and measures — above all the measure of number — and material reality. It was this process of empirical testing of *ideas* that led Bacon, and Descartes — and here they are at one — to the admiration of the measure of numbers which would be taken up by all subsequent science.[65]

For Bacon, the new science was for the purpose of the founding of a new imperial order on earth which would encompass the earth through geographical as well as mechanical discoveries and the controlled experiment. The end was human progress and betterment through dominion over nature, whose containment represented a 'heroic struggle' which would eventually lead to an artificial or mechanical utopia, which Bacon described in great detail in his last great work, the *New Atlantis*.[66]

Bacon is the originator of the new story of human salvation in the sixteenth century, which has become the dominant eschatology of the modern

63. Robert Pogue Harrison, *Forests: The Shadow of Civilisation* (Chicago: University of Chicago Press, 1992), 109.

64. James Stephens, *Francis Bacon and the Style of Science* (Chicago: University of Chicago Press, 1975), 1.

65. Whitehead argued that Bacon did not understand or appreciate the extent to which science 'was becoming, and remained, primarily quantitative'; Whitehead, *Science and the Modern World* (New York: Free Press, 1967), 45. But Bacon instructed that 'everything related to bodies and virtues in nature be set forth (as far as may be) numbered, weighed, measured, defined'; Francis Bacon, *Parasceve*, 406, cited in Dennis Desroches, *Francis Bacon and the Limits of Scientific Knowledge* (New York: Continuum, 2006), 37-38.

66. Carolyn Merchant, 'The "Violence of Impediments": Francis Bacon and the Origins of Experimentation', *Isis* 99 (2008): 731-60; this essay is a response to Pesic's 'Proteus Rebound'.

era. For Bacon, progress in dominating nature — through scientific discovery and technological invention — is the means to redeeming the original fall from Paradise.[67] The purpose of advancing scientific knowledge is not for its own sake, as it had been for Bacon's predecessors, but for its utility in advancing human happiness. Progress in human dominion over nature and progress in human flourishing are one and the same. As John Bury argues in *The Idea of Progress,* 'the principle that the proper aim of knowledge is the amelioration of human life, to increase men's happiness and mitigate their sufferings — *commodis humanis inservire* — was the guiding star of Bacon in his intellectual endeavor.'[68] The age of discovery, the age of the invention of gunpowder, the compass, and the timepiece, was a golden age, an age of the 'renovation of knowledge' in which philosophy and science were being renewed and the history of the earth and its peoples begun again. Before Bacon, the idea of progress was hardly spoken of in England, and instead history was regarded as a regress from the ages of Christ and the early Church, ancient Greece, and Rome. But for Bacon the will of men would impose a new purpose and order on history and nature, and the world would be reborn, and history begun again.

Bacon produced the definitive early modern narrative of the purpose of both government *and* science as the advancement of human dominion over the earth through scientific research and technological artifice and the recovery of the dominion over creation that had belonged to Adam and Eve in Paradise. Bacon also unfolded the politico-theological implications of this recovered dominion for the nations. In *New Atlantis* Bacon anticipates the growing alliance between governments and national scientific research institutes for asserting dominion over nature; he also with great prescience anticipates the technological fetishism which characterises a consumer culture more than ever in thrall to the fruits of that alliance:

> The scene in which the exalted scientific administrator passes in solemn pomp before the awestruck populace can be read as an allegorical representation of our contemporary situation in which citizens are the passive beneficiaries of a technological providence whose operations they neither understand nor control.[69]

67. Francis Bacon, *Novum Organum,* 247-48.

68. John B. Bury, *The Idea of Progress: An Inquiry into Its Origin and Growth* (London: Macmillan, 1921), 52.

69. Leiss, *Domination of Nature,* 70-71.

As William Leiss observes, for Bacon there is an intrinsic connection between scientific control over nature and human self-control, and hence between physical power over nature and moral and spiritual salvation.

Bacon's genius is the key source for the association between the domination of nature and Christianity which was to prove so influential in succeeding centuries, although this is not noted by Lynn White, who indicted Christianity per se, rather than its Baconian recasting, as the historic root of the ecological crisis.[70] Bacon's vision of science as salvation was also the inspiration for the founding of the Royal Society and for the Elizabethan 'age of discovery' which saw English explorers journey to the New World and the East Indies, journeys which became the basis for England's gradual transformation in the following two centuries from nation to empire. It also played a key role in the shaping of a new Tudor political theology under Henry VIII, the Cromwells, and Elizabeth I. The Tudor state was the first in modern history to take over from Church and peasants power over the land, to redistribute Church land and then common lands to landowners, and to assume responsibility for species control on the land. Scientific improvement of agriculture, the generation of more reliable harvests, was the justification to which the state resorted in this great redistribution of land upwards throughout Britain.

In Bacon we may find the root of the deep ambiguity which climate science represents in a culture guided and enriched by 'technological providence'. The power of the modern nation-state rests upon its power over nature. After agricultural enclosure and large-scale deforestation, the most visible form of the state's power over nature in the century of oil is the commissioning of fossil-fuelled roads, electric grids, and gas, water, and sewage pipes for the provision of energy and water to homes and businesses and the removal of waste from them. That the extent of human engineering interventions into the depths of the earth and the atmosphere might threaten the endurance of this provisioning challenges the godlike power of the nation as the mediator to businesses and households of technological providence. Hence the critique of the Baconian project by Merchant and others is at root a politico-theological critique. Bacon's 'technological providence' advances the mechanisation of human work with fossil-fuelled machines, a split between land and labour, and the fetishisation of commodities.

But while intending to dominate the earth, modern humanity is at risk

70. Lynn White Jr., 'The Historical Roots of Our Ecologic Crisis', *Science* 155 (1967): 1203-7.

in new ways of being enslaved by earthly powers, as C. S. Lewis warned in the 1940s.[71] And efforts to hold back climate extremes through climate engineering, while keeping the fossil-fuelled machines going, will enslave humanity to a set of technological interventions whose cost and scale will dwarf even the networks of grids and pipelines that are implicated in these other pathologies of a machine-based social order. Ultimately, as Ivan Illich argues, 'we must admit that only within limits can machines take the place of slaves; beyond these limits they lead to a new kind of serfdom'.[72] In this light the Baconian idea of salvation as progress in human fulfilment through progressive power over the exterior of nature needs now to be revised by an intentional turn toward a less asymmetric relation between scientific and technological power and the objects of scientific investigation and technological control.

Vico and the Overcoming of the Science-Ethics, Nature-Culture Divide

Against the story of human domination over nature as the core meaning of what it means to be modern, moderns need a 'new myth' of human-nature relations. Two individuals in the twentieth century who sought to advance a new myth, drawing on Anglo-Saxon and Christian mythology, are J. R. R. Tolkien and C. S. Lewis. *The Lord of the Rings, The Hobbit* and the Narnian Chronicles remain some of the most influential 'new myths' in contemporary culture; in part this is because of the enduring power of the mythological worlds created by their authors, and in part it is owing to the filmic re-creations of these worlds, especially the work of Peter Jackson.

In an essay on myth and fairy, Tolkien argues that the central feature of a successful fairy story is the 'joy of the good catastrophe' or 'eucastrophe', in which a sudden, potentially catastrophic turn in the story turns into a happy ending. This euchastrophic turn reaches its zenith in the Gospels, because in the Gospels the catastrophe of Christ's Crucifixion presages the Resurrection of Christ from the dead. In this way the Christian *evangelium* 'has not abrogated legends' but rather 'hallowed them', especially 'the happy ending':

71. C. S. Lewis, *The Abolition of Man* (London: Geoffrey Bles, 1946).

72. Ivan Illich, *Tools for Conviviality* (London: Calder and Boyars, 1973), 6.

> The Christian has still to work, with mind as well as body, to suffer, hope, and die; but he may now perceive that all his bents and faculties have a purpose, which can be redeemed. So great is the bounty with which he has been treated that he may now, perhaps, fairly dare to guess that in Fantasy he may actually assist in the effoliation and multiple enrichment of creation.[73]

This is perhaps the pithiest statement in Tolkien's oeuvre of his purpose and intention in constructing the new/old myth of Middle Earth. In it Tolkien reprises the histories of empires and peoples, the age-old struggle between good and evil. Central to his myth is the myth of the ring of power which gives to its possessor power over life and death, power over nature; promising redemption, the ring ultimately turns its possessor into the agent of destruction. Against the violent quest for power over life and death that drives the would-be possessors of the ring is the quest to redeem this struggle through the eschewal of power and the gentle care of creation through the creative and peaceable stewardship of life. As the wizard Gandalf puts it:

> "The rule of no realm is mine, neither of Gondor nor any other, great or small. But all worthy things that are in peril as the world now stands, those are my care. And for my part, I shall not wholly fail of my task, though Gondor should perish, if anything passes through this night that can still grow fair or bear fruit and flower again in days to come. For I also am a steward. Did you not know?"[74]

The ring in Tolkien, as in Wagner, is the mythic metaphor for alchemic technological power over nature, life and death, the myth of Baconian redemption through scientific knowledge as power.

Tolkien's critique of the twentieth-century turn to scientific power over nature, and violent imperial war, is anticipated in the remarkable but obscure writings of an Italian seventeenth-century humanist, Giambattista Vico. Vico is said by Isaiah Berlin to have first envisaged the split between the humanities and the sciences that has endured to this day in the modern university and in modern culture.[75] But far from proposing a split between

73. J. R. R. Tolkien, *Tree and Leaf* (London: Unwin, 1968), 63.

74. J. R. R. Tolkien, *The Return of the King: Being the Third Part of the Lord of the Rings*, book 5, i (London: George Allen and Unwin, 1955).

75. Isaiah Berlin, *Vico and Herder: Two Studies in the History of Ideas* (New York: Viking Press, 1976), xvii.

the sciences and the humanities, Vico's project was an attempt to rehumanise science, and to *resituate* scientific knowledge in human as well as natural history by recovering the mythological origins of all human cultures, including the culture of science.[76]

In contrast to Bacon, Vico realised that it was language and literature, myth and metaphor, not mechanical making and technical power over nature, which had enabled human beings to evolve from barbarism toward more peaceable, health-giving, and beautiful ways of dwelling, and not merely surviving, on earth. In Vico's mythic retelling of the origins of peoples and of European civilisation, humans first and foremost are makers, and they know most fully what they have made — their own artefacts, customs, and institutions — rather than the world of nature which God and not human beings has made; hence Vico's famous dictum *verum ipsum factum.*[77]

For Vico, the origin of human civilisation, and of humanity's stable dwelling on earth, is in human culture and history, and not in the human understanding of and control of nature. Hence the scientific, industrial, and political revolutions that birthed the modern age carry with them the threat that they will erase the true knowledge of the origins of civilisation and therefore threaten its endurance.

Climate change science suggests that this endurance is indeed threatened. Nature, of course, is not threatened. Nature, or 'Gaia', will respond to the changes human beings are wreaking on the atmosphere and the weather; nature will not end but change. Here is the crux of the potential contribution of Vico's thought to the climate crisis. The climate crisis is not a threat to nature *apart* from culture. It is a threat to human culture as it is situated *in* nature: to the cities, dwellings, fields and farms, judicial and political institutions, rituals, customs, and traditions which are the evolved forms of culture. If humanity is not to be reduced back to a barbaric struggle of all against all for diminishing fertile and habitable land on a dramatically warmed and less stable earth, then the humanities have a central role to play in mediating climate change science.

At the same time, Vico's dialogue with the new science indicates the urgent need to give up the myth of progress as power *over* nature if human

76. Joseph Mali, *The Rehabilitation of Myth: Vico's* 'New Science' (Cambridge: Cambridge University Press, 1992).

77. Elio Gianturco, 'Translator's introduction', in Giambattista Vico, *On the Study Methods of Our Time* (1709; Ithaca, NY: Cornell University Press, 1990), xli-xliv.

beings are to mitigate extreme climate change and adapt to a warming earth without descending back into global war and violence.

Vico's remythologisation of history rests upon his lifelong study of the heroic fables, myths, poetry, religion, and ritual from ancient Egypt and Israel through Classical Greece and Rome to early modern times. According to Vico, fable, myth, and religion are the ancient source of the organisation of peoples into nations, a process which 'divine providence initiated' and 'by which the fierce and violent were brought from their outlaw state to humanity and entered upon national life.'[78]

The first god of the peoples was Jove, the god of the sky, since human beings uniquely invest the things they do not understand with human properties and indicate this in language and signs. Since the sky exercises such a powerful hold on human life — whether the sun shines, the fields are watered, the wind blows, the air thunders or is lit with lightning — human beings originally envisage the sky as the home of a capricious god visiting good and ill according to divine mood. The first god of the sky from which all the others are derived is for Vico proven by philology to be Jehovah, or Yahweh, the god of the Hebrews. The first human thing that Jove commands is marriage, from which the family and hence the commonwealth of nations derives; and the second is the burial of the dead, for by reverence to the bodies of the dead human beings recognise that souls endure beyond bodily death. From these institutions the tribes, first in ancient Israel, were formed into a nation; and from that nation all the nations of the Gentiles are descended. Piety to the gods and ploughing of the land are the twin sources of the settled life of cities, and religion is the means that restrains wars between cities. The natural law of nations proceeds from the gradual emergence of agriculture and urban settlement and the accompanying Hebrew and Gentile religions. The evolution of the nations takes place in three periods: the age of the gods, in which people imagined they were ruled by direct divine oracle; the age of heroes, in which aristocrats ruled over plebs; and the age of men, 'in which all men recognised themselves as equal in human nature, and therefore there were established the first popular commonwealths and then the monarchies' as the lately evolved forms of human government.[79]

The evolution of the civilising institutions of marriage, burial, equality before the law, nations, and governments is the outcome of wonder, and

78. Giambattista Vico, *New Science,* trans. Thomas Goddard Bergin and Max Harold Fisch (Ithaca, NY: Cornell University Press, 1948), book 1, 178.

79. Vico, *New Science,* 'Idea of the Work,' 11, 31.

more especially the fabulous fables that humans first told about the gods of the sky, for 'wonder is the daughter of ignorance'.[80] People wonder most at what they do not understand, and they make sense of what they do not understand in nature by investing it with their own nature through language and signs. Hence language and mythology, not mathematics or mechanics, are the way in which human beings first give order to nature and promote civil governance and the enduring order of human institutions, laws, morals, and rituals. For Vico, this settles the Classical dispute in philosophy as to whether human beings are naturally sociable or naturally competitive and individualistic. The human ability to live together civilly, peaceably, is not natural but divine: 'this same axiom proves . . . that man has free choice, however weak, to make virtues of his passions; but that he is aided by God, naturally by divine providence and supernaturally by divine grace.'[81]

Against the utopian myth of the *New Atlantis,* in which Bacon envisaged human redemption in an engineered and mechanical future, Vico argues that history has an unfolding and providential purpose which, though sometimes hidden from view, is the true source of the movement of history from barbarism to civility, from war to peace, and from vice to virtue.[82] For Vico, the supreme form of making is poetry, and more broadly *poiesis,* which adds beauty, meaning, and order to bare life. Poetic narratives of the origins of the earth and of human history, which science was increasingly setting aside, are therefore for Vico *true myths* which carry within them the accreted 'common sense' of peoples which common sense philosophers tend to ignore or set aside. Reflection on and argument about these narratives in tradition and their celebration and enactment in myth and ritual are therefore the origin of those institutions, including government and nationhood, through which human beings have progressed from barbarism to civility.[83] This origin, however, is increasingly neglected by mechanistic scientists and rationalist philosophers alike. They forget, and Vico reminds them, that before there were craftsmen or engineers there were poets and storytellers, and it is therefore in *poiesis* — in the crafting and sustaining of artefacts, farms, gardens, institutions, myths, and rituals — that the true nobility of the human is to be found and through which humanity makes an enduring and not destructive mark on the earth.

80. Vico, *New Science,* book 1, 184.

81. Vico, *New Science,* book 1, 136.

82. Berlin, *Vico and Herder,* 74.

83. Mali, *The Rehabilitation of Myth,* 3.

Vico's new science is an attempt to rehistoricise knowledge of both the human world and the natural world, and to *reconnect* human and natural history. For Vico, mythic knowledge is more reliable and ennobling than mechanical knowledge as a way of engaging the world because it participates in the constitution of culture by nature. Vico therefore criticises the tendency he observes in the culture of early modern science to reduce all spheres of human endeavour to the measure of numbers. Mathematisation obscures the actual evolved and historical processes through which human beings acquire and adumbrate knowledge about the world, processes which evolve through *making* rather than through abstract knowing.[84]

For Vico, the *origin* of human making is not, as for Bacon, the development of tools and mechanical skill, but the generation of language. If the world becomes knowable for humanity only when it is constructed linguistically, then this brings to the fore the mediating role between matter and mind of metaphor, mythology, narrative, and rhetoric: language makes the world by reading the world. Vico here draws on the thought of the Hebrews, which he reads as the first literary record of mediation between the eternal mind of God and the history of humanity.[85] In Hebraic thought, the world is written by the speech-acts of God. Against the nominalist transformation of Christianity's conception of creation into the 'contingent acts of an omnipotent will',[86] whose fruit in philosophy is the Cartesian identification of being with subjective thought, Vico reaches back into myth and history to retrieve the Hebraic account of divine creative action as the 'speech of God' and of those who carry the image of God as those who uniquely possess speech.[87]

Against the Cartesian claim that innate ideas and the ability to reason are *a priori* to human embodied experience, Vico argues that human rationality arises and develops from the sensory experiences of children and their progress into adulthood.[88] Vico philologically resituates rational-

84. William J. Mills, 'The Relevance of Giambattista Vico', *Transactions of the Institute of British Geographers,* New Series 7 (1982): 1-14.

85. Frederik R. Marcus, 'Vico's *New Science* from the Standpoint of the Hebrews', *New Vico Studies* 27 (2009): 1-26. See also Robert C. Miner, 'The Hebrew Difference', in Miner, *Vico: Genealogist of Modernity* (South Bend, IN: University of Notre Dame Press, 2002), 106-15.

86. Sandra Rudnick Luft, *Vico's Uncanny Humanism: Reading the* New Science *between Modern and Postmodern* (Ithaca, NY: Cornell University Press, 2003), 88.

87. *Midrash Tebillium* CVII, 3, ed. Salomon Buber (New York: Om Publishing, 1947), 462, cited in Luft, *Vico's Uncanny Humanism,* 88.

88. Alfonsina Albini Grimaldi, *The Universal Humanity of Giambattista Vico* (New York: S. F. Vanni, 1958), 1.

ity in sense perception and in the language of signs and words, which are the accreted depository of the evolved *common* sense and traditions of the ancestors. Hence the origin of humanity's scientific knowledge lies in the ancient move from language to making as the symbolic representation of person-Jove, human-nature relations unfolds in the crafting of altars, burial stones, dwellings, ploughs, and tools as the constitutive medium of humanity's ordering and knowing of the natural world. For Vico, making — *factum* — is the true transcendental. But this is obscured by the Cartesian divide between mind and body and by the asymmetric scientific relationship between culture and nature. God is not a mechanical artisan but the God who calls all things into being by the divine Word.[89] *Factum* and *verbum* are therefore synonymous, and God as creator is both poet and painter, continuously holding together a world of matter and mind, which becomes more mindful through the motion of matter and through mindful making.

Against Bacon's metaphor for scientific rationality, as inductive thought from the laboratory trials of matter and organisms in contained experiments, Vico argues that the modern phase of civilisation — the 'age of men' — rests upon the prior histories of the ages of gods and heroes in which true knowledge of universal truths about humanity and the cosmos was gradually uncovered and accrued in human cultures. From this prior history arose the 'imaginative universals' and myths that guide human thought toward its humanist — and modern — phase.[90]

For the unfolding crisis of climate change Vico's ideas are powerfully relevant. Scientists continue to believe that if the profit-motivated and political suppression of climate science in the United States and beyond can be stopped, and the clear consensus of mainstream climate science can be promulgated as to the threat of uncontrolled fossil fuel emissions and deforestation to the present form of human civilisation, then governments and peoples will learn that they must change their way of life and respond accordingly. However, the language and data models that scientists use to describe the threat, together with the political claim that what the UNFCCC calls 'dangerous climate change' of two degrees Celsius can be predictably prevented by limiting carbon dioxide in the atmosphere to 450 ppm,

89. John Milbank, *The Religious Dimension in the Thought of Giambattista Vico,* vol. 1 (Lewiston, NY: Edwin Mellen Press, 1991), 120.

90. Donald Kunze, 'Giambattista Vico as a Philosopher of Place: Comments on the Recent Article by Mills', *Transactions of the Institute of British Geographers,* New Series 8 (1983): 237-48.

carry within them the Baconian myth of mechanistic control.[91] There is still a great deal that science does not know about the earth's climate because it is not made by human beings; it is more complex than science can yet, or may ever, comprehend, and far less predictable than a thermostatic control on a furnace.

Climate science and climate politics tend to be about numbers. How much cumulative atmospheric CO_2 equates to how much warming? How 'sensitive' is the climate to anthropogenic greenhouse gases? These become the definitive questions, the questions on which may even hang the future of civilisation. For Vico, it is the seminal semiotic characteristic of modern science to create a fictional world of abstract shapes and numbers through 'the point that can be drawn and the unit that can be multiplied.'[92] Through mathematical abstraction Newton first intended to explain the way in which God originally created the universe and the cosmos, and the rules which govern their subsequent motions. Against math, Vico advances myth as the way to overcome the nature-culture and science-ethics divides. He proposes an account of creation as *conatus,* in which 'nature' is a mediatory realm between God who is at rest and the physical world which is in motion: 'the works of nature are brought into being by *conatus* and brought to perfection by motion. The genesis of things presupposes motion, motion presupposes *conatus,* and *conatus* presupposes God.'[93] *Conatus,* for Vico, as Sandra Luft argues, represents 'the power of motion' which is

> ubiquitous in a physical world not composed of separate substances. The material body 'moved by a motion common to all bodies' is air, the pressure of air being 'God's perceptible hand, by which all things are moved,' causing the motion by which water rises in a siphon, fire burns, plants grow, animals frolic. In relation to the human organism, Vico appeals

91. There is accumulating evidence that even two degrees of globally averaged warming undermines ecosystem services essential to human agriculture and habitation and is therefore not 'safe'; e.g., C. H. E. Elkin, A. G. Gutierrez, et al., 'A 2° C Warmer World Is Not Safe for Ecosystem Services in the European Alps', *Global Change Biology* (2013), DOI: 10.1111/gcb.12156; Brigitte Knopf, Martin Kowasch, et al., 'The 2° Target Reconsidered', ch. 12 in O. Edenhofer et al., eds., *Climate Change, Justice and Sustainability: Linking Climate and Development Policy* (New York: Springer, 2012), 121-37.

92. Giambattista Vico, *On the Most Ancient Wisdom of the Italians: Unearthed from the Origins of the Latin Language,* trans. L. M. Palmer (Ithaca, NY: Cornell University Press, 1988), 50-52, cited in Luft, *Vico's Uncanny Humanism,* 126.

93. Vico, *On the Most Ancient Wisdom,* 109-10, cited in Luft, *Vico's Uncanny Humanism,* 127.

> once again to etymology to show that the Latins called air *anima,* meaning soul, distinguishing it from *animus,* spirit. 'Thus we may surmise that the ancient philosophers of Italy define soul and spirit by reference to the motion of air. And truly air is the vehicle of life.'[94]

The modern distinction between climate and weather is no more present in Greek and Hebrew thought than that between air and spirit, or nature and culture. In a Vichian perspective, earth system science — the science of climate change — is a newly made construct of the human-nature relationship since it excludes divine, human, and creaturely interactions which myth understands to be involved in the movements of the air and of bodies in the air.

In order for the truth of humanity's destabilisation of the atmosphere to become *true* for modern humans, Vico's thought suggests that there is an urgent necessity to translate climate science into a new myth analogous to the archaic myths which underwrote practices that distributed nature, land, and their use to families, and between social groups, in ways that situated human making and use of the earth within ecological limits and boundaried spaces and that restrained conflict.[95]

In Hebrew myth, nature and culture are not separate realms. The Hebrew Bible describes the first historically recorded arrangement for the fair sharing of use-rights to, and the usufruct of, the land and nature. The Hebrew covenant was a cosmic covenant between God and human beings in which the land and its nonhuman inhabitants were included as moral subjects, and fidelity to the covenant involved their respectful use according to the terms of the covenant. The Hebrews were required never to forget that they only occupied the land of Israel as former slaves who had been liberated from slavery through the gift of good land; they were never to forget that 'the earth belongs to the Lord', and their continued enjoyment and possession of land was therefore dependent upon their recognition and worship of its owner. It was dependent also on their continued observance of the divine commandments for living together in the land in a way that did not compromise the worship of God, and in which access to land and its fruit was equitably shared among the households of Israel.[96]

94. Luft, *Vico's Uncanny Humanism,* 128; text within inverted commas is from Vico, *On the Most Ancient Wisdom,* 85.

95. Luft, *Vico's Uncanny Humanism,* 174.

96. For fuller discussion and exposition of the Hebrew Bible on this point see further

The idea that unprecedented and extreme inequality in the United States in the late twentieth and early twenty-first centuries, as well as between the United States and other poorer nations in the Americas and beyond, is implicated in the extreme weather that has forced some to abandon houses and gardens in New Orleans or New York City, and others to abandon drought-plagued land in the American Southwest, is fanciful to those trained in the Baconian myth of nature as machine and of scientific knowledge as power over nature. But despite intermittent wars visited on European farmers by invading armies and would-be emperors, it was the development of settled post-slavery agriculture in early modern Europe which enabled households and communities to provide the crop surpluses on which the wealth, as well as the wars, of nations were built in the five hundred years since the scientific revolution. The possibility of an urban and scientific civilisation, such as grew up and overspread from early modern Europe to the New World and beyond, ultimately rests not upon technical innovation in the towns but upon the capacity of people to reliably grow more than enough food to feed their families and so make possible specialisation and the division of labour in the towns. But the quantities of greenhouse gas emissions required to sustain consumption levels sufficient to satisfy rent-seeking global economic actors in cities are beginning to destabilise food growing conditions in rural areas; by threatening food surpluses, these greenhouse gas emissions threaten through extreme weather the division of labour that makes possible the time of engineers and scientists, academics and financiers. There is therefore a relationship between extreme inequality in access to and use of fossil fuels, and the accumulation of wealth derived from their use, and current and growing threats to world food supplies.

For readers of the Hebrew Bible, that there is such a relationship comes as no surprise: Hebrew myth indicates that where the rich 'add house to house and field to field so that there is no room left' for others (Isaiah 5:8) eventually the rich will be punished and their cities and great palaces ruined: 'For Israel has forgotten his Maker, and builds temples; and Judah has multiplied fenced cities: but I will send a fire upon his cities, and it shall devour the palaces' (Hosea 8:14).

In the Hebrew Bible scarcity is not an inevitable feature of the human condition. In the myth of Eden, Adam and Eve have enough for their needs.

Michael S. Northcott, *The Environment and Christian Ethics* (Cambridge: Cambridge University Press, 1996), ch. 5.

But their desire to have more than enough, and above all to secure control over the earth from its owner by acquiring the 'knowledge of good and evil', is said to be their downfall. The Hebrew Bible tells this same story again in the history of Israel. A land of abundance 'flowing with milk and honey' is given to the people as a gift, after being violently wrested from its earlier inhabitants by divine intervention. Israel's continued possession of the land depends on their equitable sharing, so that all have enough; the rich and successful are always to ensure that they restrain the lust for power and seek instead to bring the poor, the widow, and the orphan along with them. But in Israel, as in Eden, the quest for power over nature wins out over the desire for justice and peace. As the Hebrews neglect the divine command to live equitably in the land and worship the Lord, they lose the land and are turned off it into Exile.

The history from Eden to Exile is now repeated in the climate-changing denouement of the Baconian salvation story. Copernicus displaced humanity from the centre of the earth and detached natural from human history. After Copernicus, humanity is in Exile on an earth spinning around the sun at the edge of a universe without ultimate meaning or purpose. In the new creation story of Bacon, humanity derives her own salvation by taking charge of the human-nature relation; the mediation of reason, and the powers released by science through technology, create a 'novum organum', a utopia in which species are subdued, territories deforested, oceans harvested, and mines and wells sunk into the crust of the earth as humanity strives to overcome natural contingencies, to reduce human suffering, and to defer, if not to defeat, death. But this new control over nature leads to a sundering of the relationship between land and labour, the concentration of growing numbers of people in smoky cities, and the eventual pollution of the skies not only of the cities but of the whole earth by the stored energy that drives forward the new story.

The Copernican turn comes full circle in the Anthropocene. If the radical Copernican turn decentred humanity in the destiny of the earth, now through science and technology humans are again becoming aware that they are central to the earthly networks of agency and being which stretch from the rocky substrata to the skies. But this new awareness has yet to stimulate the interior work to subdue greed, sin, and the will to power in the human heart and in the structures of society. The unique vocation of the Hebrews is that they are the first people in history to understand that this onerous struggle is one in which humanity must engage and endure if she is to bear responsibly her pivotal place below the angels. Instead, the

Baconian mediation of the Anthropocene suggests that the human tenure on earth can only be sustained, and Exile from a stable and fertile earth held back, by the further enhancement of the scientific and technological will to power over the external world of 'nature'. Hence the geoengineers propose to continue the Baconian story, to take control of the weather and the climate by mechanical carbon scrubbers, rocket-propelled sulphates, and ocean seeding with iron filings. Far from a return to the Adamic vocation of 'sons of the soil', the geoengineers respond to the threat of Exile by seeking to become 'earth masters'.[97]

Against the geoengineering vision of earth mastery, I turn in the next chapter to a view from below, as advocated by Marxian ecologists, who propose a historical origin of the climate crisis in the metabolic rift between land and labour which preceded the development of the modern global and imperious economy, whose most powerful agents have taken control of the climate by the extent of their interventions in the earth's carbon stores above and below ground. And I frame this view from below by contrasting it with the view from above of the United Nations project to respond to climate change by a global regime of markets in greenhouse gas emissions, which, the regulators intend, will ultimately price fossil fuel use in such a way as to hold back climate change below a catastrophe threshold.

97. Clive Hamilton, *Earth Masters: Playing God with the Climate* (New Haven, CT: Yale University Press, 2013).

4. Carbon Indulgences, Ecological Debt, and Metabolic Rift

The United Nations Framework Convention on Climate Change is founded on the assumption that the nations, severally and acting together, are morally and politically responsible for reducing their greenhouse gas emissions to a level commensurate with avoiding 'dangerous climate change'. Dangerous climate change was first identified in a risk analysis of the threat of climate change by the economist W. D. Nordhaus.[1] Subsequently temperature increase of two degrees Celsius was identified as a threshold by climate scientists and was written into Article 2 of the UNFCCC. However, the claim that an increase of two degrees is *not* dangerous is contested by climate scientists and others; this argument rests, in Nordhaus's case, on a cost-benefit analysis that near-term action to prevent heating of two degrees would be so costly that it is not feasible. That such a judgment is already a value judgment and not a 'natural scientific' judgment is clear.[2] That this judgment is associated with climate science, while in reality being a social scientific judgment about cost and benefit, just indicates that the fact-value distinction is not as secure as natural scientists and philosophers like to believe.[3] However, climate scientists use this uncertain number to quantify another number, which is the threshold of carbon dioxide emissions which will equate to two degrees and no more. In its Fourth Assessment Report

1. W. D. Nordhaus, 'Strategies for the Control of Carbon Dioxide', Cowles Foundation for Research in Economics, Yale University, 1977.

2. See further Michael Oppenheimer, 'Defining Dangerous Anthropogenic Interference: The Role of Science, the Limits of Science', *Risk Analysis* 25 (2005): 1399-1407; and Bron Szerszynski, 'Reading and Writing the Weather: Climate Technics and the Moment of Responsibility', *Theory, Culture and Society* 27 (2010): 9-30.

3. Carlo C. Jaeger and Julia Jaeger, 'Three Views of Two Degrees', *Regional Environmental Change* 11 (2011): S15-S26.

the IPCC judged that atmospheric CO_2 concentrations would need not to exceed 400 parts per million for there to be a prospect that global temperature rise might subsequently not exceed two to two-and-a-half degrees.[4] Four hundred parts per million of CO_2 was measured at the Mauna Loa Observatory for the first time on May 11, 2013. A more recent Global Circulation Model of the present global emissions pathway indicates that the two degree threshold will be passed around 2040.[5]

Given the already published IPCC scenario that 400 ppm equates to above two degrees of warming and that the 400 ppm limit was passed in May 2013, few climate scientists believe that the two degree temperature rise limit is now achievable. More likely is a rapid warming far beyond it because of the rapid rise, since 2000, of global greenhouse gas emissions. There is some cause for hope, in that temperature rise since 2000 has been lower than it might have been on some estimates. Some argue that this is because significant and growing atmospheric aerosols from China and India from coal burning power plants, factories, and domestic fires are presently shading the earth from the full effects of warming to be expected at present atmospheric CO_2 levels.[6]

The Cult of Carbon

The rate of deposit of CO_2 in the atmosphere rose during the first commitment period of the Kyoto Protocol, from an annual average of two parts per million in the 2000s to three parts per million since 2009. The 'Great Acceleration' of greenhouse gas emissions since 1950, far from slowing down, continues to grow in speed. At the present rate of increase, there will be a gaseous 'double glazing' on the atmosphere before 2050, with a doubling of atmospheric levels of CO_2 since the pre-industrial era.[7]

The Conferences of the Parties (CoP) and their first regulatory treaty,

4. Intergovernmental Panel on Climate Change, *Climate Change 2007: Synthesis.*

5. N. W. Arnell, J. A. Lowe, et al., 'A Global Assessment of the Effects of Climate Policy on the Impacts of Climate Change', *Nature: Climate Change* 3 (2013): 512-19; the Fifth Assessment Report of the IPCC was not available at time of writing.

6. Thomas Kuhn, Antti-Ilari Partanen, et al., 'Impact of Aerosol Emissions in China and India on Local and Global Climate', *Geophysical Research Abstracts* 15 (2013): EGU2013-10188.

7. See for example Kevin Anderson and Alice Bows, 'Beyond "Dangerous" Climate Change: Emission Scenarios for a New World', *Philosophical Transactions of the Royal Society* 369 (2011): 20-44.

known as the Kyoto Protocol, are failing to restrain fossil fuel extraction after more than fifteen years of deliberations. First among the reasons for the failure of CoP is the low target of an average of only 11 percent of emissions reduction on 1990 levels of pollution by 2012 established for the principal historic polluters — the developed countries — in the Kyoto Protocol. The European Union had sought a higher target of 40 percent, but under pressure from the United States, which ultimately did not sign the Protocol, the target was set at 11 percent. The target was far too low to affect the total of global greenhouse gas emissions because it would be more than outpaced by growth in emissions in developing countries. The CoP argued that because these countries had much lower historic emissions and much lower per capita emissions than the developed world, they should be excluded from initial efforts to rein in carbon emissions. The underlying assumption here is that 'climate justice' requires that developing countries enjoy *fossil fuel–based* economic growth in order to build the industrial infrastructure and wealth already available to developed countries. The better solution, and the only feasible solution for restraining climate heating below dangerous thresholds, is that the developing world takes an alternative pathway to development which does not involve the fossil fuel–based industrial one, with the destruction it involves not only of nature but of culture, but this path has not been taken.[8]

The Kyoto Protocol did not only fail to restrain total global carbon emissions, but even the emissions for which signatory developed countries are responsible have continued to rise. This is because the Protocol is based on national carbon accounts which only include territorial *production* of CO_2. By excluding consumption of products for which carbon is emitted in other nations, the treaty rewards those countries such as Denmark, Sweden, the UK, and the U.S. which, before the treaty came into force in 2005, had begun relocating carbon-heavy industries such as iron, steel, and cement making, ship building, and coal mining to the developing world. Products and services from these offshore industries are still consumed as imports, but the carbon involved in their production does not count against national Kyoto targets.[9] To put this another way, the Kyoto Protocol rewards Ameri-

8. See further Ivan Illich, *Tools for Conviviality* (London: Calder and Boyars, 1973); Arturo Escobar, 'Reflections on "Development": Grassroots Approaches and Alternative Politics in the Third World', *Futures* 24 (1992): 411-36; and Adrian Smith, 'Transforming Technological Regimes for Sustainable Development: A Role for Alternative Technology Niches?' *Science and Public Policy* 30 (2003): 127-35.

9. Dieter Helm, *The Carbon Crunch: How We're Getting Climate Change Wrong — and How to Fix It* (New Haven: Yale University Press, 2012), 67-72.

can and European corporations for exporting working-class industrial jobs and destroying working-class communities in the pursuit of higher profits, lower cost or 'flexible' labour markets, and weak environmental regulatory regimes in developing countries. Associating 'saved' carbon emissions in this way with the outsourcing of working class jobs, the creation of long-term structural unemployment in working class communities and the export of pollution to the skies and lungs of residents of developing countries bring climate change science and treaty making into disrepute.

There is a third reason for the failure of Kyoto, which is that, at the imperious insistence of the ultimately non-signatory party, the United States, the CoP used the Kyoto Protocol to set up an international regime of financial derivatives known as carbon 'credits' and 'permits'. Since 2005 the CoP has constructed regional and global markets in these credits and permits, the largest of which are the Clean Development Mechanism and the European Union Emissions Trading Scheme. These markets have had no impact on rising atmospheric levels of CO_2, but they permit nations and corporations to avoid genuine reductions in emissions by trading 'permits to pollute' with credits from 'saved emissions' bought in the various carbon markets now in operation.[10] In theory these markets create a carbon price which motivates companies and individuals to use less fossil fuel. But the price of carbon these markets are generating is too low to motivate changes in behaviour.

The adoption of carbon emissions trading as the primary means for restraining greenhouse gas emissions reflects a larger feature of the international negotiations on governance of the use of the atmosphere, which is competition and rivalry over property rights. The largest historic polluter — the United States — together with large oil, gas, and coal producing nations, including Australia, Canada, China, Russia, and Saudi Arabia, have resisted real physical reductions in greenhouse gas emissions as a threat to their 'right' to continuing profits from fossil fuel extraction and use. The creation of markets in carbon emissions trading reflects neoliberal prejudice against laws and taxes as means to modifying the behaviours of consumers and corporations; it also reflects the neoliberal description of the ideal society as one in which individuals realise their own good through maximal preference satisfaction via rational choice behaviours in competitive markets, rather than through cooperative and caring behaviours in

10. For a clear and detailed account of CoP approved emissions trading see Michael Grubb, Duncan Brack, and Christian Vrolijk, *Kyoto Protocol: A Guide and Assessment* (London: Earthscan, 1999), ch. 3.

families, neighbourhoods, cities, and nations. Carbon emissions markets, like many neoliberal governance procedures, are essentially anti-political devices which remove deliberation over greenhouse gas emission reduction from the arena of face to face deliberation by human beings in parliaments, law courts, local and community councils, churches, neighbourhoods, and households. Instead, they promote an ideal of autonomous mechanical markets in permits and credits of such complexity that they are only comprehensible to financial experts.

Leaving aside the problem of corruption and confusion in the global complexities of carbon markets, there is another underlying reason for their failure to restrain greenhouse emissions and this is that, while they should eventually raise the price of fossil fuel consumption, and make renewable energy, and energy efficiency measures, more price competitive, trading in carbon emissions derivatives does not restrain the *extraction* of fossil fuels. The failure to restrain extraction is the largest failure of the CoP, and the reason for this is identified in a paradox in fossil fuel energy supply first identified by William Jevons in relation to coal in the nineteenth century. Jevons found that economically justified energy efficiency savings would not result in less energy consumption. Instead, greater energy efficiency would have the same effect as cheaper energy: it would tend to promote more energy consumption, since the economic rewards of activities that would previously have been more costly are enhanced by energy efficiency, which lowers the price of energy while increasing the rewards gained from using it.[11]

It may be objected that the paradox assumes that people do not have other motives — such as concern about climate change — for reducing their energy consumption, and this is a reasonable objection. Some developed nations, including Scotland where I live, have under citizen pressure committed themselves to substantial reductions in their production of carbon dioxide in the next twenty years.[12] However, if all the displaced carbon from domestic carbon production is simply imported from other countries where the carbon is emitted instead, then national agreements only serve to move carbon production around, not reduce consumption. Even if effi-

11. William S. Jevons, 'The Coal Question: Can Britain Survive?' in A. W. Flux, ed., *The Coal Question: An Inquiry Concerning the Progress of the Nation, and the Probable Exhaustion of Our Coal-mines* (New York: Augustus M. Kelley, 1865).

12. Since 2005 Finland, Germany, New Zealand, Sweden, and the UK have all made substantial legislative commitments to decarbonise their economies, and all are in process of introducing a significant proportion of renewable energy electric power generation.

ciency gains, for example in transportation or building lighting and heating, lead to lower carbon consumption in some domains, if the global extraction of fossil fuels remains constant, these gains will simply result in a correlative reduction in demand and hence a lower price which will make fossil fuels cheaper for other uses that would previously have been uneconomic.

The other element of the Jevons paradox concerns what individuals or corporations do with the money saved from reduced energy costs. A survey in Sweden reveals that when Swedes eat less meat in response to animal welfare and climate change concerns they use the money saved to travel more.[13] Apart from cycling, walking, and sailing, travel is at least as fossil fuel intensive as raising animals for meat consumption. If money saved from energy efficiency and reduced energy consumption is used in other activities — such as foreign holidays — then the net impact of energy efficiency improvements in building use or domestic transportation is still neutral. Analogously, a corporation that saved money by becoming more energy efficient would be able to use the money saved to pass on greater rewards to shareholders, or higher salaries to employees, or both. If these translate into consumption activities, the Jevons paradox holds true again. As Jevons argues, 'it is wholly a confusion of ideas to suppose that the economical use of fuel is equivalent to a diminished consumption. The very contrary is the truth.'[14]

The Jevons paradox occurs in a particular fashion in present terrain-based cap and trade systems, as the following example indicates. Germany has made a significant commitment in the last twenty years to a transition to renewable energy, now deriving 15 percent of its electricity from renewable sources including biomass, solar, and wind. However, the increased availability of renewable energy has allowed German power producers to trade more of their permits to pollute beyond Germany in the European Union Emissions Trading Scheme (EUETS). Hence the emissions saved by the availability of renewable power in Germany are sold as permits to pollute, which enable power producers in other domains to pollute beyond agreed caps.[15] In the absence of international supply side controls on the

13. Joseph A. Tainter, 'Foreword', in John Polimeni, Kozo Mayumi, et al., eds., *The Jevons Paradox and the Myth of Resource Efficiency Improvements* (London: Earthscan, 2008), ix.

14. Jevons, 'The Coal Question', 123.

15. Alfred Waidermann, 'Green Energy Not Cutting Europe's Carbon', *Der Spiegel*, 10 February 2009. See also Christoph Bohringer, Tim Hoffmann, and Casiano Manrique-de-Lara-Penate, 'The Efficiency Costs of Separating Carbon Markets under the EU Emissions Trading Scheme: A Quantitative Assessment for Germany', *Energy Economics* 28 (2006): 44-61.

extraction of fossil fuels, no form of carbon pricing will reduce the global use of fossil fuels. But no nation, and not even the United Nations' IPCC, envisages restraining fossil fuel corporations from extracting and marketing fossil fuels.[16]

New coal is being commissioned on a weekly basis by developing countries which are not yet included in the emissions reduction regime established under the Kyoto Protocol. China and India are at time of writing firing up two new coal-fired power stations a month. Emissions from these new plants more than wipe out saved emissions in Europe under the Kyoto Protocol. Worse, the saved emissions in Europe promote the sense that Europeans are 'leading' the world to a solution on climate change when in reality they continue to consume carbon embedded in imported products even while 'green' politicians pat themselves on the back for promoting renewable energy and marginally reducing domestic production of carbon, while adding to foreign production of the same climate-heating gases.[17]

If the Jevons paradox is right — and it is based on common sense and simple mathematics — then the only effective way to reduce greenhouse gas emissions is to limit fossil fuel *extraction,* since with a continuous supply there is no limit to the potential of energy efficiency and savings in better regulated economies to contribute to emissions increases in less regulated economies. Helm argues that this problem can only be addressed if developed countries begin a practice of taxing carbon on carbon-intense imported goods such as aluminium, cement, iron, and steel (and I would add tropical timber, animal feeds, and plywood).[18] Carbon taxes on domestic carbon are also essential, since carbon emissions trading has hardly affected energy prices. Carbon taxes are much more efficient than markets in carbon emissions offsets and may have other morally good effects since, if they are imposed equitably on domestic and imported goods, they will end the export of carbon and other kinds of pollution to less well regulated economies. This might even repair some of the structural unemployment which has severely affected many industrial communities since the 1970s in North America and Europe. This approach will also assist in the development of a global price for carbon. But only when carbon is priced at equivalent levels across the *world* will the price of carbon motivate reduction in global *extraction* of fossil fuels. This is because a global carbon tax on internationally traded goods will provide a mo-

16. Helm, *Carbon Crunch,* 127.
17. Helm, *Carbon Crunch,* 136-38.
18. Helm, *Carbon Crunch,* 192-94.

tive in every country to reduce fossil fuel extraction and carbon consumption as well as domestic carbon emissions.

Helm is in a minority of economists in proposing a carbon tax as the key device for mitigating climate change. Vast amounts of political energy and time have been and will be invested in global treaty talks and in the related construction of carbon markets, because they do not challenge continued growth in fossil fuel extraction. Fossil fuel companies like carbon emission trading because governments let them have 'grandfathered' emissions permits for free, based on their fossil fuel production levels before the carbon markets were inaugurated. Governments like carbon emissions trading because it creates the impression that governments are doing something about climate change even though the net effect of their actions has been to hide an increasing rate of greenhouse gas emissions rather than genuinely to reduce carbon consumption in Europe, where the main architects of the Kyoto Protocol are located.

The proposal that environmental pollution should be treated as a cost that ought to be internalised by markets was first made by the economist Ronald Coase, who argues that since markets are more efficient at resolving social costs or 'market externalities' than government regulation, economists need to develop procedures for incorporating costs within market pricing mechanisms.[19] Building on Coase, John Dales argues that environmental pollution is a clear candidate for mitigation through markets in pollution credits and permits.[20] The first domestic trial of trade in pollutants took the form of a 1990 amendment to the United States' Clean Air Act, which inaugurated a new acid rain programme in the United States. The programme created a market in permits to emit sulphur from coal and oil-burning power stations and motivated annual sulphur emissions reductions of 4 million tons per annum.[21]

Emissions trading is said to be more efficient than command and control source-specific approaches to pollution because choices on how to cut emissions are left to market actors to achieve for themselves on the lowest cost basis.[22] Claims for its success are based on its early use in the United States. However, studies in the outcomes of sulphur emissions trading in

19. R. Coase, 'The Problem of Social Cost', *Journal of Law and Economics* 3 (1960): 1-44.

20. J. H. Dales, *Pollution, Property and Prices* (Toronto: Toronto University Press, 1968).

21. R. Schmalansee, L. Joskow, et al., 'An Interim Evaluation of Sulfur Dioxide Emissions Trading', *The Journal of Economic Perspectives* 12 (1998): 53-68.

22. A. D. Ellerman, A. Denny, et al., *Markets for Clean Air: The U.S. Acid Rain Program* (Cambridge: Cambridge University Press, 2000), 253.

American power utilities do not compare its effects with regulatory approaches in other domains but rather with previous command and control abatement schemes within the United States, which achieved lower rates of compliance and higher costs from litigation. The United States is a uniquely litigious jurisdiction, in which corporations regularly behave along 'rational choice' lines, estimating the relative costs of compliance with pollution regulation against the cost of payment of fines for noncompliance should government agencies discover law breaking and exact financial punishment through the courts. For it to be proven that market solutions are always cheaper it would have to hold true not just in the U.S. but also in countries such as Germany and Sweden, where corporate compliance with environmental legislation is much greater.

There is an additional problem with emissions trading, which is that this approach had never been tried on an international basis before the Kyoto Protocol. Carbon markets within and between nations involve the creation of several levels of concealment between market actors. The 'black box' of the carbon market trade reveals no information about the quality of efforts to conserve energy or reduce reliance on fossil fuels for which tradable emission credits are granted.[23] As with markets in other kinds of derivatives — such as the 'credit default swaps' based on mortgages and which precipitated the 2008 global financial crisis — the potential for fraud in these paper devices is very great.

The principle behind carbon emissions trading is known as 'cap and trade'. The idea is that power-producing companies in countries that participate in the Kyoto Protocol and successor treaties are assigned permits to produce carbon up to an assigned cap, set in relation to their previous emissions and gradually reduced. If they emit more than their cap they must go into the carbon market and purchase carbon permits from countries and corporations that have earned credits from activities that putatively reduce atmospheric emissions of greenhouse gases. Billions of dollars are invested in the creation and maintenance of new markets in carbon, rather than in activities that would reduce fossil fuel *extraction*. The result of this considerable diversion of human creativity and wealth into an economic activity has been to grow the shared perception, amongst many environmentalists and policy makers as well as scientists and political theorists, that carbon emissions trading is the best approach to climate mitigation, even although,

23. Donald MacKenzie, 'Making Things the Same: Gases, Emission Rights and the Politics of Carbon Markets', *Accounting, Organisations, and Society* 34 (2009): 440-55.

as Michael Grubb points out, the structure of emissions trading inaugurated by Kyoto 'makes emissions higher than they would be in the absence of trading.'[24] It is, in other words, a classic example of the fallacy of misplaced concreteness, and at least some natural scientists are now naming this vast misdirection of efforts to mitigate carbon pollution.[25]

Fraud has characterised carbon markets from the outset. Twenty-five percent of funds in the principal Kyoto Protocol emissions trading scheme — the Clean Development Mechanism — went to Chinese hydroelectric plants which the Chinese government was building or had planned before the inauguration of the CDM. Chinese factories also began producing a greenhouse gas known as HCFC-23, despite the availability of cheaper substitutes, solely because it was profitable under the CDM to produce the gas and then destroy it.[26] Investigation of the large market in carbon offsetting reveals that the majority of offset projects are unreliable as carbon sinks, and are sometimes just paper exercises.[27] The fiction of the carbon market begins with the procedure itself, which requires that the one who acquires the credit has simply to affirm the intention that the activity for which the credit is acquired is undertaken in part to offset carbon. Such an intention can be claimed without any risk of prosecution for fraud and hence carbon trades cannot be legally verified. Added to the concealing nature of the carbon market is the suppression of local knowledge of, and hence power over, energy, forest, and land resources where these markets are operative.[28] Tribal peoples who manage their forest habitats through common property practices lose knowledge and power over their forests when they are drawn into carbon emissions trading schemes, such as Reduced Emissions from Deforestation and Forest Degradation in Developing Countries, or 'REDD', in which corporately owned plantations are considered tradable and attract carbon credits while old growth forests lived in sustainably by indigenous peoples do not.[29]

24. Michael Grubb, 'International Emissions Trading under the Kyoto Protocol: Core Issues in Implementation', *Implementing Trading Mechanisms* 7 (1998): 140-46.

25. William H. Schlesinger, 'Carbon Trading', *Science* 314 (24 November 2006): 1217.

26. Jacqueline M. Drew and Michael E. Drew, 'Establishing Additionality: Fraud Vulnerabilities in the Clean Development Mechanism', *Accounting Research Journal* 23 (2010): 243-53.

27. Barbara Happya, *Failed Mechanism: How the CDM Is Subsidizing Hydro Development and Harming the Kyoto Protocol* (London: Rivers International, 2007).

28. Larry Lohmann, 'Carbon Trading, Climate Justice and the Production of Ignorance: Ten Examples', *Development* 51 (2008): 359-65.

29. L. Lebel, A. Contreras, S. Pasong, et al., 'Nobody Knows Best: Alternative Perspectives on Forest Management in Southeast Asia', *International Environmental Agreements: Politics, Law and Economics* 4 (2004): 111-27.

The advancement of carbon emissions trading over restraint on fossil fuel production — the principal geophysical cause of anthropogenic climate change — indicates a number of problematic philosophical premises rarely discussed in the social scientific literature on emissions trading. The first is that in modernity 'values' in the earth system are only recognised when they become part of the human economy of making and production and are given monetary equivalence. If there is a value to the avoided harms that a proportionate reduction in greenhouse gas emissions represents, this value is said to have more *productive* power when it is translated into monetary terms and becomes tradable against other values. This premise goes back to the political theology of John Locke, for whom values in the earth are not intrinsic, or derived from a divine Creator, but instead arise from human activities in making the earth fertile and productive. Locke enunciates the first theological justification for the dominance of money values in modern political economy when he proposes that money is the means by which work is preserved from the decay that afflicts the fallen natural order since it is 'a lasting thing which man may keep without despoiling'.[30]

The Lockean character of carbon emissions markets is expounded in a recent essay by a moral philosopher in defence of carbon markets, and more especially of 'grandfathering' of emissions permits to those with the longest history of substantial serial pollution. Luc Bovens argues that the atmosphere is analogous to unmanaged and unproductive land of the kind that Europeans settled in the New World and to which Locke argued they had acquired rights provided they made the land more productive and they did not own so much land that they wasted it. Emission rights reflect the Lockean principle that those who 'usefully' turn a polluted atmosphere into a productive resource by deriving wealth from the burning of fossil fuels ought to continue to enjoy at least a proportion of the property rights they had so acquired.[31]

Private Property and the Governance of the Commons

Emissions trading reflects an understanding of natural right which emanates from late medieval theology, whose first application to common en-

30. J. Locke, *Second Treatise of Civil Government,* section 46.

31. Luc Bovens, 'A Lockean Defense of Grandfathering Emission Rights', in Denis G. Arnold, ed., *The Ethics of Global Climate Change* (Cambridge: Cambridge University Press, 2011), 124-44.

vironmental goods was proposed by Hugo Grotius. Grotius wrote a treatise on prize and booty in response to the piracy of Dutch ships by Portugal, Spain, and Britain on their way back to Holland with goods from the East Indies.[32] He asked how property rights might be asserted, and legally affirmed, in domains that do not fall within the physical terrain of any one country, such as the ocean. Grotius locates property rights in the 'law of nature' according to which 'it shall be permissible to acquire for oneself, and to retain, those things which are useful for life.' Further, he proposes that 'every man is the governor and arbiter of affairs relative to his own property.'[33] Defence of individual and inalienable property rights from those who would detract from them is for Grotius a cause of just war. Thus 'the proper aim of belligerents is nothing more nor less than the repulsion of injury, or (and this, in the end, amounts to the same thing) the attainment of rights, not only one's own, but also, at times, the rights of others.'[34]

For Grotius, property rights originate in a divinely created order from the first days of creation and such rights are as much a part of the state of nature as human embodiment or other features of the physical environment. In making this claim Grotius repristinates the Roman legal understanding of property as *dominium,* and in so doing he departs from the Augustinian and Thomist account of property as an extension, under the guidance of reason, of revealed accounts of justice and the common good. Instead, Grotius builds on a new account of natural rights whose roots go back, as Brian Tierney shows, to medieval canon law, and the nominalist account of the unencumbered individual as propounded by William of Ockham.[35] This account erodes the traditional association between the use of property and duties to God and other persons. Thus for Grotius the right of use of that which is ' "seized as his own" by each person' is moderated not by divine providence in originally sharing the fruits of creation equitably in the Garden of Eden, but only by the duty to have regard for the safety of other individuals, for 'one created being should have regard for the welfare of his fellow beings, in such a way that all might be linked in mutual harmony as if by an everlasting covenant.'[36] Property for Grotius is part of the

32. Hugo Grotius, *Commentary on the Law of Prize and Booty* (1603), ed. Martine Julia van Ittersum (Indianapolis: Liberty Fund, 2006), II, Appendix A: 'Table of Rules and Laws Compiled.'

33. Grotius, *Law of Prize and Booty,* III, 23.

34. Grotius, *Law of Prize and Booty,* IX, 185.

35. Brian Tierney, *The Idea of Natural Rights* (Grand Rapids: Eerdmans, 2001).

36. Grotius, *Law of Prize and Booty,* II, 24.

postlapsarian world, in which 'as soon as living in common was no longer approved of' each man appropriated to himself 'by Right of first Possession, what could not have been divided'. Hence he sides with Cicero against Aquinas in asserting that 'whatever we possess by lawful Title can never, without Injustice, be taken from us.'[37]

Grotius's account of the nature and defence of individual property rights is accompanied by an account of the way in which sovereign nations are to claim rights over 'common property' such as the sea that may not be subjected to title. Grotius affirms the rights of each nation to use the sea as it wills for passage of its ships and for fishing beyond coastal waters without let or hindrance.[38] But this sets up an opposition between private interest and the common good which is made explicit in Thomas Hobbes's *Leviathan:* whereas for creatures such as bees and ants 'the common good differeth not from the private', men by virtue of their innate competitive and warring tendencies rarely see a contiguity between their own interests and those of others.[39] Hence, according to Hobbes, only the artificial leviathan of the social contract can keep human beings from war. The State — Leviathan — requires submission to its sovereign power which is exercised above all in a war with nature, for the 'state of nature' is a war of all against all, and can only be restrained by the sovereign power of the one over the many, and the State *over* nature.[40]

The crucial innovations that Grotius and Hobbes bequeath to modernity are the concept of autonomy implicit in the idea of the sovereign individual and sovereign State, and the concept of natural rights as deriving from property which originates in the 'state of nature' rather than in an originally peaceable creation.[41] These rights confer an independence from contingent relations of persons with God, with other persons, and with creation that none before Ockham had conceded. The result is a view of prop-

37. Hugo Grotius, *Of the Rights of War and Peace,* trans. J. Barbeyne, 2nd ed. (Clark, NJ: Law Book Exchange, 2004), Book II, ch. 1.4.

38. Thomas Hobbes likewise argues that rights over property arise from human control over nature, for everything cannot be available to all, and therefore 'the "right of every man to every thing" is limited by his ability to get hold of and keep it.' Thomas Hobbes, *Leviathan,* I, 14, 4.

39. Hobbes, *Leviathan,* II, 17, 113.

40. See the insightful account of the politico-theological implications of Leviathan in Jürgen Moltmann, 'Covenant or Leviathan? Political Theology for Modern Times', *Scottish Journal of Theology* 47 (1994): 19-42.

41. Tierney, *Idea of Natural Rights.*

erty and of rights that subverts the earlier theological priority of justice over sovereignty. This priority emanated from a view of property very different from that of the medieval voluntarists, who inaugurated the theological move from which the natural rights tradition takes its rise. As Joan O'Donovan argues, in the early Church Cyprian and Zeno of Verona viewed 'the communal sharing of things according to need as a true human imitation of God's equal beneficence to all men in bestowing natural gifts of, e.g., sunshine and rain, of moon, stars, and wind.'[42] In this Christian view the pattern of use of that which is divisible — land, housing, clothing, food — is shaped by the divine provision of goods such as sunshine, wind, and rain that are *indivisible*. By contrast, the modern approach to shared goods such as climate, oceans, and rivers involves the attempt — as in the case of emissions trading — to model access rights or rights of use to indivisible goods after a property regime in divisible goods.

The contemporary inability of state or corporate actors to think outside the discourses of private property and territorial rights in relation to climate pollution reflects the Grotian shaping of international relations and its subsequent unfolding in political theory. Hobbes builds on Grotius when he suggests that, like the individual, the state has a right to use whatever means may be needful for its self-preservation, since it is the states severally that judge on what belongs to whom 'by the right of nature';[43] property rights therefore necessitate the coercive policing of property law within states, and of territorial borders between them. So although the first Law of Nature with regard to nations is the pursuit of peace, where borders or property are infringed nations will necessarily seek to defend themselves through war. Grotius does not differ from Hobbes in this respect when he suggests that the first imperative of international law is that 'foreign property is respected'; the analogy in both Hobbes and Grotius is between that of a private individual and her property and that of a state and its territory.[44] The environment becomes an 'object of ownership in itself' and limits on its use increasingly described as positivist rules imposed by the State on prior natural rights rather than emanating from its ordering to God, and to communal justice.[45]

42. Joan Lockwood O'Donovan, 'Natural Law and Perfect Community: Contributions of Christian Platonism to Political Theory', *Modern Theology* 14 (1998): 19-46, 24.

43. Thomas Hobbes, *De Cive* I-8, 46-48.

44. A. Carty, *The Decay of International Law?* (Manchester: Manchester University Press, 1986).

45. S. Coyle and K. Morrow, *The Philosophical Foundations of Environmental Law* (Oxford: Hart, 2004), 213.

At the Enlightenment this account of property as representing the unfettered natural right of the individual and the State, who can claim legal ownership, becomes more fully entrenched in European and American political thought and legal practice. Consequently social constraints on the enjoyment of property — for the sake of the common good or collective welfare — come to be seen as infringements of private rights by the State. This reflects a new split between private law, derived from the doctrine of the natural right of the individual to property, and public law, which has no doctrinal basis but is instead arbitrary and instrumental.[46] Public law at this time is therefore increasingly defined in terms of the positive action of the State, and State parliaments, rather than as arising from and regulating the multiple agencies of networks of individual householders, farmers, craftspersons, and others. So whereas property rights are written on the land through title deeds, which show a genealogy of ownership, public law, of which environmental law is a part, is less evidentially based and lacks a philosophical rationale. Instead, it appears as the arbitrary interference by statute of the state into private property rights.[47] Environmental law and regulation is therefore resisted by economic corporations and individuals as unreasonable interference in the use of private property.

Holmes Rolston III and others propose the idea of the intrinsic value of the environment, and of particular species, to resolve the lack of attention in modern political conceptions of sovereign power to other creatures, and to commons such as air and ocean that cannot be territorially divided.[48] But this does not repair the philosophical emptiness of the positivist legal conception of private property, nor the tensions this conception presents between the dominion of individuals and corporations and the common good. The formulation of a better doctrinal foundation for environmental law, of the kind offered above in the exposition of Whitehead, is therefore an urgent necessity. In its absence restrictions on personal and corporate property rights arising from carbon reduction treaties or other environmental laws will always be hard to justify, representing as they seem to do an arbitrary invasion of propertied dominion. It is this lack of justification which in part explains the preference for forms of pollution control that allocate 'rights to pollute' and minimise the impact of pollution controls

46. Coyle and Morrow, *Philosophical Foundations of Environmental Law*, 160-61.

47. Coyle and Morrow, *Philosophical Foundations of Environmental Law*, 161.

48. Holmes Rolston, *Environmental Ethics: Duties to and Values in Nature* (Philadelphia: Temple University Press, 1988).

on the propertied dominion of individuals and corporations; the refusal of governments to restrain fossil fuel extraction to the capacity of the climate to absorb carbon is not therefore simply a sign of moral corruption, as Gardiner has it, but of the deep roots of the dominion of private property in modern political arrangements.[49] Once property is defined in such a way as to detach it from its setting in a network of creaturely being under God as the ultimate owner, regulations that restrict the enjoyment of property — such as a restriction on the speed of a private vehicle or the amount of energy used in running a house — are described and resisted as infringements on individual and private 'rights.'

The construction of markets in greenhouse gas emission credits and permits has the aim of promoting a market price for atmospheric pollution while not challenging the modern 'factish' cult of private property. The markets in credits and permits are intended both to prevent pollution — by raising the price and so discouraging pollution — but also to permit those who wish to go on polluting so to do provided they are prepared to pay the price. But markets in carbon indulgences acquire an independent, *idolatrous* existence from what they signify: they 'make the same' the activity of pollution and the payment of a price for permission to continue that activity.[50] Trades in the paper that 'makes the same' become a source of wealth creation that bears no relation to the actual physical state of ecosystems whose pollution is justified by these paper 'indulgences.'

Learning from the Commons to Resist the Babylonian Captivity of Global Capitalism

Although lawmaking and political theory increasingly reflect the Grotian account, we can discern in long-enduring common property regimes of the kind that still persist in Swiss summer pastures, or in residual common lands in parts of England, the persistence of the earlier Christian idea of common property in land use arrangements. Alpine pastures in Switzerland are still managed under a property regimen which manifests a complex array of rule-governed customary behaviours that indicate a range of responsibilities that fall upon owners of private houses and fields in alpine

49. Stephen Gardiner, *A Perfect Moral Storm: The Ethical Tragedy of Climate Change* (Oxford: Oxford University Press, 2011), 302-5.

50. MacKenzie, 'Making Things the Same.'

areas. Such arrangements have endured for centuries in Swiss cantons where 'alpine grazing meadows, the forests, the "waste" lands, the irrigation systems, and the paths and roads connecting privately and commonly owned properties' have been collectively managed since the thirteenth century.[51] Only resident citizens have use rights and common use is managed in such a way as to balance the needs of present users with the need to conserve common property for future users. Four-fifths of alpine terrain is managed in this way, and 'overuse of alpine meadows is rarely reported'.[52] Time spent in governing such arrangements has been shown to be a benefit rather than a burden. Participation in face to face arrangements that build community and promote a sentiment of collective justice around collectively used spaces has been shown to promote well-being whose levels are reportedly higher in Swiss cantons than in many other settings.[53] Typically such arrangements involve days of individually assigned as well as shared work, as well as meetings in which work is assigned, and then also days of shared festivity in which the fruits of shared work are themselves shared. Participation in such arrangements, moreover, fosters the practice of justice as a virtue, which is not honoured in a non-participatory bureaucratic procedure such as a redistributive tax but is agentially engaged in. Sharing in the praxis of justice enhances rather than diminishes individuals' sense of freedom as agents since the praxis is consensual and not imposed from outside by a bureaucratic agency or top-down regulation or treaty. Thus, political and economic freedom and the quality of social institutions are intrinsically related. Where democratic participation is high so also is well-being and the sense of being engaged in a shared set of just and participatory institutions.[54]

The continued existence of commons governance regimes in Swiss cantons or Somerset forests is an enduring mark of a Christian understanding of non-proprietary use of divine creation that is not ultimately owned by humans. The failure of international negotiations to produce a common commitment to the recovery of a safe atmosphere for human and other than human life indicates the urgency of the repristination of the concept

51. Elinor Ostrom, *Governing the Commons: The Evolution of Institutions for Collective Action* (Cambridge: Cambridge University Press, 1990), 62.

52. Ostrom, *Governing the Commons*, 64.

53. Bruno Frey and Alois Stutzer, *Happiness and Economics: How the Economy and Institutions Affect Human Well-Being* (Princeton, NJ: Princeton University Press, 2002).

54. Frey and Stutzer, *Happiness and Economics*. See also John Atherton, Ian Steadman, and Elaine Graham, eds., *The Practices of Happiness* (London: Routledge, 2010).

of the common good, and of creation as the metaphysical background of human affairs in accounts of propertied relations and of sovereignty.[55] Such an account resists the modern secular account of agency as being primarily vested in individual human agents and economic corporations, reflecting instead the larger networks of agency present in the environment and in social arrangements from the canton to constitutional democracies and beyond.[56]

To begin to learn from the canton how to respect the global commons requires not just a recovery of an earlier natural law account of property but a revision of the Hobbesian conception of the state as first and foremost a protector of private property. As we have seen, the current global carbon reduction regime measures territorial carbon production but does not acknowledge the underlying importance of terrain-based fossil fuel extraction. The mediating device that distributes rights to consumption is carbon emissions trading. This fails to restrain or reform a global market economy in which environmental obligations are regularly evaded or avoided by the principal property rights claimants in this economy — large public and privately owned fossil fuel companies which trade across national borders. These corporations evade national regulatory and taxation regimes by a variety of devices, such as multi-location of their offices and accounting processes, including the use of tax havens, transfer prices, and other devices.[57] As I discuss more fully in chapter six below, the increasingly borderless nature of the global economy undermines political governance. The operations of the global economy are the contemporary form of Roman *dominium*, and hence global borderless capitalism is responsible for the 'Babylonian captivity' of the earth in the Anthropocene.

Karl Marx argues that modern capitalism really begins with the importing of large quantities of gold and silver from the Americas into Europe in

55. See further David J. Hollenbach, *The Common Good and Christian Ethics* (Cambridge: Cambridge University Press, 2002); and Timothy J. Gorringe, *The Common Good and the Global Emergency: God and the Built Environment* (Cambridge: Cambridge University Press, 2011), 18-39.

56. W. Neil Adger, 'Social Capital, Collective Action, and Adaptation to Climate Change', *Economic Geography* 79 (October 2003): 387-404.

57. Nicholas Shaxson's research reveals that half of the monetary transactions in the global economy take place 'off-shore' and are therefore not susceptible to the national carbon accounting procedures of the UNFCCC. He also reveals that offshore banks are the principal source of investment loans for the development of new and more heavily polluting fossil fuel deposits; Shaxson, *Treasure Islands: Tax Havens and the Men Who Stole the World* (London: Bodley Head, 2011).

the sixteenth century.[58] This expanded Europe's capital stock while enabling many of its beneficiaries — especially Spain — to live off wealth which had been obtained at the expense of the desecration of the lands and cultural artefacts, and ultimately the genocide and enslavement, of Latin and Central American indigenous peoples. Andrew Simms makes an analogy between the sixteenth-century expropriation of gold and silver and fossil fuel–based wealth accumulation in the twentieth century. He argues that in both cases an 'ecological debt' was built up between Europe and the developing world which requires reparation.[59] However, the standard position on colonial expropriation was established by John Locke, who argued that colonial expropriation was justified by the more productive uses to which settlers put lands, precious metals, and other natural resources than did native peoples. For Locke, productivity is redemptive and the storing of productivity in magnetised form represents therefore a kind of pecuniary redemption. Locke's position was widely embraced in Europe and the Americas as justification of the colonial project, even though it was the sin of theft that Locke had justified. Climate indulgences in the form of carbon credits or offsets have an analogous effect.

Locke's justification of colonial expropriation took the form of a new theology of money, which rejected classical and Christian suspicion of monetised relationships, including usury. One root of this suspicion was Aristotle, who argued that money is sterile and should not be given 'life' by being lent at interest.[60] But Christian opposition to usury had a more ancient root in the Hebrew concept of *chesed*, variously translated righteousness or justice. Old Testament law included both a proscription against usury and a range of devices to repair situations where individuals and their households had lost land — and hence access to the fruits of creation — through debt. The Jubilee Law in particular was designed to return forfeited lands to their original owners after fifty years, so limiting the ability of rentiers and landowners to amass landed wealth at the expense of unsuccessful farmers. The principle of debt forgiveness was at the root of the practice of Jubilee and of other laws concerning debt bondage.

58. Karl Marx, *Capital: A Critique of Political Economy*, vol. 1, trans. Samuel Moore and Edward Aveling (Moscow: Progress Publishers, 1954), 738.

59. Andrew Simms, 'Ecological Debt — The Economic Possibilities for Our Grandchildren', in Chris Jochnick and Fraser A. Preston, eds., *Sovereign Debt at the Crossroads: Challenges and Proposals for Resolving the Third World Debt Crisis* (Oxford: Oxford University Press, 2006), 84-108.

60. B. N. Nelson, *The Idea of Usury: From Tribal Brotherhood to Universal Otherhood* (Chicago: University of Chicago Press, 1969), 73-82.

For the Hebrews, the power of debt to distort relationships among the people of God was a manifestation of sin against God, against persons made in the divine image, and against the land as the gift of God. For some to own too much of the land while others had no land at all, and hence became debt slaves or bonded labourers, was dishonourable and contrary to natural justice since God had conferred creation on all persons, and the land of Israel on all Israelites.[61] The rich, who acquired more land and wealth than the poor through industry and luck, were supposed to practice generosity toward the poor, as well as offer a tenth of their produce to God, and so redeem their abundance from the danger of unrighteousness. However, in the history of Israel the ideal of natural justice and the actual practices of Israelite elites were two very different things, and the Hebrew prophets rail against those who had taken over the land and left many poor and landless without reparative generosity.

Early Christian teaching on usury universalised the Old Testament suspicion of debt and debt bondage. Whereas the Jews were permitted to charge interest to strangers under Old Testament law, the Christians were not to lend money to anyone. This rule stood for over a thousand years, but was relaxed in the late Middle Ages. The papacy ran up huge debts in pursuit of its growing geopolitical ambitions, and these grew particularly large in the two centuries that it took to build St. Peter's Basilica in Rome. To relieve his debts the pope offered to grant individuals 'time off' purgatorial punishment for their sins in return for the monetary purchase of an indulgence.[62] The commodification of grace involved in indulgences was part of a larger spiritual corruption that infected the Catholic Church in the late Middle Ages and was the occasion for the first major historical break with Rome in the West in the form of Martin Luther's *Disputation on the Power and Efficacy of Indulgences,* commonly known as his Ninety-Five Theses against the Roman Catholic Church, with whose publication as an open letter to the pope the Reformation is generally said to have begun.[63] Luther's argument against indulgences turned on his understanding of justification. The sinner for Luther was justified, and her sins forgiven, by faith in Christ, not because of good works or monies given to the Church. The commodification of forgiveness

61. I give a fuller account of Hebrew land practices in Michael S. Northcott, *The Environment and Christian Ethics* (Cambridge: Cambridge University Press, 1996), ch. 5.

62. Robert W. Shaffern, 'Indulgences and Saintly Devotionalisms in the Middle Ages', *Catholic Historical Review* 84 (1998): 643-61.

63. Kurt Aland, ed., *Martin Luther's 95 Theses: With the Pertinent Documents from the History of the Reformation* (London: Westminster John Knox Press, 1967).

was wrong, for Luther, because it undermined the essence, as he perceived it, of the Christian religion. In *On Trade and Usury*, Luther developed the implications of his critique of indulgences for the growing mercantile economy. In it he argued that the problem with usury was not so much the Aristotelian account of money as lifeless but rather that the one who lends usuriously is seeking to secure his income and wealth independently of the risks taken by those to whom the money is lent. This for Luther was wrong because creation itself is a sphere of shared risk, and it is not possible 'to equate a fixed sum of money with a fixed land value' because natural events could change the product of the land from one year to the next. For Luther, no loan that does not include within its terms the shared interest of the lender and the borrower in the productivity of the land is legitimate. Usury is a way of making 'safe, certain, and continual profit out of unsafe, uncertain, and perishable goods' and this is contrary to justice.[64] Usury for Luther is a form of idolatry in which the lender encroaches on the work of God and puts himself in place of God.[65] Idolatry is not only a sin against God but against the people and the land. Luther argued elsewhere that in practising idolatry and usury, the Jews had rendered their culture 'inorganic' and that this was reflected in the despoliation of the soil of ancient Israel.[66]

The Roots of Climate Change in the Rift Between Land and Labour

In the light of Luther's critique of indulgences and usury, the core philosophical error involved in the Lockean monetisation of nature — of which carbon emissions trading is an egregious example — is that it underwrites the claim that it is possible to guarantee a return on the earth, including pollution, regardless of the created contingencies of climate, land, rocks, soil, species, and water that determine the productivity of the earth. In other words, it posits a relationship between human well-being and the earth that excludes the networks of being and exchange between humans, creatures,

64. Martin Luther, *On Trade and Usury*, in *Luther's Works*, vol. 45, ed. Walter Brandt (Philadelphia: Muhlenberg Press, 1962), 271. I am grateful to my graduate student Chad Rimmer for this reference.

65. Martin Luther, *On Trade and Usury*, in *Luther's Works*, 45:258.

66. Martin Luther, *On the Jews and Their Lies*, in *Luther's Works*, vol. 47, ed. Franklin Sherman (Minneapolis: Fortress Press, 1971), 211; see also Bret Stephenson and Susan Power Bratton, 'Martin Luther's Understanding of Sin's Impact on Nature and the Unlanding of the Jew', *Ecotheology* 9 (2000): 84-102.

cosmic agencies, and the Creator. The fruitful working of these networks and exchanges are often described by theologians as 'providence', although this term, which is increasingly pressed into service after the Reformation, obscures more than it reveals of the transactions between humans, other creatures, and cosmic powers which are more complexly and analogically described in the biblical sources.

Luther's economic thinking reflects patristic linkage of the ownership and use of property with a divine and transcendent conception of justice according to which God originally gave to each sufficient to meet their needs. If creation is used in such a way as to deny the sufficiency of others, then its original ordering to every individual by providence is undermined, and this is sin. By contrast, for Locke, land is redeemed from the curse of original sin by being productive and hence by being privately owned, for private owners are more likely to use the land productively than those, such as indigenous people, who use land in common property regimes. Moreover, for Locke the product of land can be legitimately stored as money, wealth secured in a bank or translated into gold.

Locke's invention represents a clear repudiation of the Christian understanding of redemption, and turns an occasion for sin — the exclusion of the poor from lands they had inherited — into the *means* of salvation, much as does the contemporary practice of carbon emissions trading. As with colonial expropriation, carbon emissions trading also produces ecological exclusion. Under the United Nations' REDD programme, the forests of indigenous people are claimed by sovereign governments and often sold to private corporations, as sinks for carbon emissions that are tradable as credits on carbon markets.[67] In the process, these forests are often turned from old growth forests, in which carbon was locked down, into fast-growing plantations. In the process, watersheds as well as the habitats of myriad species and indigenous peoples are destroyed, and forested lands are turned from carbon sinks into carbon sources.

While the Reformation Church resisted the infection of divine-human relations with the mercantile revival of propertied dominion, the reformers on the whole failed to challenge the growing commodification of land and of human-nature relations. Thus, for example, Luther sided with the rising merchant and bourgeois classes of Germany and declared that the peasants were unjustified in revolting against the theft of their ancestral

67. UN REDD Programme Strategy 2011-2015, http://www.un-redd.org/ (accessed 23 February 2013).

rights to land and livelihood. The difference between Luther's response to the commodification of *grace* and the commodification of *land* has marked Christianity's response to the rise of imperialism and capitalism from the sixteenth century until the present. Here at the origins of the Reformation, and of the modern age, we have the beginnings of a split between creation and grace, nature and culture, which has characterised mainstream Christian responses to the rise of mercantilism and capitalism since that time.

John Milbank argues that the modern split between nature and culture may be traced back to the medieval scholastic dualism between being and creation, philosophy and theology, reason and revelation, which influenced not only Descartes but Luther.[68] But the roots of this split in medieval theology go further back. As Franciso Benzoni demonstrates, it is Thomas Aquinas who first argues that Creation, while it has goods internal to itself, is not morally or spiritually valuable apart from its instrumental use to human beings.[69] In essence, 'beastly' theology begins here and produces a growing sense of the earth as an objective gathering of 'things' available for human use both above and below ground. If nature is primarily a realm of things that are not ensouled and have no destiny beyond human history, then the metaphysical ground is laid for the subsequent recovery of the Roman and pagan account of absolute ownership and the rise of private property as the dominant institution in human-nature relations after the Reformation.

Fossil fuel pollution on a scale sufficient to alter the climate did not and could not have occurred before the rise of a new form of property relations in the early history of capitalism. Mining of coal on a large scale required considerable investment in machinery of the kind that was only made possible by joint stock holding in deep mines, as we have seen.[70] At the same time investment in machinery driven by coal was only possible when sufficient monetary power had been accumulated to make the development and purchase of such machines feasible.[71] Prior to these events, both in Germany and Britain, capitalism had its origins in an original

68. John Milbank, 'Knowledge: The Theological Critique of Philosophy in Hamann and Jacobi', in John Milbank, Catherine Pickstock, and Graham Ward, eds., *Radical Orthodoxy: A New Theology* (London: Routledge, 1999), 21-37.

69. Francisco Benzoni, *Ecological Ethics and the Human Soul: Aquinas, Whitehead and the Metaphysics of Value* (South Bend, IN: Notre Dame University Press, 2007).

70. Lewis Mumford, *Technics and Civilisation* (New York: Harcourt, 1934), 75-76.

71. Karl Polanyi, *The Great Transformation: The Political and Economic Origins of Our Time* (New York: Farrar and Reinhart, 1944).

sundering of relationships between people and land, and between town and country, which fomented both the German Peasants' Revolt and the English Revolution. Neither revolution was successful, although outlier groups — for example, the Diggers and other radical dissenters in England and Anabaptists in Germany — continued the struggle over the land question. Some from both groups eventually refounded a land-based economy of labour, the remnants of which can be seen on the Amish farms which persist until today in Pennsylvania and the prairies of the United States and southern Canada, though of course the ancestors of these farmers still participated in the theft of land, approved by Locke, from its original native inhabitants.[72]

If it is true that the fossil fuel economy was not possible without capitalism, is it possible for there to be capitalism without fossil fuels? This question is at the heart of the problem of climate change but strangely rarely posed, perhaps because the end of capitalism is harder to imagine than the end of the world.[73] For technological optimists such as Amory Lovins, the failure to restrain greenhouse gas emissions is not intrinsic to capitalism but arises principally from state subsidies to fossil fuels and fossil fuel–dependent infrastructure. It is not capitalism but state rigging and regulation of energy supply and use which is the problem. If national governments shift subsidies and infrastructure toward renewable energy and energy conservation, including renewable energy supply and reductions in peak energy demand through smart technologies, then a post–fossil fuel capitalism is possible.[74]

More radical ecological thinkers argue that a deeper turn from present-day capitalism is needed before global atmospheric greenhouse gas emissions will begin to come down. In particular, the problems of remoteness and scale in a global economy will have to be addressed because, so long as nations severally can choose to continue to sell goods and services to other nations in a world trading system without regard to climate change,

72. On the Diggers and the Levelers see Christopher Hill, *Intellectual Origins of the English Revolution* (Oxford: Clarendon Press, 1980); Chris Rowland, 'The Common People and the Bible: Winstanley, Blake and Liberation Theology', *Prose Studies: History, Theory, Criticism* 2 (1999): 149-60; and Andrew Bradstock, *Winstanley and the Diggers, 1649-1999* (Portland: Frank Cass, 2000).

73. Frederic Jameson, *Archeologies of the Future: The Desire Called Utopia and Other Science Fictions* (London: Verso, 2005), 199.

74. Amory Lovins, *Reinventing Fire: Bold Business Solutions for the New Energy Era* (White River Junction, VT: Chelsea Green Publishers, 2011).

cheating will always undermine efforts by individual nations to restrain national emissions, even where such efforts include border taxes on carbon. Similarly, international treaties such as the Kyoto Protocol will not work if significant fossil fuel–producing countries fail to restrain fossil fuel production for overseas use while claiming to reduce their own territorial emissions. As Helm argues, border leakage and international cheating are intrinsic to the ongoing failure of current emissions reduction regimes to restrain global fossil fuel supply. Absent a global agreement to limit the rate of extraction of fossil fuels to the point that annual global carbon emissions are reduced, the policy of restricting territorial carbon production in Europe under the Kyoto Protocol has *zero* effect on atmospheric chemistry. Yet, as Hans-Werner Sinn points out, environmental policymakers have no plan 'to induce resource owners to modify their supply behaviour' and have not made 'the slightest attempt to get the resource owners on board.'[75] A carbon restriction regime which merely exports the carbon-intense jobs of Europeans is already giving rise to long-term structural unemployment in Europe, which in turn is a source of grave injustice and potential political destabilisation. The rise of fascism in Europe in the 1930s was triggered by just such long-term structural unemployment. It is unlikely that European governments will indefinitely be able to neglect the costs to jobs in Europe of their carbon policies, even if such neglect were not unjust.

The intrinsic relationship between capitalism and fossil fuels is indicated in the extent to which reserves of fossil fuels, though under the ground, are counted as the stock of wealth owned by some of the largest economic corporations traded in European and American stock exchanges. The threat of future restrictions on carbon production is stimulating owners of fossil fuel resources to extract more of them in an effort to realise their wealth before regulation makes it unrealisable; hence the considerable growth in coal extraction since 2000. It makes no difference whether stock exchange computers and screens in Frankfurt and London are fuelled by wind or coal-generated electric power if traded stock values depend, as they do in large part, on unused deposits of coal, oil, and gas. That those values are considered as *owned,* and extractable, drives the disposition of the owners of the resource to continue to extract it, and to accelerate its extraction as they have done during the first commitment period of the Kyoto Protocol. To call this 'greed' is too simplistic. The market

75. Hans-Werner Sinn, *The Green Paradox: A Supply-Side Approach to Global Warming* (Cambridge, MA: MIT Press, 2012), 187.

value of oil reserves alone is roughly equivalent to the combined value of all the houses and factories on the planet.[76] If it were written off as unusable, as Bill McKibben and others argue it should be, the collapse of pension funds and stock markets would trigger a world economic depression far worse than the global financial crisis of 2008.[77] Billions of people have public and private pensions invested in part in energy companies. The jobs and livelihoods as well as pensions of billions therefore depend at least in part on quoted fossil fuel reserves.

The Red and the Green Contradictions of Capitalism

If the problem of climate change is at root ownership of, and access to, the means of production — of which fossil fuels are the most valuable — it might be thought that a Marxist approach would suggest a solution. For Marx the core contradiction in capitalism arises from the class conflicts which emanate from the appropriation (or theft) of the *ownership* of the factors of production by capital. This theft — which originates in the ongoing depeasantisation of agriculture, first begun in England — is the source of alienation in workers who ultimately are motivated to wrest back control over the means of production. There is, however, a second, *ecological* contradiction in the new capitalist relations of production, with which Marxist thought is less often associated: just as there are *social* limits to capitalism arising from the centralisation of ownership of the means of production, there are also *ecological* limits arising from the intrinsic metabolic relationship between natural capital and human capital.[78] Whereas the social limits to capital did not result in collapse because of the success, at least until 1979, of the social market model in equitably distributing the fruits of production, the ecological limits to capital pose the longer term threat to capitalist relations of production.

The ecological roots of Marx's political economy are submerged for the most part in his work by the fact that his philosophy of history was developed as a response not only to the Hegelian defence of bourgeois society as the fruit of the Spirit in history, but also to Malthusian conservatives and

76. Sinn, *The Green Paradox,* 164.

77. Bill McKibben, 'Global Warming's Terrifying New Math', *Rolling Stone,* 19 July 2012, at http://www.rollingstone.com/politics/news/global-warmings-terrifying-new-math-20120719 (accessed 4 February 2013).

78. See further James O'Connor, 'The Second Contradiction of Capitalism', in O'Connor, *Natural Causes: Essays in Ecological Marxism* (New York: Guilford Press, 1988), 158-77.

sociobologists. Marx wrote at a time when the average life expectancy of the industrial labouring classes was sixteen years, as contrasted with the sixty-seven-year life expectancy of the upper classes.[79] Malthus, Spencer, and others argued that the immiseration and pauperisation of the masses, of the kind that characterised the three-hundred-year history of the Enclosures and the rise of polluted industrial cities, was a 'natural' problem arising from scarcity of land and natural resources relative to human reproduction. Against this view Marx argues that a scientific explanation for the increase in suffering of the industrial masses, and of rural labouring and servant classes, lay in the history of humanity's relation to nature and the history of tools, a history which is more significant in understanding the human condition than Darwin's histories of fossils.[80] Unlike other animals, Marx contends, 'man' adds stature to his being by his use of nature: nature is both 'one of the organs of his activity, one that he annexes to his own bodily organs' and man's 'original store house' and 'tool house'.[81] Labour is intrinsically work which transforms nature. Before capitalism this transformation took place within the limits of an evolved and emplaced relationship between labourers and land in which nature was not only consumed but sustained and stewarded on peasant farms. For Marx, pre-capitalist production is an ideal since the independent yeoman with four acres attached to his cottage, as the law required in sixteenth- and seventeenth-century England, had a self-sufficient life far superior to that of the industrial worker.[82] In the condition of the peasant farmer,

> Man lives from nature, i.e., nature is his body, and he must maintain a continuing dialogue with it if he is not to die. To say that man's physical and mental life is linked to nature simply means that nature is linked to itself, for man is a part of nature.[83]

Capitalism was the first historical form of economy that — through land expropriation, the criminalisation of vagrancy, and steam power — sys-

79. George Davey Smith, Douglas Carroll, et al., 'Socioeconomic Differentials in Mortality: Evidence from Glasgow Graveyards', *British Medical Journal* 305 (1992): 1554-57.

80. Karl Marx, *Capital: A Critique of Political Economy,* vol. 1, trans. Samuel Moore and Edward Aveling (Moscow: Progress Publishers, 1954), 352, n. 2.

81. Marx, *Capital,* 1:175.

82. Marx, *Capital,* 1:674-76.

83. Karl Marx, *Economic and Philosophical Manuscripts,* in David McLellan, ed., *Karl Marx: Selected Writings* (Oxford: Oxford University Press, 1977), 81.

tematically sundered labour from land, and hence production from consumption, town from country. Marx protests the results of this sundering in the suffering and criminalisation of the landless, and in the terrible living conditions of their descendants in unsanitary and polluted industrial cities. City air was toxic from the wastes of chimneys and factories; it was so polluted that it denied residents sunlight and led to an enduring epidemic of rickets; and city waters were fouled with human waste. The countryside, by contrast, was being emptied of people and the vitality of the soil was gradually leaching out of the land as its product in meat and cash crops was traded far from its source:

> Capitalist production, by collecting the population in great centres, and causing an ever-increasing preponderance of town population, on the one hand concentrates the historical motive power of society; on the other hand, it disturbs the circulation of matter between man and the soil, i.e., prevents the return to the soil of its elements consumed by man in the form of food and clothing; it therefore violates the conditions necessary to lasting fertility of the soil. By this action it destroys at the same time the health of the town labourer and the intellectual life of the rural labourer.[84]

In the third volume of *Capital* Marx describes how large-scale property ownership and remote trading arrangements enhance this break between land and labour, and the loss of nutrients to soils, which leads to their exhaustion:

> Large landed property reduces the agricultural population to an ever decreasing minimum and confronts it with an ever growing industrial population crammed together in large towns; in this way it produces conditions that provoke an irreparable rift in the interdependent processes of social metabolism, a metabolism prescribed by the natural laws of life itself. The result of this is a squandering of the vitality of the soil, which is carried by trade far beyond the bounds of a single country.[85]

John Bellamy Foster documents Marx's reliance in this crucial account of 'metabolic rift' on the work of soil chemist Justus von Liebig.[86] Liebig ar-

84. Marx, *Capital,* 1:474.

85. Karl Marx, *Capital,* vol. 3: *The Process of Capitalist Production as a Whole,* ed. Friedrich Engels (Moscow: Institute of Marxism-Leninism, 1959), 949-50.

86. John Bellamy Foster, *Marx's Ecology: Materialism and Nature* (New York: Monthly Review Press, 2000), 149-58.

gued that industrial civilisation, because it forces the majority of the population off the land, depletes soils of essential nutrients. In one study he observed that London's Thames River was being polluted systematically with human waste, which was subsequently washed into the ocean. This resulted in a loss of essential nutrients — especially nitrogen, potassium, and phosphorus — to agricultural soils where the city's food had been grown. Without the return of these nutrients to the soil, the soil's metabolic capacity to nourish crops was reduced and therefore there was a need to replace these lost nutrients with imported nutrients.

Despite the recognition of ecological problems in *Capital,* Marx was resistant to the idea that nature presents absolute limits to economic growth. He believed science would overcome limits to soil fertility, even given the tendency of industrial agriculture to fail to return nutrients to soils. The import of Peruvian guano and Chilean nitrates countered the loss of nutrients from English farms in the nineteenth century.[87] In the twentieth century farmers turned to fossil fuel–derived fertilisers for the same purpose, underlining Marx's expectation that science would trounce Malthusian accounts of biological limits.

Marx resisted the Malthusian claim of ecological limits to production not only because it was premised on bad science but because it underwrote the immiseration of the inhabitants of industrial cities. For Marx the key to a meaningful human life, and a purposive and progressive society, is the *transforming* potential of human labour when applied to matter. Use values, and their maximisation for social progress, arise from the application of human labour and tools to natural resources. The key to progress is the appropriation of these use values for collective uplift rather than private gain. The exclusion of workers from participation in the fruits of their own labour for Marx therefore has *spiritual* consequences and should be resisted on these grounds above all. If being a person becomes meaningful through agential participation in production, then the deprivation of this is the definitive source of alienation.

For most Christians, Marx's theory of agency and alienation places too much emphasis on material transformation and production relationships as the sources of sin and redemption. For green political theorists, Marx places too much emphasis on the human economy and insufficient emphasis on the economy of the land; hence his dismissal of ecological limits.

87. Justus von Liebig, *Letters on the Subject of the Utilization of the Metropolitan Sewage* (London: W. H. Collingridge, 1865), cited in Foster, *Marx's Ecology,* 154.

These criticisms are brought into relationship through a deep reading of the role of land and the priority given to equitable use rights in the Jewish and Christian traditions. As Hilaire Belloc, G. K. Chesterton, and the distributists argue, the unfolding impact of the Christian ideal of salvation in the history of Christendom led to the end of slavery in a large-scale civilisation for the first time in history in the late Middle Ages. Guild associations relating to crafts and technology also played crucial rules in the emergence of the economy of the late Middle Ages, and sustained new kinds of post-feudal citizenship and cultural and economic agency in towns and cities. It is therefore not irrelevant that the most productive country in Europe to this day, Germany, is a nation where more than two thousand guilds are still recognised and where these guilds participate in co-responsible relationships with factory or capital owners. From a distributist perspective, the Marxist critique of capitalism is too simplistic, not only in its neglect of ecological limits but also in its neglect of societal conditions. Relatively equitable access to use rights and genuine participation in production relationships are at least as possible under capitalism as under other economic arrangements, but arguably only under the influence of Christian ideals of community, freedom, and respect for persons.

The extent to which Germany remains a place where guilds and small farms are still a strong feature of the economy may also help to explain the extent to which the German state has been prepared to regulate electricity supply in such a way as to promote community and household renewable energy production as a high priority. The promotion of locally produced solar power has been a central plank of Germany's *Energieweinde* (energy turning). But absent a larger turn toward the local in the global energy economy, the Jevons paradox has ensured that this great turning in Germany has not translated into significantly reduced emissions in or beyond Germany. Furthermore, the installation cost over 30 billion Euros, and the ongoing and very costly feed-in tariffs have resulted in the German government looking to cheap domestic coal to provide for base load power. It brought two lignite-fuelled coal plants on stream in 2012, and more are planned.[88] At the same time the growing cost of carbon permits in Germany is driving carbon-intense industries such as cement and steel making out of Germany into domains where carbon is unpriced.

At first glance, the outcome of green technology in promoting coal looks perverse. Doing the right thing makes bad things happen. The prob-

88. Helm, *Carbon Crunch*, 127.

lem here is ultimately metaphysical rather than pragmatic. Green politicians, drawing on a romantic and ecological critique of modernity, reject the modern rationalist divide between humans as morally considerable subjects and nature as an assemblage of morally inconsiderable mechanical objects. If biological limits to economic growth are not to threaten human welfare, and even survival, nature as well as humanity should be brought into the domain of the morally considerable. But mainstream green political thought, despite adopting some 'red' hues in recent years, has not addressed the larger problem of ownership and power in a global economy. Here the relevance of Marx's critique of the accumulation of ownership of capital in fewer hands is crucial in understanding the failure of the nations in a global capitalist economy to restrain fossil fuel extraction. The nations regard their ownership of fossil fuels as a crucial source of wealth and capital accumulation, and such accumulation is pursued without regard to ecological limits. This is equally true of communist countries. Indeed, communist China is now the largest global producer of greenhouse gas emissions, while the cities of China are also sites of life-shortening levels of sulphur dioxide and particulate pollution. Marx and Marxism share with historic Christianity a deep consciousness of the capacity of unequal ownership of land and wealth to distort human relationships. But Marxism is no less Baconian than capitalism, and it is therefore equally promethean. Both ideologies are premised on the redemption of the human condition through the growing control over nature which is the salvific promise of modern science.

In an attempted ecological repair of Marxian materialism, Ted Benton argues that Marx places too much emphasis on the promethean and transformative powers of human beings, and insufficiently considers the context-dependent nature of the human economy.[89] Much human labour is about optimising and sustaining the conditions for the kinds of intentional transformations through labour of the craftsman or the engineer which Marx sees as the spiritual essence of all meaningful work: mining coal is not in itself a transformative experience, but the use of coal in creating an optical lens or a beautiful glass jug is.[90] In highlighting the social contradictions of the labour process under capitalism, Marx neglects the ecological contradictions and their equivalent potential to give rise to crises and conflicts,

89. Ted Benton, 'Marxism and Natural Limits: An Ecological Critique and Reconstruction', in Ted Benton, ed., *The Greening of Marxism* (New York: Guildford Press, 1996), 157-86.

90. Benton, 'Marxism and Natural Limits', 161.

including either natural limits on raw materials entering the production process, or on the capacity of natural systems to absorb residues:

> Once the dependence of productive labor processes on contextual conditions is explicitly recognized, then the possibility that they may be undermined by their own naturally mediated unintended consequences is open to investigation. Among these unintended consequences may be the effects of accessory raw materials and their residues as well as unutilized energy releases upon water supplies, atmospheric conditions, climatic variables and so on.[91]

Marx's failure to recognise created order as something that is truly inherited and not made is related to his ascription of ultimate metaphysical force to productive relations as the meaningful centre of human existence and of capital creation. In so doing Marx also clearly rejects theism of any kind; the one true transcendent becomes the human spirit either alienated or realised fully in the process of production.

Marx's overt position on religion arose in large part from his critique of idealism. For Marx, idealism is the last gasp of the religious impulse, which he, like Feuerbach, conceives as cultural projection. Religion reflects and obscures the alienated relations between rich and poor, capital and labour, which characterise the capitalist form of production. The construction of a transcendent realm of meaning and spirit beyond the material undermines the desired transition of material production from its oppressive and alienated form to its post-capitalist phase. Instead, this transformation is assured by the reassertion of a pagan naturalism which would redeem the essence of the human. For Marx, this essence finds intrinsic worth and meaning in the transformative work of humanity on nature and matter. The recovery of that transformative possibility after capitalism is conditional on the unfolding of the incipient antagonisms of class in the transition to the last phase of historical materialism, that of communist naturalism. As the associated producers take back control of the machinery and the land that capital had wrested from the workers, human beings would restate their labour and sociality. The recovered harmony with nature is said to heal the historic and metabolic rift between land and labour which was the condition for the evolution from pre-capitalist to capitalist modes of production.

91. Benton, 'Marxism and Natural Limits,' 167.

In Marx's vision, associated producers do not act alone. Reflecting the legacy of Hegel, the state would play a determinative role in this transition once violent revolution had drawn the state and civil society together, so enabling the workers to recover the means of production from the capital owners. Marx's account of the redemptive potential of the state is the hidden religion of Marxism and Marxian humanism.[92] This is writ large in the development of communist state cults which took the form of a religious mythos around communist struggles, 'heroic' leaders and workers, and the sacrifices of individual freedoms for the collective. Really existing communist states such as those of the Soviet Union and Maoist China, and their mass persecution of religion, as well as genocidal slaughter of perceived enemies of the state, are therefore not so easy to set apart from the legacy of Marx as some contemporary Marxists propose.

Marxist atheism does not eviscerate the need for redemption, nor even, as it turns out, for religion, which reappears in the guise of the cult of the state or civil religion. For Marx, the need for redemption arises originally from the age-old conflict between nature and humanity. But when humans, after Bacon, learn both nature's laws and how to control nature more effectively for their needs and purposes, they achieve this on a scale, and under an ownership system, which creates metabolic rift between land and labour, which is another way of describing the nature-culture divide. Marx, like Darwin, views the evolutionary struggle for survival as the original source of conflict in the human condition, which is resolved under capitalism by a new conflict between humans and nature in the form of an alienation between land and labour. Commodity fetishism, alienating forms of work, and degraded and polluted living conditions for workers are all the result of the split between land and labour, which turns workers into alienated producers who participate in the fetishism of what they and other workers make. The *end* of production under capitalism is the commodity form; it produces a new alienation between human beings and nature, including their own nature as embodied beings as well as mindful spirits. Hence in Marxist thought alienation between consumption and production is the root also of the *ecological* limits to capitalism. Large-scale production, combined with alienation and commodity fetishism, produces a situation in which the inner life of the worker is directed toward the monetised exchange of commodities which are wrested from land and labour. This new set of conflicts is resolvable only through a novel, utopian, and harmonious

92. John Milbank, *Theology and Social Theory* (Oxford: Blackwell, 1991), 183.

relationship between land and labour which, though never realised in history, Marx argues is realisable *after* capitalism.

Marx is essentially an Epicurean, having written his doctoral thesis on Epicurus at the outset of his career, who also came under the influence of Darwin.[93] That the path to the materialist utopia Marx envisages involves large-scale suffering and violent revolution suggests that the Darwinian revision of Epicurus is thoroughgoing. The struggle for existence which Darwin takes as his guiding metaphor for species history gives to Marx's account of social struggle a biological imperative. But it also raises the question of the end or goal of the redemption *through* suffering that Marx envisages.

One end of Marx's envisaged utopian future is the recovery of an agrarian society where small scale farmers and associated producers recover a harmonious relationship between production and consumption, return to the soil what they take, and hence leave it in a good condition for the next generation. Whereas capitalist agriculture is guided primarily by fluctuations in prices, what Marx calls 'rational agriculture' 'has to concern itself with the whole gamut of permanent conditions of life required by the chain of human generations'.[94] This agrarian vision of stewardship for future generations is matched with an awareness of the waste of the past heritage of the earth involved in capitalist industrial production. Thus Engels, in a letter to Marx which anticipates the problem of climate change, observes that

> the working individual is not only a stabiliser of the present but also, and to a far greater extent, a squanderer of the past, solar heat. As to what we have done in the way of squandering our reserves of energy, our coal, our ore, forests, etc., you are better informed than I.[95]

Recognition of the foreshortening of time by capitalism is a central theme in Marx's work, and in subsequent Marxism. It is also illustrative of the central role of history in Marx's philosophy of political economy. Whereas Classical and neoclassical economists describe economic relationships and exchanges almost entirely without regard to the historical events that precede the present order of such exchanges, for Marx the prehistory of capitalism in the theft of land from labour is the definitive historical horizon which

93. Foster, *Marx's Ecology,* 56.

94. Marx, *Capital,* 3:754, cited in Foster, *Marx's Ecology,* 164.

95. Karl Marx and Friedrich Engels, *Collected Works,* 46:411, cited in Foster, *Marx's Ecology,* 166.

sets up the horizon of the future beyond capitalism where labour — with the state — takes back the land and the other means of production from capital. However, the Marxian dream of a liberated agrarian future is given the lie by the history of really existing communism. Communist countries that have enforced 'back to the land' movements on 'liberated' peoples have had a very poor record indeed of actually improving agriculture and reducing waste. Indeed, in the history of the twentieth century fewer major agricultural disasters were precipitated by capitalism than by communism and state socialism. Soviet and Chinese collectivised farms had terrible records of production and land care compared to the peasant agriculture that preceded them and compared to their capitalist counterparts in Europe and the Americas.[96] There were similar problems in the disastrous collectivisation of agriculture in socialist Tanzania.[97]

There are, however, instances of small-scale communist governance which have been less disastrous. The communist government of the South Indian state of Kerala enacted a process of radical land reform in 1964. Large land holdings were distributed from high caste to low caste Keralans, and the procedure had exceptionally good outcomes. Kerala subsequently enjoyed improved agricultural productivity as well as educational, health, and mortality outcomes equivalent to those of the United States and far in advance of the rest of India.[98] Communist Cuba initially nationalised vast foreign-owned sugar plantations, but then was forced in the 1990s to organise an ultimately successful agrarian revolution under the stimulus of the Reagan-era trade and oil embargo and the collapse of the Soviet Union. These two events forced Cuba to find ways to grow food, and to manufacture and repair clothing construction materials and machinery, without significant inputs of oil. While acknowledging the dominant role of the state in Cuba, and related suppressions of freedoms of association and expression, nonetheless the Cuban case shows that a transition to a post-carbon agriculture is possible within a short time scale. It also had relatively good outcomes in terms of reduced infant mortality and rising adult health outcomes because it increased dietary quality and consumption of locally grown organic foods. It led to a greening of urban areas with more space devoted to food growing and less space reserved

96. On environmental outcomes in the Soviet Union see Philip R. Pryde, *Environmental Management in the Soviet Union* (Cambridge: Cambridge University Press, 1991).

97. See further James C. Scott, *Seeing Like a State: How Certain Schemes to Improve the Human Condition Have Failed* (New Haven, CT: Yale University Press, 1998).

98. G. Parayil, 'The "Kerala Model" of International Development: Development and Sustainability in the Third World', *Third World Quarterly* 17 (1996): 941-57.

for the movement of vehicles; it promoted decentralisation of food production and supply and hence a more equitable distribution of property — or at least use — rights; it improved community relationships as people invested time in shared management of urban allotments; and it improved urban-rural relationships as people in the towns got more involved in farming in rural areas.[99] Cuba and Kerala are different cases, and Cuba is marred still today by a relative lack of political and religious freedoms. But both examples reveal the potential that Marx had envisaged when a multigenerational history of land theft is reversed, and farming and the land are put back in the hands of small family farmers and urban residents. Both cases also indicate that there can be a synergy between the emancipatory project of Marxism and socialism and the ecological project of bringing the economy back into relation to biophysical limits.

The heart of the problem with Marx remains, however, his overt refusal of transcendence and of any teleological shaping of human life, and natural life, toward moral and spiritual ends beyond that of the transforming power of work. Underlying this refusal of teleology there is a deep ontological philosophy of violence which argues for the necessity of suffering and violent revolution as the means to societal redemption. Green political theorists argue analogously that there is antagonism between biological limits and human development and that the ecological crisis is therefore an inevitable result of an inbuilt contradiction between industrial and technological development and its contribution to human population growth on the one hand, and ecological limits on the other. Ecological collapse and consequent human suffering and even 'die-off' play the same role in green political thought that class conflict and social crisis play in Marxism. Hence it is not surprising that the two contradictions of capitalism, the red and the green, perversely coalesce in the green political drive for carbon caps, which have the effect of exporting working-class jobs from Europe while leaving fossil fuel corporations in charge of ongoing fossil fuel extraction.

From Political Economy to Political Theology

At the turn of the twentieth century, the Russian political economist and theologian Sergei Bulgakov saw the limits of Marx's theoretical account of

99. Julia Wright, *Sustainable Agriculture and Food Security in an Era of Oil Scarcity: Lessons from Cuba* (London: Earthscan, 2009).

agriculture and made a definitive break with Marx in a major two-volume study entitled *Capitalism and Agriculture.*[100] For Bulgakov, the key problem with Marx's account of capitalist agriculture is a metaphysical one. Whereas in industry human beings are masters of their machinery and its products, in agriculture they remain dependent on the earth and its products, and this dependence has a metaphysical root. Humans are co-creatures with the rest of creation as well as bearers of the divine image. Land is the form in which creation enters into their agrarian activities, and use rights to land, and their economic form of land rent, constitute a constraint on capitalist accumulation through agriculture which does not exist to the same degree in industry.[101] Bulgakov is therefore critical of the American style of mechanised farming, which, although it had applied capitalist industrial economics to farming, was mistaken since it represented 'not technical progress but a barbarian means of violating the virgin soil.'[102] Bulgakov is also critical of Marx's account of English agriculture as the basis of a universal pattern of agricultural and industrial exploitation. Few countries followed England in its near complete separation of the peasantry from the land, and in some, including Germany and France, there remained a viable peasant agriculture into the twentieth century.

A third strand of Bulgakov's critique of Marxian political economy concerns creativity. For Bulgakov, human beings, as Marx asserts, require an appropriate level of material security which peasants have when they own sufficient land on which to make a living. But once they have material security other kinds of creative and contemplative forms of life become possible. These forms of life are not extrinsic to the human condition for Bulgakov; they are not merely cultural extrusions of the material base of existence, as Marx had it, but they are rather the means for the ennobling and transcendence of the conditions of struggle with nature and mortality that

100. Sergei Bulgakov, *Kapitalism e zemledelie* (Capitalism and agriculture), 2 vols. (St. Petersburg, 1900).

101. Bulgakov's two-volume work has never been translated into English. I am following here the summary of Bulgakov's argument in Catherine Evtuhov, *The Cross and the Sickle: The Fate of Russian Religious Thought* (Ithaca, NY: Cornell University Press, 1999), 33-36. Earlier treatments and responses to Bulgakov's views on capitalism and agriculture include an unsympathetic reading by Lenin in his *Capitalism in Agriculture: Kautsky's Book and Mr Bulgakov's Article,* in V. I. Lenin, *Collected Works,* vol. 4, trans. Joe Fineberg and George Hanna (Moscow: Progress Publishers, 1960), 105-59.

102. Bulgakov, *Kapitalism e zemledelie,* 2:49, cited in Evtuhov, *The Cross and the Sickle,* 35.

reduce the life of the poorest in every age to a diminished existence shorn of the possibilities of moral and spiritual transcendence.

For Bulgakov, freedom from bondage to material conditions is not true freedom, but it is the material precondition of that moral and spiritual freedom which is the true end of the human condition:

> The spiritual life requires one condition, a negative condition, but invaluable and irreplaceable: *freedom*. Ethical self-determination cannot but be free, and, conversely only free self-determination can have an ethical goal. It would be superfluous to try to demonstrate this self-evident truth. But perfect freedom belongs only to pure spirit, which is free from all external influence and open only to interior motivation. Man exists in *bodily* form and so is bound to the external and material world in which mechanical necessity dominates. The freedom of the human spirit is thus necessarily subject to external limitation, not to mention internal limitation as well; complete and purely spiritual freedom is an unattainable ideal for empirical man. Yet it is still his ideal. The nearer man comes to this ideal and the more his spiritual life is autonomous, the more fully can he express his spiritual self, his spiritual 'I'. This contradicts Hegel's formulation of the matter: Hegel sees the structure of history simply in the development of the human spirit toward liberty and self-awareness.[103]

The activities of production and exchange that constitute the economy have a guiding goal and purpose beyond the economy: to enable humanity to draw the material and the natural character and constitution of created order into an aesthetic, moral, and spiritual purpose which apart from the human spirit they cannot sustain. This is why for Bulgakov 'poverty creates the kind of suffering that degrades man and excludes the possibility of a properly human and spiritual life' and hence 'the battle against poverty is a battle for the rights of the human spirit'.[104]

Modern technology and political economy endlessly expand the ecological invasiveness of human needs to the point where mastery over nature becomes an end in itself. Human identity and purposiveness are also at risk of the poverty of spirit that is the defining feature of modern political economy. Hedonism for Bulgakov is the characteristic sin of modern political

103. Sergei Bulgakov, 'The Economic Ideal', in Rowan Williams, ed., *Sergei Bulgakov: Towards a Russian Political Theology* (Edinburgh: T&T Clark, 1999), 23-54, 41.

104. Bulgakov, 'The Economic Ideal', 49.

economy, just as asceticism — particularly *ascesis* for the poor — was the besetting sin of the pre-capitalist era. The answer for Bulgakov to the resolution of the sins of modern capitalism is not, however, violent overthrow of the existing order, but 'spiritual combat' against hedonic luxury and excess which represent the 'triumph of sensuality over spirit, of mammon over God whether in the individual soul or in the whole of society':

> Once the cult of gratification, aesthetic or non-aesthetic, has become a guiding principle, we have luxury. Luxury is the reverse side and the constant peril of wealth. Just as in a state of poverty the spirit's liberty is negated by external limitations, so in wealth it falls victim to internal temptation. Luxury and poverty are equally the enemies of culture: there may be as much spiritual poverty in the nobleman's palace as in the pauper's hovel. The spiritual decline that accompanies luxury sooner or later leads to economic decline as well, so that luxury is self-condemned even from an economic point of view. The spiritual state of a nation is very far from being a matter of indifference for its economic life.[105]

This state of economic excess and hedonism has not only poor spiritual and ultimately economic outcomes. It also undermines the possibility of a genuinely moral life:

> The modern mind, with its excessive mechanisation of life, is particularly given to designing ideal social systems in which people will be virtuous automatically, as it were, without any kind of struggle with themselves. No: virtue is always bestowed and created only by moral conflict; and if our destiny is above all to struggle against poverty and all kinds of tyranny the destiny of humanity in the future will be, along with other kinds of battle, the struggle against luxury and wealth.[106]

Here is the kernel of the disagreement between those who would discount that there is any issue about limits to consumption arising from climate change, and the larger ecological crisis, and those who argue that luxury and excess can no longer be justified in the face of the coming emergency, and that consumerism is a symptom of a deeper disease in capitalist (and communist) industrial societies. It is not so much an argument about cap-

105. Bulgakov, 'The Economic Ideal', 49.
106. Bulgakov, 'The Economic Ideal', 49.

italism and communism, since both forms of political economy prioritise economy as an end in itself. It is instead an argument between political economy and political *theology,* but not of the Marxist kind.

The climate emergency is a genuine emergency; it is not a hoax. It is an emergency in particular for those people who might otherwise have expected finally to be delivered from the drudgery and material want of the poor peasant farmer or the urban squatter. But it is an emergency that also represents an opportunity to rethink the apparent antagonism between the socialist project of class emancipation from capital ownership and the green political project of bringing the economy back into scale to ecological limits.[107] The crisis is a genuinely transformatory historical moment, which reveals that the socialist quest for emancipation of farmers and industrial workers from immiseration, and the green political quest for forms of human economy that resist scientific prometheanism, are intertwined, rather than in opposition, as demand-side approaches to climate change seem to indicate. A green politics that does not destroy working-class jobs requires a critique of patterns of capitalist ownership in which giant private corporations continue to value and extract fossil fuels, and other natural resources, without regard to ecological limits or justice. The quest for social emancipation through an emancipatory response to climate change requires the refusal that fossil fuels in the crust of the earth are 'private property' belonging to economic corporations whose guiding responsibility is to maximise their fiduciary obligations to their shareholders. Hence the Christian account of use rights to the earth as limited by social obligations and obligations to the Creator, which I discussed above, is essential to a resolution of the blind spot in green politics which embraces 'natural' ways of generating energy while neglecting to critique the institution of private property. It is this conception which underwrites the corporate claim of right to continue to burn fossil fuels at rates that will bring on four or even six degrees of global warming and render much of the planet uninhabitable by human beings by the end of this century.

The climate crisis reveals the moral and spiritual limitations of Marxism *and* capitalism. Against the stultifying materialist excess promoted by the illusory freedoms of modern political economy, whether in communist or capitalist guise, climate change presents a moment of crisis, and opportunity. It both uncovers and reveals the hedonic moral and spiritual depravity of the present 'advanced' state of industrial civilisation. This admission runs

107. Benton, 'Marxism and Natural Limits', 172-73.

counter to the established liberal view that personal choice in the economy is to be limited only when precise particular harms to other proximate persons result from an individual's consumptive activities, such as death from dangerous driving. The linkage of the ecological crisis with the moral critique of hedonism is resisted by those who sustain the right of the rich to drive and fly and generally continue to pollute the atmosphere without let or hindrance. The consumption of air travel, cars, meat, and other luxury goods by around one billion people are activities that are judged by the global market to be more economically *valuable* than the subsistence activities of more than two billion hunter gatherers and peasant farmers whose livelihoods are being eroded by drought, flood, and rising temperatures. Planetary limits on the activities of the relatively wealthy go unrecognised in the dominant valuing devices of market economics. This refusal of planetary limits also reflects the dominant philosophy of economic liberalism according to which activities in the economic sphere are better governed by commercial interest and personal choice than by collective deliberation.

Grotius, Locke, and Smith all played crucial roles in the emergence of modern liberal political economy. All took up the politico-theological account of human freedom which in the late Middle Ages ended slavery, and they attempted to apply it to the emergence of a Christian-originated global economy. For all three, the rights of the individual European householder and business owner were the crucial restraint on the growing power of empires to draw individuals into new kinds of collectivity, which placed more emphasis on sovereign power than on individual rights and therefore underwrote a new dominion of ownership of nature. But in so doing, the architects of modern liberalism and political economy ironically and unintentionally colluded with the merchants, the papacy, and the nascent imperial nation-states of England, Holland, Portugal, and Spain in undoing more than a millennium of moral and customary restraints on ownership and wealth accumulation in Christian history.

Marx's Hegelian-influenced repair to economic liberalism drew on the resources of the Jewish and Christian traditions in re-emphasising the role of land and property regimes, and unequal access to the fruits of nature and economy, in an attempted repair to liberal capitalism and imperialism. His emphasis on the potential of social crisis to create the conditions for the emergence of a new and more egalitarian society also recalls the core insights of Christian eschatology and the nature of apocalyptic literature in the Old and New Testaments. However, Marx overemphasised the power of social and structural change to redeem social and individual sin, and he therefore advanced a collectivist statist doctrine whose totalitarian outcomes were a cause ultimately of greater suffering than liberal capitalism.

If Marx's critique of imperialism and economism represents an attempt to repair the growing instrumentalisation of persons in liberal regimes of modern capital accumulation through a new account of Christian moral and political collectivism, then the philosophy of Immanuel Kant can be understood as an equivalent attempted repair to modern liberalism from the side of a restatement of Christian moral individualism. Kant also developed a powerful account of cosmopolitanism between persons and nations which was highly influential in the emergence of the United Nations, and hence the UN project to limit dangerous climate change. It is to Kant's attempted repair of liberalism through cosmopolitan reason, and his related account of the *a priori* moral law, that I turn next in tracing the reasons for the failure of the UN project to limit climate-damaging fossil fuel emissions.

5. The Crisis of Cosmopolitan Reason

The eighteenth Conference of the Parties to the UNFCCC met in Doha, Qatar, in December 2012. Qatar's wealth rests principally on its exports of natural gas and oil, and the nation supplies energy free to all its citizens. It was symbolically fitting, given the low expectations of the conference, that the gathering took place in the country with the highest per capita emissions of greenhouse gases in the world, and a country that has not acknowledged a need to restrain its greenhouse gas emissions in the light of climate change. The conference ended in extra time during which the 'industrialised nations' agreed to a second commitment period for the Kyoto Protocol from January 2013, albeit with very modest further cuts in emissions than those already achieved in the first period.[1] However, the conference also decided that nations such as Poland, which had lower emissions than permitted in the first commitment period due to economic recession, would be allowed to raise their emissions correlatively in the second commitment period. These 'hot air' allowances negate much of the potential of the second commitment period to reduce total greenhouse gas emissions from signatory countries. In any case, more than two-thirds of current global greenhouse gas emissions emanate from countries — including Brazil, Canada, China, Indonesia, India, Russia, Saudi Arabia, Singapore, South Korea, Taiwan, and the United States — who have either refused to ratify the Protocol or are not included in the Annex 1 list of nations which are required to reduce their emissions under the Protocol.

The UNFCCC has so far promoted two kinds of agreed actions by the nations as ways of responding to the threat of dangerous climate change.

1. 'What Doha Did: No Progress Today, but a Slightly Better Chance of Progress Tomorrow', *The Economist*, 15 December 2012.

The first is the establishment of agreed targets for national terrestrial emissions of greenhouse gases, the intention of which is to promote mitigation of climate change. These targets are related to national inventories of greenhouse gas emissions, which include electricity generating plants above 2 megawatts, cement works, factories and refineries, and fossil fuels supplied to businesses, homes, offices, and transportation vehicles excluding planes and ships. The Clean Development Mechanism, discussed above, also comes under the heading of mitigation arrangements, and is used, among other things, to fund renewable energy projects in developing countries.

Second is the establishment of a range of climate funds to enable developing countries to adapt to growing extreme weather events, for example by strengthening flood and sea defences. These include four funds: the Least Developed Countries Fund, the Special Climate Change Fund, the Global Environment Facility, and the Adaptation Fund. Given the costs of adaptation already being experienced by developing countries, none of these funds represent sufficient compensation for the climate change–related damage already being experienced in developing countries.[2]

The existence of a significant gap between damage and reparations was acknowledged at Doha and led to a new third type of envisaged international activity under the UNFCCC in the form of compensation for 'loss and damage'. This introduces into the UNFCCC for the first time the 'polluter pays' principle, creating an important link between present failure to mitigate climate change and present and future damages from extreme weather because of this failure. The United States, as the heaviest historic polluter, strongly resisted use of the word 'compensation', and no actual funds were committed at Doha, but it is possible that this third strand may turn out to be the most significant of the three types of activity mandated by the UNFCCC. It may help to underwrite legal claims for damages in national or international courts against governmental and corporate agents of fossil fuel extraction, supply, and marketing.

The Doha conference underlined that there is no prospect of the nations agreeing to reduce global greenhouse gas emissions in the next five years. The Conference of the Parties agreed in 2011 to launch a treaty negotiation process in 2015, which will see all nations, and not just developed countries, agree to emission caps under a new treaty to commence in 2020.

2. Donald A. Brown, *Climate Change Ethics: Navigating the Perfect Moral Storm* (New York: Routledge, 2012), 184.

But at present rates of growth in global greenhouse gas emissions of 3 percent a year, which the Kyoto Protocol's second commitment period will not affect, many believe the earth will be locked in to a future of at least three degrees of warming by 2020. This is in part because the investment over the next seven years in fossil fuel energy provision far outweighs investment in renewables, and hence commits the nations long beyond 2020 to ongoing growth in fossil fuel energy use.

Climate denialists argue that the IPCCC has overestimated climate heating, particularly in the period since 1998, when an exceptionally strong El Niño effect prompted a record global temperature high, but IPCCC predictions of climate heating have actually tracked real world measurements well.[3] The most glaring *inaccuracy* in IPCCC reports is consistent underestimation of growth in greenhouse gas emissions. For this reason there is a growing gap between estimates of future temperature rise in IPCCC reports and the likely rise given growth in emissions. Hence the International Energy Authority in 2012 finds that on present emissions trajectories the planet could heat by six degrees Celsius by 2100:

> We cannot afford to delay further action to tackle climate change if the long-term target of limiting the global average temperature increase to 2°C, as analysed in the 450 Scenario, is to be achieved at reasonable cost. In the New Policies Scenario, the world is on a trajectory that results in a level of emissions consistent with a long-term average temperature increase of more than 3.5°C. Without these new policies, we are on an even more dangerous track, for a temperature increase of 6°C or more.[4]

A more conservative 2012 report on the development implications of climate change from the World Bank envisages a four degree warmer world as increasingly likely and argues that such an outcome would be catastrophic, as it would involve

> the inundation of coastal cities; increasing risks for food production potentially leading to higher malnutrition rates; many dry regions becoming dryer, wet regions wetter; unprecedented heat waves in many regions, especially in the tropics; substantially exacerbated water scarcity

3. Myles R. Allen, John F. B. Mitchell, et al., 'Test of a Decadal Climate Forecast,' *Nature Geoscience* 6 (2013): 243-44.

4. International Energy Authority, *2011 World Energy Outlook* (Paris: IEA, 2011), 2.

> in many regions; increased frequency of high-intensity tropical cyclones; and irreversible loss of biodiversity, including coral reef systems.[5]

Impacts of a four degree warmer world in the tropics will be much more extreme, with higher sea level rise and higher temperature rises, rendering these areas prone to extreme drought, flooding, and storms. So, for example, the populous and highly developed tropical region of South and Southeast Asia will have greatly reduced availability of food, since even a two degree temperature rise will significantly reduce rice production.[6] Temperate and semi-temperate areas will also be dramatically affected by a four degree temperature rise, since 'the coolest months are likely to be substantially warmer than the warmest months at the end of the 20th century.' In the Mediterranean region July will likely be nine degrees warmer than presently. In all but the far north of Europe and Canada, food production will be seriously compromised as recent science indicates crop outputs decline with higher local daily temperatures. Combined with severe water shortages, heat waves, and extreme storms, four degrees will likely see the collapse of many existing social institutions and systems. Because such collapses would affect the poor unequally, in part because there are more poor people in the most at-risk areas, 'the global community could become more fractured and unequal than today.' The World Bank report's authors therefore urge that 'the heat must be turned down' and that only 'early, cooperative, international actions can make that happen.'[7]

A peer reviewed current best scientific guess for the timing of a four degree temperature rise is 2070.[8] The world has not been that warm for three million years, and never in the time of *Homo sapiens.* According to climate scientist Kevin Anderson, a four degree warmer world is 'incompatible with organised global community, is likely to be beyond "adaptation", is devastating to the majority of ecosystems, and has a high probability of not being stable (i.e., 4°C would be an interim temperature on the way to a much higher equilibrium level).'[9]

5. Jim Yong Kim, 'Foreword', *Turn Down the Heat: Why a 4° C Warmer World Must Be Avoided* (Washington, DC: World Bank, 2012), v.

6. R. Wassmann, S. V. K. Jagadish, et al., 'Climate Change Affecting Rice Production: The Physiological and Agronomic Basis for Possible Adaptation Strategies', *Advances in Economy* 101 (2009): 59-122.

7. *Turn Down the Heat,* 8.

8. Richard A. Betts, Matthew Collins, Deborah L. Hemming, et al., 'When Could Global Warming Reach 4° C?' *Philosophical Transactions of the Royal Society* 369 (2011): 67-84.

9. Kevin Anderson, 'Going Beyond Dangerous Climate Change: Exploring the Void be-

Despite the growing risks of an outcome of climate change of a three, four, or even six degree warmer world, greenhouse gas emissions continue to grow. The nations express an 'intention' in UNFCCC conferences to limit climate change to two degrees. But their actions do not match this intention. Global greenhouse gas emissions in 2011 were thirty-five billion metric tons of CO_2, a record high, and 54 percent higher than in 1990, which is the Kyoto Protocol reference year. Coal was the biggest source at 43 percent, followed by oil at 34 percent, gas at 18 percent, and cement at 5 percent. The biggest contributors were China at 28 percent, the United States at 16 percent, the European Union at 11 percent, and India at 7 percent. EU and U.S. emissions were down 2 and 3 percent respectively in the previous year, while the other countries more than made up for these reductions. China and Europe had reached near parity on per person emissions by 2012.[10]

Both the IEA and the World Bank argue that subsidies to fossil fuel production should end in order to remove perverse incentives for climate-changing levels of fossil fuel extraction. However, even Kyoto Protocol Annex 1 countries, including Australia, Canada, and the United States, increased subsidies to coal, shale gas, and oil in the period of the Protocol. The UK government in 2012 announced that it would provide tax breaks for shale gas exploration. As discussed above, Indonesia, which is not in Annex 1, has increased its coal exports to China tenfold since 2000. In sum, the UNFCCC treaty process is not working to reduce fossil fuel extraction and related greenhouse gas emissions, and shows no prospect of so doing.

According to paleoclimate data, the last time the earth was four degrees Celsius warmer than the pre-industrial global average temperature of 14 degrees Celsius was 3 million years ago in the Pliocene geological era.[11] The Northern Hemisphere was then entirely free of ice, and sea levels were 25 metres above the present. Besides inundating multiple world cities, and the settlements of the more than three billion people who live at or below 25 metres above current sea level, average temperature rise of this magnitude would provoke endur-

---

tween Rhetoric and Reality in Reducing Carbon Emissions', Public Lecture, London School of Economics, 21 October 2011, at http://www2.lse.ac.uk/newsAndMedia/videoAndAudio/channels/publicLecturesAndEvents/player.aspx?id=1208 (accessed 10 December 2012).

10. Global Carbon Project, *Global Carbon Budget 2011,* at www.globalcarbonproject.org/carbonbudget (accessed 10 December 2012).

11. James E. Hansen and Makiko Sato, 'Paleoclimate Implications for Human-made Climate Change', in A. Berger, F. Mesinger, and D. Sijaci, eds., *Climate Change at the Eve of the Second Decade of the Century: Inferences from Paleoclimate and Regional Aspects,* Proceedings of Milutin Milankovitch 130th Anniversary Symposium (New York: Springer, in press).

ing drought not just in present drought-prone areas but in tropical forests. At three degrees the UK Hadley Centre estimates that the Amazon forest, which represents 10 percent of global photosynthesis, would dry out and turn to savannah, releasing its multi-million-year-old store of carbon and triggering a cycle of spontaneous terrestrial carbon emissions, regardless of anthropogenic emissions, which would set in train a runaway warming event.[12] Three degrees of warming would see the desertification of most of the western United States, Mexico, North and South Africa, southern Europe, and all of Australia apart from the Northern Territories, rendering these areas uninhabitable and useless for agriculture.[13] Given the potential for such extreme climatic outcomes in the present century, it is perhaps unsurprising that the growing focus of national and international risk assessment in relation to climate change includes the frame of national security and violent conflict as indicated in chapter 1.

Modernity was not supposed to end like this. In 1989 Francis Fukuyama announced that the ideals of equality, liberty, democracy, and the rule of law, the principal fruits of the European Enlightenment, had triumphed in the victory of liberal capitalism over totalitarian communism in the late twentieth century.[14] For Fukuyama, the promise of the Enlightenment, and of classical political economy, came together in the collapse of the Berlin Wall in 1989, the end of the Cold War, and the breakup of the Soviet Empire. Together these crushed the idea that there is an alternate end to history other than the progressive trajectory of modern liberal democratic capitalism; the political left's hope in socialism was finally revealed as an illusion. With the subsequent adoption by China of a state-sponsored form of neoliberal capitalism, there is no alternative reality.[15] But the earth herself now contests the claim that there is no alternative to neoliberal capitalism; hence Fukuyama's Hegelian 'end of history' is an apocalyptic delusion.

12. Jorge L. Sarmiento and Nicolas Gruber, 'Sinks for Atmospheric Carbon', *Physics Today* 55 (2002): 30-36.

13. On the spread of deserts in the American West see John D. Marshall, John M. Blair, et al., 'Predicting and Understanding Ecosystem Responses to Climate Change at Continental Scales', *Frontiers in Ecology and Environment* 6 (2008): 273-80; on southern Africa see David S. G. Thomas, Melanie Knight, et al., 'Remobilization of Southern Africa Desert Dune Systems by Twenty-first Century Global Warming', *Nature* 435 (2005): 1218-21.

14. Francis Fukuyama, *The End of History and the Last Man* (London: Hamish Hamilton, 1992).

15. The neoliberal, or what is sometimes called the 'post-political', condition is governed and promoted by nation-states — in partnership with economic corporations — in all domains where this new form of economic libertarianism holds sway; see further David Harvey, *Neoliberalism: A Brief History* (Oxford: Oxford University Press, 2005).

If climate change is not only a scientific datum but a shaper of social and political experience, then liberal democratic capitalism is itself built on an illusion: the illusion that the corporately sustained engine of economic growth can spread freedom and material prosperity to all seven billion humans on the planet provided they abandon their old gods and magical attachments to fields, fisheries, forests, and folk tales and acknowledge the supremacy of Enlightenment reason, and in particular economic rationality, as the means to progress the human condition. Climate change may underline what Zygmunt Bauman identifies as a 'postmodern divide' between Europe (and other advanced industrialised countries) and the rest of the world.[16] If modern European identity is conceived in essence as political and economic liberty, then this identity is increasingly threatened by ecological crisis and resource constraints. Instead of a Hegelian and progressive *end* or goal of history, of the kind announced by Fukuyama, history may be moving toward a different kind of end, in which nature herself calls time on the freedom of the human species to continue to raid the planet for the resources to sustain industrial civilisation, while relations between assemblages of humans grow more conflicted in a struggle for access to diminishing food and water sources.[17]

The project to limit fossil fuel emissions to prevent dangerous climate change was envisaged by the framers and signatories to the United Nations Framework Convention on Climate Change as an ethical one. The 'ultimate aim' of the convention is

> to achieve, in accordance with the relevant provisions of the Convention, stabilization of greenhouse gas concentrations in the atmosphere at a level that would prevent dangerous anthropogenic [originating in human activity] interference with the climate system. This objective is qualified in that it 'should be achieved within a time frame sufficient to allow ecosystems to adapt naturally to climate change, to ensure that food production is not threatened, and to enable economic development to proceed in a sustainable manner.'[18]

The Convention states that present humans, and in particular the agents who are parties to the convention which are *national governments,* have a

16. Zygmunt Bauman, *Postmodernity and Its Discontents* (Cambridge: Polity Press, 1997); see also Leigh Glover, *Postmodern Climate Change* (New York: Routledge, 2006).

17. Slavoj Žižek, *First as Tragedy, Then as Farce* (London: Verso, 2009), 63.

18. *United Nations Framework Convention on Climate Change* (Geneva: UN, 2002), and at http://unfccc.int/resource/docs/convkp/conveng.pdf (accessed 22 August 2012).

duty to future generations to act prudentially to limit the future damage of present atmospheric pollution. The moral claim that present human beings have duties to future human beings in relation to ecological degradation has not, however, translated into agreed collective actions by the nations, and their citizens and corporations, to reduce greenhouse gas emissions. Instead of cooperating to reduce greenhouse gas emissions, the nations continue to compete for remaining fossil fuel sources, including new potential sources opened up by melting ice in the Arctic. This competition remains violent, as is evidenced in the conflicts that continue in oil- and gas-rich lands, including the Middle East, the Caucasus, and the Niger Delta.

Kant, Cosmopolitanism, and the Nature-Culture Divide

Climate change represents a challenge to Enlightenment modernity in two ways. First, it challenges the Enlightenment ideals of equality, liberty, and fraternity in civil society, and the spread of constitutional government within and between nations as the means to the achievement of cosmopolitan peace and order. Second, it challenges the nature-culture and science-ethics divides which underwrote and accompanied the Enlightenment accounts of reason, of the dignity or intrinsic worth of the human person, and of moral values.

The European ideal of transnational government, ordered by constitutional law among nations at peace that for centuries had been at war, was first clearly enunciated in Kant's *Idea for a Universal History with a Cosmopolitan Purpose.*[19] In this late essay in political theory, Kant argued that there was a tide in human, and more especially European, history whose ultimate end would be peace and an international, law-governed, cosmopolitan society. Kant's political idealism had long historical roots. Europe's present-day borders mirror those of the Holy Roman Empire at the time of Charlemagne, and Europe's legal codes draw on a tradition of law first developed in the Roman Empire. But for Kant there is a crucial difference between the rise of enlightened nation-states in Europe and the preceding empire: the form of constitutional government in which the rights of the person are legally recognised and protected. At the foundation of Kant's

19. Immanuel Kant, 'Idea for a Universal History with a Cosmopolitan Purpose', in *Kant: Political Writings*, ed. H. S. Reiss, trans. H. B. Nisbet (Cambridge: Cambridge University Press, 1991), 41-53.

ethics is the claim that all persons are due equal respect and that this is true *a priori*, and is the core maxim of the moral law, which is comprehensible by all reasonable persons. From this principle derive the claims that persons are to be treated equally before the law and that they should be free from coercion to the extent that their own behaviour does not involve the coercion or harm of others. For Kant, this meant that no national civil society could any longer be built on slavery or serfdom. All born within the borders of a civil state ought to have the rights of free citizens.

The Enlightenment project involved a break with the past, while at the same time Enlightenment philosophers recognised that it was built on the foundations, including more especially the Christian foundations, of the present in the past. As well as the moral law, the ideas of progress and of a universal society all have Christian roots for Kant. But it is education and the development of reason in individual citizens, and no longer primarily the Church or revealed religion, which become the means for the realisation in the enlightened nations of a civil society characterised by freedom and equality before the law.

Kant envisages a new condition of international relations as the outcome of the emergence of enlightened civil societies. This new condition will spread through the exchange of goods and ideas, so that the nations will subject their relations to one another, as well as within their borders, to the rule of law and not war. Wars will also be undermined by the spread of public education and by the growing cost of technologically sophisticated weapons and warring strategies. For Kant, humanity is capable of moral progress which requires 'a long, perhaps incalculable series of generations, each passing on its enlightenment to the next.' The ultimate end will be the formation of a cosmopolitan civil constitution — a law-governed state of relations between the nations — which ensures that persons live in a state of international peace. At the same time he envisages the emergence of 'a great political body of the future, without precedent in the past'; this expectation Kant calls a '*chiliastic* expectation', invoking the Christian millennial hope, though he emphasises that this hope is not primarily a religious but a philosophical hope.[20]

Kant's account of cosmopolitan reason could not have been more wrong as a description of the history of Europe from the eighteenth to the twentieth centuries. The Napoleonic Wars reignited border wars and national rivalry in Europe, which eventually drew the whole world into their

20. Kant, 'Idea for a Universal History', 32.

compass in the global wars of the twentieth century. However, in the dying embers of the fires of total war that Napoleon first ignited, a new project to revive Kant's vision of a cosmopolitan order, in the form of the United Nations, came to fruition. But while the UN has arguably played an important role in limiting war since its founding, the failure of the UNFCCC to restrain greenhouse gas emissions indicates that climate change challenges the ideal of a law-governed international order where nations do not harm other nations intentionally.

So dominant is the hold of Kantian cosmopolitanism on modern political theorists and philosophers that few are prepared to acknowledge that climate change is a test which Kantian rationalism and cosmopolitanism fails, and instead seek to find alternative explanations for the failure. In an extensive Kantian treatment of the problem of climate change, *A Perfect Moral Storm,* Stephen Gardiner narrates the failure of climate change politics in a Kantian rationalist frame as an example of 'moral corruption'. He describes the failure to act now as a classic case of intergenerational 'buck passing'. Those who have the responsibility and the capacity to act now are wealthy consumers of fossil fuels. In refusing to reduce their uses of fossil fuels — many of them for luxuries such as foreign holidays or unnecessary car journeys — present consumers pass on the necessity to suffer the effects of climate change to those who are not responsible for atmospheric pollution, who include the poor, future generations, and nature.[21] Gardiner, like Kant, assumes that progress in reason and law will over time generate a cosmopolitan world order in which rule-governed, equitable behaviours between the nations in relation to the use of climate sinks will lead to the end of the subjugation of poor nations by rich ones and the exploitation of the weak by the strong.

However, it is not only climate change or the two hundred year post-Kantian history of global war in Europe which reveal the limits of Kantian cosmopolitanism. The treatment by Europe and the United States of the formerly colonised nations on the continents of Asia, Africa, and Latin America — of which more below — also runs directly counter to Kantian cosmopolitan chiliasm. Yet despite the deep structural injustices in the laws and practices which govern the global economy, cosmopolitans still contend that the cosmopolitan frame stands against all other alternatives.[22]

21. Stephen Gardiner, *A Perfect Moral Storm: The Ethical Tragedy of Climate Change* (Oxford: Oxford University Press, 2011), 32-47.

22. See for example Martha C. Nussbaum, *Frontiers of Justice: Disability, Nationality,*

Gardiner defends cosmopolitanism as the only game in town when it comes to negotiating internationally agreed greenhouse gas emissions reductions. The result is that Gardiner assumes that the failure of enlightened nations to reduce their greenhouse gas emissions is unrelated to flaws in the cosmopolitan frame but is instead a particular feature of the 'moral storm' of climate change which, because of its global scale and long-run consequences, is uniquely exposed to the problem of moral corruption.[23] Having established why the frame has so far failed, Gardiner turns to a solution grounded in game theory, in which he attempts to persuade his readers that, provided rational choices are made by all parties — particularly concerning discount rates of costs borne by future generations — climate change is amenable to a cosmopolitan solution. He does not, in other words, consider the possibility that cosmopolitanism is not the principal motive of constitutional nations in their relations with other nations, despite the extent to which his own nation of residence and employment, the United States, continues to use violence, and the threat of violence, as a default mode in many of its interactions with other nations, and despite the fact that it refuses to ratify international treaties not just in relation to climate change but in relation to many other matters of international justice and law.

Gardiner also does not consider that among the most powerful actors in the 'game' of negotiating or refusing to negotiate emissions reductions are economic corporations, which are not rational choosers in the Kantian sense, since corporations are not real persons nor constitutionally governed assemblies of persons. Large corporations such as Exxon, Shell, and Shenhua Coal use their considerable funds and influence over the media and politicians to stall action on climate change — and other ecological problems — and to mislead the public as to the nature of these problems.[24] Corporations act rationally only in the narrowest of utilitarian frames, since they are primarily driven by narrow and statistical — and hence impersonal and often extralegal — considerations of market share, profit, and shareholder value.[25] When a corporation does not like a national law governing,

Species Membership (Cambridge, MA: Belknap Press of Harvard University Press, 2006); and for a critique see Michael S. Northcott, *A Moral Climate: The Ethics of Global Warming* (London: Darton, Longman and Todd, 2007), 155-59.

23. Gardiner, *A Perfect Moral Storm*, 301-38.

24. See further Eric Pooley, *The Climate War: True Believers, Power Brokers, and the Fight to Save the Earth* (New York: Hyperion Books, 2010).

25. For a fuller discussion of the lack of respect for persons, species, and ecosystems intrinsic to cost-benefit analysis, which is the culturally dominant form of utilitarianism, see

say, employment of workers, or atmospheric pollution, it will often 'rationally' choose to outsource its activities to domains where such laws either do not exist or are not enforced. Given that the economic power of most nations is smaller than that of the largest corporations, this alone ought to give Gardiner and other defenders of the cosmopolitan frame pause for thought.[26]

Cosmopolitans argue that ecological crisis, like other global crises, such as that occasioned by the 2001 attack on the United States, 'terrorism', and conflict in the Middle East, do not so much reveal the weaknesses of Kantian cosmopolitanism as require its extension. Against Samuel Huntington's claim that civilisational conflict is an underlying feature of all, including modern, history, Ulrich Beck constructs an account of what he calls 'realist cosmopolitanism', which addresses the question of 'how "societies" deal with "difference" and "borders" under conditions of global interdependence crises'.[27] With Peter Berger, Beck suggests that what is needed is a secular account of relations between nations which eschews particularisms of creed or ethnic identity, and which promotes an international public dialogue between all 'neighbours'. The distinctive feature of the late modern situation is the extent to which such neighbours include 'transnational' actors and influences which are no longer limited to bordered neighbourhoods. Once this is recognized, it should become possible to draw all actors into a new international form of public reason and communication in which shared solutions can be negotiated.[28]

For Bruno Latour, this extended account of transnational rational communication still fails to include nature, since it excludes agents, such as nonhuman species, that do not have a voice in human assemblies.[29] But Mike Hulme argues that climate change aids in the resolution of this problem, for it reveals that modern cosmopolitanism reaches not only across all human cultures but into the earth itself. As the climate increasingly reflects the imprint of industrial civilisation, 'climate change, by dissolving different

---

Mark Sagoff, *The Economy of the Earth: Philosophy, Law, and the Environment*, 2nd ed. (Cambridge: Cambridge University Press, 2008).

26. See further Michael Northcott, 'Artificial Persons against Nature: Environmental Governmentality, Economic Corporations and Ecological Ethics', *Annals of the New York Academy of Sciences* 1249 (2011): 104-17 .

27. Ulrich Beck, *Cosmopolitan Vision* (Cambridge: Polity Press, 2006).

28. Beck, *Cosmopolitan Vision*, 48-71.

29. Bruno Latour, *Politics of Nature: How to Bring the Sciences into Democracy*, trans. Catherine Porter (Cambridge, MA: Harvard University Press, 2004).

types of boundaries, is performing significant work in extending and deepening the cosmopolitan perspective.'[30] Hulme is right that climate change challenges the nature-culture divide, as I have argued elsewhere.[31] But the implications of this for cosmopolitanism are more profound than he acknowledges, since this divide reaches back to the origins of Enlightenment thought in the Newtonian revolution and the subsequent division of labour erected by Kant between theoretical and practical reason, of which more below. Before I address this, I will describe more fully two problematic areas of the Kantian frame which conventional defences of cosmopolitan justice do not acknowledge. The first of these is best summarised as the 'white man's burden', and the second as rationalist individualism.

Climate Change and Postcolonial Burdens

Cass Sunstein argues that the Montreal Protocol reduced atmospheric pollution from CFCs, whereas the Kyoto Protocol did not reduce greenhouse gas emissions, because the advocated behaviour of the latter was not in the self-interest of the United States.[32] But whose interests is Sunstein referring to here? Is it the interests of fuel poor citizens who cannot afford adequately to heat or cool their homes in the midst of growing climate extremes? Or of farmers suffering the growing 'dustbowlification' of the American Southwest? What Sunstein does not openly acknowledge, though it might be said to be implicit in his analysis, is that the 'interests' that are served by the United States' refusal to mitigate climate change are principally those of its economic corporations, especially its all-powerful fossil fuel corporations. Furthermore, the refusal of the United States to agree to mitigate climate change by global treaty suggests that the United States is not prepared to join global agreements which promote a fair sharing of climate burdens between developed and developing countries. Nor is it prepared to join those which acknowledge that developed countries bear historic responsibility for the causation of climate change, and in many cases continue to see the

30. Mike Hulme, 'Cosmopolitan Climates: Hybridity, Foresight and Meaning', *Theory, Culture and Society* 27 (2010): 267-76, 268. For a more conventional defence of cosmopolitanism in relation to climate change see Simon Caney, 'Cosmopolitan Justice, Responsibility and Global Climate Change', *Leiden Journal of International Law* 18 (2005): 747-75.

31. See further Northcott, *A Moral Climate.*

32. Cass R. Sunstein, 'Montreal vs. Kyoto: A Tale of Two Protocols', *Harvard Environmental Law Review* 31 (2007): 1-65.

developing world as a source of resource harvesting and for offloading or making less visible the social costs of unrestrained fossil fuel consumption. In other words, the United States remains in the position, long after theoretical independence has been granted to formerly colonised and subjugated nations, of managing its relations with these nations in such a way as to sustain the markedly unequal access to natural resources and economic wealth that characterised those relations in the colonial era.

The United States is not alone in taking this stance. The European Commission, while it more often accedes to international laws and treaties, is also committed to pursuing a range of policies whose effect is not to improve the balance of economic power and the fairness of the terms of trade between European and former colonial countries but to worsen them. In concert with Europe's largest economic corporations, the Commission has strenuously pursued a series of bilateral agreements between Europe and individual developing countries in the South known as Economic Partnership Agreements. These agreements are opposed by civil societies in North and South as unjust and as generators of poverty and not wealth in the developing world. One such agreement permitted European trawlers to destroy African fishing grounds off the west coast of Africa and the livelihoods of millions of fisher folk in the region. When Senegalese young men subsequently turn up off the coast of Italy looking for work, they are mostly sent back to Africa as 'economic migrants'. Having destroyed their local economies and natural resources, Europe still treats them as pariahs.[33]

The problem we are up against here is the postcolonial form of the 'white man's burden'. For three centuries Europeans, both within Europe and in other terrains that they have permanently settled, have convinced themselves that in their relations with nonwhite and native peoples they have been advancing the civilisation, and even the salvation, of these peoples and their lands. Yet before European colonisation, most of the peoples in Africa, Asia, and the Americas were in a better state in terms of their capacity to feed themselves and to enjoy a richly biodiverse natural environment, fertile soils, well stocked forests, unpolluted rivers and lakes full of fish, and extensive coastal fisheries, than they were at the end of European colonisation. Climate change and its unequal effects on Africa, far from encouraging the powerful former colonisers to reduce their resource extraction from African soils and waters, is leading to a new land grab in Af-

33. Vlad M. Kaczynski and David L. Fluharty, 'European Policies in West Africa: Who Benefits from Fisheries Agreements?' *Marine Policy* 26 (2002): 75-93.

rica, with large areas of arable land being stolen from native peoples — with the license of corrupt politicians — and sold to American, Chinese, and European investors, who use the land for cash crops such as animal feeds, and for flowers and vegetables for sale in developed world supermarkets.[34]

Kant, it must be said, is sensitive to this problem of the 'white man's burden', and far more so than Adam Smith or John Locke. Kant is sharply critical of the 'commercial states' — especially 'America', which he names, though the other principal imperial powers of England, Holland, and Spain are also likely indicated — who in visiting foreign countries enslave their populations or treat foreign territories as ownerless and purloin the land for themselves.[35] But even Kant envisages that the path for primitive peoples from their precolonial condition to constitutional democracy and freedom is one that will be opened up by the spread of constitutional democracy from the 'enlightened' nations to the rest. For only as law-governed peoples, and subjects of constitutional states, can tribes or other assemblages of people have agency and be granted a measure of sovereignty over their own lands in a cosmopolitan order.[36]

Continuing advocacy of Kant's restatement of the 'white man's burden' represents an ideological cover for the international state of affairs in which former colonising nations continue to utilise for the maintenance of their standards of living an ecological footprint which represents a land area three to five times those of their own borders.[37] That these nations also choose to purloin — and to persist in purloining — an inequitable amount of atmospheric space for their greenhouse gas emissions therefore only requires resolution as a *unique* problem of 'moral corruption' if one ignores these continuing elements of the 'white man's burden'. In this perspective it is not only the different outcomes for the self-interest of the United States and its corporations that led to the creation of an effective environmental treaty to control CFCs (the Montreal Protocol) and an ineffective treaty to control GHGs (the Kyoto Protocol). It is also the different way in which the

34. See further Cecilie Friis and Anette Reenberg, *Land Grab in Africa: Emerging Land System Drivers in a Teleconnected World* (Copenhagen: Global Land Project, University of Copenhagen, 2010); and Bertram Zagema, *Land and Power: The Growing Scandal Surrounding the New Wave of Investments in Land,* Oxfam Briefing Paper 151 (Oxford: Oxfam, 2011).

35. Kant, *Perpetual Peace,* in *Kant: Political Writings,* 106-7.

36. Kant, *The Metaphysics of Morals,* in *Kant: Political Writings,* 143-47.

37. C. Monfreda, M. Wackernagel, et al., 'Establishing National Natural Capital Accounts Based on Detailed Ecological Footprint and Bbiological Capacity Assessments', *Land Use Policy* 21 (2004): 231-46.

burdens of the ozone hole and anthropogenic climate change fall on white and nonwhite people. In the case of the ozone hole, it is white people — the descendants of Europeans who had settled in the temperate countries north of the southern ocean — who are experiencing sharp rises in skin cancer rates from increased ultraviolet light. In the case of GHGs, it is largely nonwhite people whose lands will be, and are already being, rendered less habitable and less water and food secure by climate change.

Securing Persons from Scientific Power via the Nature-Culture Divide

The second problem with the Kantian underpinnings of cosmopolitanism is the modelling of human moral behaviour on *a priori* rationalism. Kant's account of law-governed moral reasoning between persons and between nations rests upon his foundational distinction between theoretical and practical reason. The need for this distinction is envisaged in Kant's first published work on the philosophical implications of the mechanistic cosmology of Isaac Newton. Kant argues that Newton had, through the posits of theoretical reason, described the laws which govern the behaviour of objects moving in space. Nature after Newton can no longer be said only to be governed directly by divine providence but is also the outcome of laws set into the creation by the Creator at its origin.[38] This account of causation and laws of nature forces Kant to find another framework within which to secure human freedom from determinism and instrumentalism. Hence in Kant's later critical philosophy he envisages an unbridgeable chasm between the animalistic, instinctual, and mechanistic nature of the cosmos, which humanity participates in as embodied and sensory beings, and the rational capacities of humanity, which permit persons to act according to a universal law which transcends the mechanical laws which govern material phenomena.

This chasm produces two domains, governed by two distinct forms of reason: theoretical reason, which pertains to the causally determined earth and heavens, including the animalistic nature and interior instincts of sensory bodies, including those of humans; and practical reason, which pertains to free, law governed, rational deliberation by persons. The possi-

38. Immanuel Kant, *Universal Natural History and Theory of the Heavens* (1775), trans. Stanley L. Jaki (Edinburgh: Scottish Academic Press, 1981).

bility of ethical life for Kant rests upon recognition of the universal claim of the moral law on every reasonable person. The reasonable person ought to recognise that she has a duty to obey a moral law, defined as a maxim that is universalisable, which Kant calls the duty of self-preservation; and by extension she has the duty to promote, or at least not to subvert, the preservation of other selves. Respect for persons as ends and not means is foundational to the moral law as Kant describes it. That there are persons who do not respect other persons is for Kant what is meant by the problem of evil. The origin of this problem is in the will. The person who refuses to act according to her rational duty to respect her own self and other selves is a person whose desires are malformed and who therefore has a bad will.[39]

There are a number of advantages to Kant's post-religious foundations for ethics — hence its continuing authority — and there are also a number of problems. Principal among the advantages is that it appears to safeguard the sphere of the human from subjection to scientific manipulation, since it sets apart mechanistic relations from personal relations by privileging the latter. Second, Kant's account is apparently amenable to being understood by all reasonable people regardless of cultural background. It does not require — or at least it does not appear to require — adherence to any prior set of beliefs or stories about God or gods, or revealed religion. It is therefore at least potentially a transnational as well as a transpersonal morality. Third, it is a morality for the non-magical and ultimately secular era inaugurated by the mechanistic cosmology of Newton and his heirs. It distinguishes the human good from its former moorings in the natural law of a divinely created universe and yet sets a standard of equality of treatment of persons and freedom from harm that is higher than that offered by the other dominant moral paradigm of the post-Enlightenment era, which is utilitarianism and its cost-benefit progeny. This is particularly helpful when considering the issue of harms from greenhouse gas emissions because it suggests that no individual, and no community, ought to be subjected to disproportionate harms from these emissions, and that in principle every individual ought to have the same access to carbon sinks no matter the part of the world they inhabit or the level of development their nation has reached. Fourth, it sets moral duty apart from contextual judgements of consequence: what is right is transcendent and prior to utility and may not be set aside for individual or collective advantage. Fifth, it emphasises not

39. Immanuel Kant, *Groundwork for the Metaphysics of Morals*, trans. Mary Gregor (Cambridge: Cambridge University Press, 1998).

only that there is a transcendent good that reasonable persons ought to recognise but also that there can be failure to recognise this good, or to act upon it, and that such failure has to do with an inexplicable original failure to desire or will the good. Kant's approach can then account for, if not quite explain, not only the existence of the good and the fact that reasonable people are drawn to the good, but also that some people may fail to be so drawn to it because they did not will it.

There are, however, a number of problems with Kant's account. These were identified almost immediately by his successors, including Hegel. First, Kant's account of reason and the will is highly individualistic. Kant envisages the good as an *a priori* moral law knowable in the consciousness, and conscience, of each rational individual concerning the respect that is due to their person and hence to other persons. While he does talk of education as playing a role in the development of rational capacities and a good will, nonetheless he does not acknowledge the seminal part played by the nurture of children and young people in the formation of moral adults. He strongly argues that imitation plays no role in truly moral action. This failure is in part a consequence of Kant's overemphasis on the *a priori* moral law as a category universally available to personal consciousness rather than one that develops as the body and mind of the person develop in interaction with those around them *and* with the natural environment. The second and related problem is the account of the individual self as an autonomous agent. This account entails a failure to appreciate the extent to which consciousness is a feature of sensory, animalistic bodies which exist through their relations to other persons and other creatures.

The sense of an individual self is a function of a distinctively modern consciousness, but it is illusory if it leads individuals to believe that they author their own life, or that they alone are responsible for the experience of selfhood, as Charles Taylor, echoing Hegel's critique of Kant, has shown.[40] The autonomous rational willing being who is Kant's idealised person is dependent, even when she sits alone in her study, on the atmosphere that gives oxygen to her lungs, on the rains, rivers, and lakes that meet her thirst and the agents that bring it to her in potable form, and on the fields and farmers and marketers who sustain her with food. The self is situated in communities of other selves, and in ecological communities. The sense of duty that a person experiences arises from these relational and embodied

40. Charles Taylor, *Sources of the Self: The Making of the Modern Identity* (Cambridge: Cambridge University Press, 1992), 182-96.

contexts, and is not therefore solely the product of rational recognition of *a priori* categories.

A third problem with Kant's account is that his emphasis on the rational will leads him to downplay the role of desire and emotion — and not just of instinct — in the recognition and performance of the good. Desires and emotions for Kant are analogous to instinct: they are closer to the animal, mechanistic, and sensory nature of bodies than they are to the unique rational faculties which make persons of transcendent and intrinsic value.[41] Kant here is markedly at odds with his premodern predecessors, including Aristotle and Augustine, for both of whom the failure to perform the good is a defect which is evidence of a lack of desire for or emotional attraction to the good, and hence a lack of virtue. Kant's account, however, requires a conception of radical evil as an expression of our freedom to be unfree, and the fact that we fall into this unfreedom.[42] This trend toward radical evil (which is quite different from the traditional doctrine of original sin) points to the larger problem of what John Hare calls the 'moral gap' in Kant's account between the duty to act according to the good and the capacity of individuals so to act.[43] For Gardiner the failure of individuals and collections of individuals — or governments — to honour the rational duty to limit greenhouse gas emissions is a unique problem of moral corruption arising from the peculiar nature of climate change and issues in what he calls an 'atmosphere of evil'.[44] But for Hare, as for Augustine, and to a lesser extent Aristotle, the problem of moral corruption is intrinsic to the human condition and not confined to uniquely long run chains of moral consequences such as those associated with climate change.

Kant's resolution of the 'moral gap' is his appeal to the moral law, which has an irresistible attraction for all reasonable people. If the law guarantees a person's freedom, and hence her own good, reasonable people will recognise that a law that requires others to respect them requires them to respect others.[45] But this approach leads Gardiner and others down the road of

41. See further Immanuel Kant, *Religion within the Boundaries of Mere Reason and Other Writings*, ed. Allen Wood and George di Giovanni (Cambridge: Cambridge University Press, 1998).

42. Peter Dews, *The Idea of Evil* (Oxford: Blackwell, 2008), 22-23.

43. John E. Hare, *The Moral Gap: Kantian Ethics, Human Limits, and God's Assistance* (Oxford: Oxford University Press, 1996).

44. Gardiner, *A Perfect Moral Storm*, 390.

45. The emphasis on freedom is more explicit in *The Metaphysics of Morals* than in the more widely quoted *Foundations*. Thus, 'Every action which by itself or by its maxim enables

game theory, because it represents the problem of doing one's duty as an issue of calculation. If doing one's duty means that others do their duty to one, then all ought to be well among rational people in law-governed societies. All that is needed in relation to a long-run, large-scale problem like climate change — though of course this has proven the trickiest part of the problem — is evidence to balance up the costs and harms between spatially and temporally near and distant, or present and future, peoples. Once such costs and harms are factored into appropriate game-theoretic models of transnational and transgenerational relations, it should be possible for reasonable people to agree on measures to reduce emissions appropriately so as to minimise harms.[46]

As Peter Dews has shown, Kant's account of the moral imperative is not as reductionist as that of his successors; it arises from Kant's concern to preserve the integrity of the will, so as to be able to hold humans accountable for their actions, while acknowledging that humans seem to display a propensity for evil *(Hang zum Bösen).* It is not so much a matter of calculation as of an inner contradiction within the free will.[47] Nonetheless, Kant's account decontextualises the ways in which persons come to recognise, and to be capable of and desirous of performing, the good.[48] The consequent decontextualisation of moral action is reductionist, since neither embodied environmental relations nor interpersonal relations are intrinsic to the ways in which persons apprehend the moral law.

The reductionist descriptions of the moral life sustained by Kant reflect the desire to describe the moral law in a way analogous to Newton's descriptions of physical laws. The consequent reduction of morality to calculation, which was also felt powerfully by Mill, reaches its zenith in the mathematics of a game theorist like Martin Nowak, who reduces an argument for the evolutionary advantages of cooperation over competition to five mathematical rules.[49] Such approaches depart significantly from Kant, whose entire moral theory rests on an account of freedom, though they have a genealogical root

---

the freedom of each individual's will to co-exist with the freedom of everyone else in accordance with a universal law is *right*'; Immanuel Kant, *The Metaphysics of Morals,* in *Kant: Political Writings,* 133.

46. Gardiner, *A Perfect Moral Storm,* 51-88.

47. Dews, *The Idea of Evil,* 19-27.

48. Bernd Wannenwetsch, *Political Worship: Ethics for Christian Citizens,* trans. Margaret Kohl (Oxford: Oxford University Press, 2004), 1-2.

49. Michael Nowak, 'Five Rules for the Evolution of Cooperation', *Science* 314 (2006): 1560-63.

in his quest for an analogy between natural and moral laws. But game theoretic arguments rest upon a *scientific* account of rationality, that is, what Kant would have called 'theoretical' rather than 'practical' reason.[50] They therefore discount *normative* discussion of what is or is not desirable in human affairs and are in danger of the naturalistic fallacy: the fallacy of suggesting that because nature at times favours cooperation over competition then this is sufficient reason for humans to favour cooperation over competition. Such arguments are also problematic because they obscure the essential role of nurturing narratives, practices, and rituals in the induction of people into a habitus in which the performance of the good becomes possible as well as desirable. It seemed obvious to Aristotle and Augustine that the community in which a person is educated and habituated plays a central role in the capacity of that person to realise moral excellences, or the virtues. Kant, too, recognised that this is so, but it did not play a central role in his theory because it ran counter to his emphasis on autonomy. He no doubt did have wonderful parents and grandparents and great teachers, but he does not sufficiently acknowledge the significance of these relational experiences of community for his own capacity to identify and desire the good. Game theory goes further and undermines Kant's concern with freedom through its privileging of calculation.

Resort to game theory also raises a cross-cultural issue which is crucial to understanding the continuing failure of international climate negotiations. Non-Western philosophers argue that modern scientific accounts of rationality — such as cost-benefit analysis and rational choice behaviour — are forms of 'cultural violence' that justify continuing postcolonial destruction of the non-Western world by excess consumption backed by military power, unfair trade rules, and transnational corporations.[51] Scientific descriptions of rationality, on this account, are ethnically and culturally bound up with the Newtonian world of mechanistic science and technology. This framing of human rationality is alien to cultures where language discourse and familial and ecological relations are still understood to be deeply interconnected.[52] As Edward Said argues, when Europeans lay claim

50. Murray Code, *Process, Reality, and the Power of Symbols: Thinking with A. N. Whitehead* (Basingstoke: Palgrave Macmillan, 2007), 2.

51. Ashis Nandy, ed., *Science, Hegemony and Violence: A Requiem for Modernity* (New Delhi: Oxford University Press, 1988); see also Murray Code, 'Reason and Violence', in *The Encyclopaedia of Peace, Violence and Conflict*, 2nd ed. (San Diego: Elsevier, 2009).

52. See further Michael R. Dove, Marina T. Campos, et al., 'The Global Mobilization of Environmental Concepts: Re-thinking the Western/Non-Western Divide', in Helaine Stein and Anre Kaland, eds., *Nature across Cultures: Views of Nature and the Environment*

to ideas that are said to be universally applicable and when they take over other territories as 'empty' and unowned — including atmospheric carbon sinks as well as other lands — these are 'different sides of the same essentially constitutive activity, which had the prestige, and the authority of *science*'; it is a mistake to underestimate imperialism's power to redefine reality and hence 'to treat reality appropriately'.[53] Against the claim of science to universal knowledge and rationality, the concepts of indigenous and local knowledge offer important alternative imaginaries for describing the actual constitution of relations in particular — and especially particular non-Western — places and languages.[54]

The resort of Gardiner and other modern analytic philosophers to game theory to explain failures in the balance between 'selfist' and 'altruist' behaviours in relation to climate change is deeply intertwined with the scientific shaping of the philosophical and technological cultures of the United States and other imperial powers, including Kant's Germany and Locke's Britain. This is evidenced in the characterisation of the good by Kant and Kantians as a universal law set into human relations, and analogous to the mathematically predictable laws of motion set into mechanistic relations by the divine Creator or First Cause. Kant believed that by describing human freedom as a matter of assenting to the moral law, rather than as the reception of a tradition, he would be able to give something approaching the same degree of universal validity that applied to the laws of motion. His core metaphor of rationality is, however, not empirical, but an imaginary in the form of 'an island, enclosed by nature itself within unalterable limits', 'the land of truth', 'surrounded by a wide and stormy ocean, the native home of illusion, where many a fog bank and many a swiftly melting iceberg give the deceptive appearance of farther shores'.[55] Rationality on this account — though it is naturalistic — is con-

---

in Non-Western Cultures (Dordrecht, the Netherlands: Kluwer Academic, 2003), 19-46; and Cornel W. Du Toit, 'The Environmental Integrity of African Indigenous Knowledge Systems: Probing the Roots of African Rationality', *Indilinga: African Journal of Indigenous Knowledge Systems* 4.1 (2005): 55-73.

53. Edward Said, 'Zionism from the Standpoint of Its Victims', in *The Said Reader* (New York: Vintage Press, 2000), 131, cited in Code, *Process, Reality, and the Power of Symbols*, 2.

54. The most influential first report of the countervailing truths of indigenous cultures to enlightened rationality is Clifford Geertz, *Local Knowledge: Further Essays in Interpretative Anthropology* (New York: Basic Books, 1983); see also David Turnbull, 'Rationality and the Disunity of the Sciences', in Helaine Stein and Ubiratan D'Ambrosio, eds., *Mathematics across Cultures: The History of Non-Western Mathematics* (Dordrecht: Kluwer Academic, 2001), 37-54.

55. Immanuel Kant, *Critique of Pure Reason*, trans. Norman Kemp Smith (New York:

ceived over against desires, loves, passions, and relationships between embodied persons in communities, and between persons and other organisms. The island logic — the thinker as Robinson Crusoe — gives to Kant's influential account of rationality a peculiarly unsituated and contextless character, apart, that is, from its metaphorical location in the geographical imaginary of the foggy northeastern European coastline.

Kant's metaphor of an unsituated observer marooned on a fog-bound island promotes a dualistic divide between the animal instincts, emotions, and sensory motors of human bodies, which participate in the mechanistic nature of the cosmos, and the capacity of the rational mind to know the *a priori* transcendent ideals of the moral law. This requires that the moral law is not empirical and does not arise from reflection on experience, but rather is transparent to the reasoning mind of each individual self. On this account, relations of dependence between bodies, such as those which all humans experience as children and when they are sick or infirm, are not crucial to the ability of the human mind to know rationally and respond appropriately to the duties that arise from the moral law.[56] Since the transcendent good of the person is not given in the embodied material substrate of the mechanical universe, and the ability to apprehend it does not depend upon biological relations of nurture between kin, or upon the senses, it becomes a logical but a-cosmic *a priori,* and because it is a matter of radical freedom, Kant believed, it is not amenable to or reducible to empirical scientific investigation — though the game theorists are now trying to prove otherwise! The difficulty is that in order to redeem the moral law from the danger of infection by mechanistic science, the moral law is only knowable through the pure light of individual reason. Hence dualisms between cosmos and cosmopolis, desire and duty, emotion and reason, individual and community, instinct and intention are unbridgeable within the terms of Kant's grand schema, or without unwelcome innovations such as the invention of radical or 'horrendous' evils.[57]

For Kant and Kantian cosmopolitans, there is a clear trajectory to human history, which is realised by the freedom of enlightened persons from religious tutelage and unscientific superstition to perform what the moral

---

Macmillan, 1965), 245; see also John Sallis, *Spacings — of Reason and Imagination: In the Texts of Kant, Fichte and Hegel* (Chicago: University of Chicago Press, 1987).

56. Alasdair MacIntyre, *Dependent Rational Animals* (Chicago and La Salle, IL: Open Court, 1999).

57. Marilyn McCord Adams, *Horrendous Evils and the Goodness of God* (New York: Cornell University Press, 1999).

law requires by the light of reason. Hence for Kant the 'guardians' of the modern age are those who have thrown off the 'immaturity' of dogma and metaphysical speculation and can 'disseminate the spirit of rational respect for personal value and for the duty of all men to think for themselves': 'For enlightenment of this kind, all that is needed is *freedom.* And the freedom in question is the most innocuous form of all — freedom to make *public use* of one's reason in all matters.'[58] Peace on earth is the goal of enlightenment because rational persons will be content to join their wills to constitutional societies where communicative discourse and law-governed behaviour rise to the fore, while the irrationality and the costs of war, together with the exchange of people and goods between nations, will ultimately ensure that such societies seek law-governed ways of interacting with their neighbours. But this enlightenment vision has not been realised in economic relationships between independently governed rich and poor countries. It has similarly not restrained the growing gap in life chances between rich and poor within even developed nations since 1979 under neoliberal modifications to conceptions of the common good and distributive justice.

The Cult of Facts and the Culture of the Enlightenment

That public reason has not triumphed in relation to climate change, or in relation to other aspects of the ecological crisis, is not only because of the uniquely global or long-run character of the interspecies as well as interpersonal nature of the crisis. Granted, the voices of those most affected by climate change are not heard in the boardrooms and parliaments of the rich, and the voices of future generations and other species are also absent. But even if they were present, it is unlikely the outcome would be different from what it is on the Kantian account of cosmopolitanism since it rests upon a misdescription of the nature of the autonomous individual and of the formation of morally sensitive persons, and in particular lacks an adequate account of the role of history, memory, and tradition in the formation of persons and peoples. This misdescription arises not only from an effort to model the moral life after mathematical and mechanistic cosmology but from the related nature-culture divide which is fundamental to Kant's epistemology, and to his moral ontology. This divide sets the life of the mind,

58. Immanuel Kant, *An Answer to the Question: What Is Enlightenment?* in *Kant: Political Writings*, 54-60, 55.

and human freedom and culture, in a realm of law and value which is independent of the life of the cosmos and 'nature'. While the intention is to construct a division of labour between science and ethics, so that ethics is not manipulable by or reducible to science, the resultant nature-culture divide is not bridgeable by the procedure of inclusion of subdominant voices — the voices of indigenous minorities or peasant farmers — in international domains of public reason such as the United Nations, the Conferences of the Parties, or the European Parliament. The problem is that the definition of the domain of cosmopolitan public reason is itself already denatured and cannot do justice to the relations of minds and mindful communication to bodies and embodied exchanges, either organic or personal.

Kant's account of moral and political reason is not only at odds with key figures in Western moral philosophy before Kant — including Aristotle, Augustine, and Aquinas. It is also at odds with the thought of other Enlightenment philosophers, including David Hume and J. G. Herder, and with subsequent discoveries in modern moral psychology. Kant and Kantians cannot explain the failure of individuals to desire the good other than by an account of 'radical' evil, or a 'propensity to evil', in which the free rational will is preserved utterly intact. Hence the failure of moderns to respond to the ecological crisis, and to climate change more especially, and the harms it is visiting and will visit on those innocent of its causation in present and future generations is understood by a Kantian such as Gardiner as evidence of a force in human history — moral corruption — which is at odds with the progressive unfolding of reason described by Kant and his heirs. This alternate force is as empirically exceptional as the force of the moral law since it also cannot be explained experientially, historically, or narratively. It acquires a mysterious power that in an odd way replicates the superstitious fear of devils and demons that rationalists claim to have left behind.[59] It also explicitly claims that conflict between these two forces of good and evil is an inbuilt feature of the human condition from the beginning and that it is fear of conflict and violence that drives people to organise their affairs constitutionally and collectively so as to get to peace. It suggests, in other words, that at the back of peace there is always conflict and violence and fear.[60]

59. See further Northcott, 'The "War on Terror", the Liberalism of Fear, and the Love of Peace in St Augustine's *City of God*', *New Blackfriars* 88 (2007): 522-38.

60. Like Hobbes, Kant argues that conflict between groups of individuals over natural resources is the state of nature and the natural cause of the spread of people across the planet and the creation of tribes and ultimately nations, and that a desire to reduce conflict is what

The turn of public discussion of climate change toward apocalypse and denial that I reviewed in the first chapter is entirely explicable on this analysis. The politics of denial, fear, and scapegoating are perverse attempts to respond to the failure of public reason concerning this imminent threat. The resultant 'climate of fear' produces an increasingly conflictual and antirational shaping of public discussion; emotive arguments and denial of the science consequently move from the margins to the mainstream of public reason in arguments about climate change in the United States, the United Kingdom, and beyond. Kantians would, of course, have argued that this is why there is no way forward. Public reason works only when superstition and religion are excluded, they used to say, although more recently figures such as John Rawls and Jürgen Habermas seem less sure, as their accounts of 'comprehensive doctrines' (Rawls) or 'religious languages' (Habermas) indicate.[61]

As John Milbank argues, the exclusion from descriptions of the character of public reason of nature *and* the spirit as non-rational, and analogously of embodied organic life *and* the life of the emotions, has a perverse outcome:

> if one restricts reason to the formal and insists that it operates only within knowable boundaries, one will encourage entirely irrational and purely emotive political movements to take centre-stage, even though they are playing by the rules.[62]

It is precisely because the cosmopolitan ideal of reason is socially and ecologically unsituated that it is also emotionally and spiritually insensitive. Its claim to exclude belief, cultus, religion from public life closes off discussion about different conceptions of substantive goods, especially when these goods are related to religious beliefs, or, in the case of climate change, beliefs about empirical science. Yet Kantianism merely constructs another cult: the cult of the 'factish gods' as Latour calls it.[63] These new factish gods are unlike the old gods: they have no past or future, they are not revelations in

---

drives people towards constitutional government, and that the escalating cost of war between constitutionally governed nations will ultimately lead to the end of war between them; Kant, *The Metaphysics of Morals*, 165-70; and Kant, *Perpetual Peace*, 95-96.

61. John Rawls, *Political Liberalism* (New York: Columbia University Press, 2005), 175; Jürgen Habermas, *Between Naturalism and Religion* (Cambridge: Polity Press, 2008), 129-30.

62. John Milbank, 'Hume *Versus* Kant: Faith, Reason and Feeling', *Modern Theology* 27 (2011): 276-97, 277.

63. Bruno Latour, *On the Modern Cult of the Factish Gods* (Durham, NC: Duke University Press, 2010).

history, but rather apprehensions of individual minds moment by moment which guide a fragmented presentism that is unconscious of the origins of the present in the past. As Jacob Taubes puts it, channelling Herder, for Kant 'every event bears within itself the measure of its time within itself.'[64] The scythe of Kantian reason slices history into moments of individual apprehension. At the same time it proclaims enlightened reason as the lord of history, the uniquely conscious maker of histories both human and natural.

Climate change, and the ecological crisis more generally, reveal that Kant's announcement of the triumph of enlightened humanity over natural and cultural history was premature. Kant reinvented morality as the law in order to save humanity from the fated physicality of the mechanistic cosmos described by modern science. In so doing he not only denied the founding insight of the Christian tradition, which is that the law made by God cannot save but only the Messianic intervention of the Incarnate One. He also put enlightened humanity in the place of the Creator when he reinvented the law, revealed commands and natural law, as the supreme work of human reason. As Pierre Manent argues, Kant announced not only the rise of enlightened humanity but the 'end of nature', for it is in reasoned respect generated by the law alone that humanity realises the highest good that is possible in physical existence; respect for the earth, and her Maker, no longer forms part of the moral law.[65]

The Kantian account of constitutional cosmopolitanism is the most influential philosophical expression of the Baconian idea that the human domination of nature through empirical science is the source of human flourishing. But, paradoxically, Kant excludes nonhuman nature from the construction of this new constitutional sphere. Nature is instinctual and mechanical; humans are intuitive and rational. As Latour argues, the Enlightenment constitution is an artifice, a willed rational and subjective construct whose makers deny that human societies are hybrids of the human and the natural.[66] By the same token, the empirical domain investigated and translated by climate scientists — the fabric of the earth from ocean floor to upper atmosphere — is said not to be human. Scientists translate

64. Jacob Taubes, 'On the Current State of Polytheism', in Jacob Taubes, *From Cult to Culture: Fragments Toward a Critique of Historical Reason,* ed. C. E. Fonrobert and A. Engel (Stanford, CA: Stanford University Press, 2010), 302-14, 311.

65. Pierre Manent, *The City of Man,* trans. Marc A. LePain (Princeton, NJ: Princeton University Press, 1998), 188-89.

66. Bruno Latour, *We Have Never Been Modern,* trans. Catherine Porter (Cambridge, MA: Harvard University Press, 1992), 13-33.

natural facts that are for them objectively real, even if the data they use to reveal and represent these facts are constructed by scientists and reside in computer models and 'climate proxies' such as tree ring data sets or Antarctic ice cores stored and sliced up in laboratory freezers. But that the climate is changing for the scientist is not a social construct.

The foundational Enlightenment separation between nature and culture, and hence between natural history and the history of the earth, is the core conundrum of climate change, as I first argued in *A Moral Climate.* In the 'anthropological matrix' of those who lived before — or who now live without knowing — Bacon, Locke, Kant, and Marx, the climate interacted with human morality. For the Hebrew poets and prophets, weather and weather extremes carried messages from the Creator about human activities: the rain comes to water the crops and the sun shines on the fields to give good harvests when the people and their ruler treat the poor with justice and when they worship the Lord (Psalm 72). But when the rich neglect the terms of the Mosaic covenant and pile up lands and houses for themselves while enslaving their fellow Hebrews and neglecting to care for orphans, the poor, strangers, and widows, and when they worship idols instead of the Creator, then the weather is said to turn against them, the rains dry up, the ground is baked hard, and the vineyard and the field no longer give their increase.[67] In these and other ways the earth is said to 'cry out' and bear witness against the Hebrews' unfaithfulness to the everlasting covenant. This is because the society of the Hebrews was founded not on an artificial constitution but on a divine and cosmic covenant between God, human beings, and the land. The original covenant with Noah was sealed with the natural sign of the rainbow; similarly, when Moses ascended Mount Sinai to renew the covenant and receive the divine commands, the cloud enfolded him as he met with his maker.[68] And hence the earth, and more especially the *climate,* also bore witness to the failure of the covenant.

Moderns are moderns precisely to the extent that they reject that God judges the nations and their rulers through climatic effects on crop productivity. When harvests fail from drought, or snow piles high for months in the fields, this is no longer because God disapproves of a king's divorce, or

67. I give a fuller account of the prophetic interpretation of natural disaster in Michael S. Northcott, *The Environment and Christian Ethics* (Cambridge: Cambridge University Press, 1996), ch. 5.

68. Robert Murray, *The Cosmic Covenant: Biblical Themes of Justice, Peace and the Integrity of Creation* (London: Sheed and Ward, 1992); and Northcott, *The Environment and Christian Ethics,* ch. 5.

resists a would-be emperor in his attempted conquest of Russia. As Herbert Butterfield argues, Christians and Jews until modernity shared the Old Testament view that God in history judges prideful humanity through the course of events both natural and political: 'what was unique about the ancient Hebrews was their historiography rather than their history — the fact that their finer spirits saw the hand of God in events.'[69] For Jews and Christians, history is not a series of 'freeze-frames' of the kind offered in scientific papers with their graphs of data snapshots of heat in the ocean or the atmosphere.[70] The factish gods of these scientific freeze-frames are ahistorical, detached from the histories of peoples, lands, and species, and they do not in the end move us in the way the fables of the gods of old moved our forebears. They do not speak truthfully of the human condition because their 'facts' are frozen in time, moments of gnosis that hide more than they reveal. But this new cult of facts, and of enlightened worshippers who set their days and hopes by them, displaces the old gods, including the angel of history, who 'rides the whirlwind and directs the storm', and who sends forth rain and sun in due season to water and warm the crops where a righteous ruler judges for the poor and the needy.

If there is a place for God in the modern constitution it is, as William James saw clearly, in the movements of the mind and the soul, perhaps even in familial relations; but it is not in the climate, the economy, or the land.[71] As Latour argues, modernity therefore had to reinvent spirituality so that 'the all-powerful God could descend into men's heart of hearts without intervening in any way in their external affairs.'[72]

The interiorisation of God, the exclusion of God from nature, and the new duality between natural and human history are necessary corollaries of the Baconian scientific revolution. Empirical science objectifies the natural world in order to subject it to the artifice of the repeatable laboratory experiment and the hypothesis-confirming data freeze-frame. The scientist studies objects in a laboratory, 'the field', and in computer models: the atmosphere, climate, forests, oceans, species are such objects. They behave as they do because of the laws of nature; they are not changed by human minds nor by divine spirits. As Latour puts it, 'no one is truly modern who

69. Herbert Butterfield, *Christianity and History* (London: Fontana Books, 1949), 98.

70. Latour, 'Thou Shalt Not Freeze-frame, or How Not to Misunderstand the Science and Religion Debate', in Latour, *On the Modern Cult of the Factish Gods,* 99-123.

71. William James, *The Varities of Religious Experience: A Study in Human Nature* (New York: Longmans, 1902).

72. Latour, *We Have Never Been Modern,* 33.

does not agree to keep God from interfering with Natural Law as well as with the laws of the Republic.'[73]

But climate scientists invoke natural objects as witnesses to the fact that the modern republic in its energetic mobility, technical productiveness, and wealth accumulation rests upon a natural foundation which includes fossil fuels and forests. The anthropic mobilisation of these objects into the atmosphere and the ocean is said to undermine the stability or fruitfulness of other objects — coastlines, croplands, deltas, icebergs, soils, species — which provides the stable backdrop to the human cultural economy. Paradoxically, these 'nonhuman testimonies' contradict the modern constitutional claim that society is an artifice made by rational subjects while nature is an unmade stable backdrop. The manner of their testimony — data sets, computer models, scientific papers — contradicts the claim that nature is an object and not an artifice. Climate science therefore involves a politics as well as natural science. It too is a cult — 'the cult of the factish gods'. It makes claims about the human constitution, and not only the biosphere in which constitutions are constructed. And hence climate scientists find themselves accused of being 'liberals' or 'progressives', or even 'lobbyists', because their science has political implications and interferes with the neat division of politics and economy from science on which the modern constitution is founded.[74]

Logos, Sovereign Power, and the Angel of History

The modern world is often described as humanist, but its announcement by scientists and philosophers required 'the simultaneous birth of "nonhumanity" — things, or objects or beasts — and the equally strange beginning of a "crossed-out God" '.[75] The modern constitution, precisely because it is a rational artifice, requires the separation of nature and society. But the founding fathers of this modern constitutionalism leave out two fundamental features of this artificial political and social order, namely, 'scientific power and the work of hybrids'.[76] The secret that can hardly be uttered by

73. Latour, *We Have Never Been Modern,* 33.

74. This point was made in Bruno Latour's University of Edinburgh Gifford Lectures, 'Facing Gaia: A New Enquiry concerning Natural Religion', February 2013, at http://www.ed.ac.uk/schools-departments/humanities-soc-sci/news-events/lectures/gifford-lectures/series-2012-2013/lecture-four (accessed 13 March 2013).

75. Latour, *We Have Never Been Modern,* 13.

76. Latour, *We Have Never Been Modern,* 14.

humanists is that the Enlightenment constitution is and has always been a hybrid of nature and culture even although its framers envisaged the constitution, the economy, and politics as the purely cultural products of human reason.[77] At the same time, the secret that cannot be told by scientists is that science not only reveals the workings of natural laws; it also exercises power over them, even in the act of revealing them. And as soon as we become aware that the constitution rests upon the production of hybrids, such as the fossil fuelled economy, and that the technological construction of these hybrids is also the work of science, 'we then discover that we have never been modern' and that modernity, as a form of life that is premised on the separation of the history of nature from the histories of peoples, is a form of 'false consciousness'.[78]

Kant's epistemological divide, on which modern constitutional cosmopolitanism rests, is, as Latour argues, the 'canonical formulation' of nature-culture dualism; in Kant's splitting of theoretical and practical reason, 'things-in-themselves become inaccessible, while, symmetrically, the transcendental subject becomes infinitely remote from the world'.[79] Attempts by philosophers from Hegel to Merleau-Ponty to repair the divide fail because they do not acknowledge the extent to which scientific power over objective nature sustains the subjectively human realm, nor the extent to which 'quasi-objects' proliferate in the realm of the subject and produce the subject, through the 'industrial revolution(s)'.[80] For Latour, we have never really been dualists; we have always been pre-Copernicans, pre-Cartesians, pre-Kantians. The difficulty is in recognising that this is so, and in facing up to its tremendous implications for how moderns conventionally think of and describe nature and society as separate realms of objective fact and subjective creation. The answer for Latour, as for Whitehead, is to challenge the Copernican revolution: nature is greater than and enfolds society; society grows in scale and complexity because of scientific power over nature. Humans are still at the centre of the cosmos; they are there with the Being of beings, and they are surrounded by the multiple agencies of angels, crea-

77. David W. Ehrenfeld also makes this point in his *The Arrogance of Humanism* (Oxford: Oxford University Press, 1981).

78. Latour, *We Have Never Been Modern*, 46.

79. Latour, *We Have Never Been Modern*, 56.

80. Latour, *We Have Never Been Modern*, 57; on the ways in which technologies such as computers, keyboards, and mobile phones increasingly produce subjects see Andy Clark, *Natural Born Cyborgs: Minds, Technologies and the Future of Human Intelligence* (Oxford: Oxford University Press, 2003).

tures, and spirits. In this sense the 'Anthropocene', which scientists of the Royal Society have announced as the new geological epoch of humanity's rising power over nature as revealed by climate science, is not a new epoch at all.[81] It was the Copernican invention of the Heliocene which was the illusion.

The sun may seem since Copernicus to dominate history, but since the invention of agriculture, and the journey of Adam out of the hunter-gatherer Paradise of the forest of Eden, humans have been, and are still becoming, central to the destiny of life, as they nurture children, domesticate animals, hybridise crops, build cities and drains, mine fossil fuels, and split atoms. The Jewish and Christian story of redemption is the story of the redeeming of these from the heightened powers, and violence, that they release into human history. The Copernican revolution, however, challenged this story of redemption by destroying 'man's dwelling in the cosmos' and threatening 'the analogy of earthly and heavenly order'.[82] But it is only where such a correspondence is sustained that it is possible to understand that the heavens and the earth interact and reflect one another. Without the analogy of being there is only sovereign power, unmediated, unrepresented, by priesthood, sacrifice, and worship. A merely immanent actor network of scientific facts, or around carbon emissions, cannot recover mediation. As Taubes realises, from Copernicanism to Baconianism is a short step in the realm of ideas. The Baconian instrumentalisation of nature results in the instrumentalisation of humans by the technologically expanded domination of sovereign power. The Kantian borderline between 'science' and 'religion' underwrites the rise of the science-informed technological cult of sovereign power. Hence Protestant theology since Kant denies that redemption is possible within the cosmos, or in the relations of power that are the histories of peoples and nations. After Kant's rejection of the Messianic intervention of the Logos as the source of the redemption of the nations, God becomes the deists' outsider to being and time.

Like Vico, Latour refuses to divide human history from natural history, and like Vico he refuses to divide either of these histories from the history of religion. The human science that helps reconnect these divided histories is anthropology, the human science Vico is said to have founded, because anthropologists still access what Lévi-Strauss calls the 'savage mind'

81. Jan Zalasiewicz, Mark Williams, et al., 'The Anthropocene: A New Epoch of Geological Time?' *Philosophical Transactions of the Royal Society* 369 (2011): 835-41.

82. Taubes, 'Dialectic and Analogy', in Taubes, *From Cult to Culture,* 170-71.

in which nature, society, and the gods form one sacred universe. But the challenge climate change raises is how, in the midst of the vast technological scale of human power over nature and the unintended enhancement of natural powers *over* humans (rising oceans, strengthening storms, enduring droughts), the primitive sense of an interconnection between nature, society, and the sacred might be recovered.

For Latour, the answer to this question is in the mediation that takes place in actor-networks. Climate science is an actor-network which includes scientists, animals, and plants that observers record responding to a changing climate, and robotic data sensors such as the thousands of floats used to measure ocean temperatures. Climate politics is another actor-network, which includes climate activists and climate scientists, bureaucrats, legislators, democratic representatives, heads of state, diplomats and conference negotiators, the UN and the IPCCC, nongovernmental organisations, fossil fuel companies, miners, mining and drilling machines, divers, tunnelling robots, computer-controlled well heads, electric utility companies, car drivers, householders, office managers, energy market makers and investors, and the builders of windmills and coal-fired power stations. To name the actor-network in all its diversity, locality, and globality is to indicate that this network can address climate change only through the mediatorial discourse of politics, by becoming a 'parliament of things' as well as of subjects. In this parliament,

> There are no more naked truths, but there are no more naked citizens either. The mediators have the whole space to themselves. The Enlightenment has a dwelling-place at last. Natures are present, but with their representatives, scientists who speak in their name. Societies are present, but with the objects that have been serving as their ballast from time immemorial. Let one of the representatives talk, for instance, about the ozone hole, another represent Monsanto chemical industry, a third the workers of the same chemical industry, another the voters of New Hampshire, a fifth the meteorology of the polar regions; let still another speak in the name of the State; what does it matter so long as they are all talking about the same thing, about a quasi-object they have all created, the object-discourse-nature-society whose new properties astound us all and whose network extends from my refrigerator to the Antarctic by way of chemistry, law, the State, the economy, and satellites.[83]

83. Latour, *We Have Never Been Modern*, 144.

Latour's 'parliament of all things' is reminiscent of John Seed and Joanna Macy's 'council of all beings'.[84] The council is a ritual process in which participants rediscover their connectedness to nature by acting out the voices of other beings, places, and species, including those threatened by climate change and ecological crisis. The rituals are designed to re-create the networks which sustain human and other than human life but which modern rituals of consumption and production obscure and fail to honour. Both Latour's 'parliament of all things' and Seed and Macy's council of all beings involve an attempt to overcome the nature-culture divide, sustained by science and political economy, through mediation and participation in rituals which honour the hybrid actor-networks that sustain human dwelling on a more than human earth.

But both metaphors — parliament and council — partake of another shared feature that makes them incommensurable with the global actor-network involved in climate change: this is that they are local and face-to-face meetings. In the case of parliaments, the house may be a semicircular arrangement of chairs around a dais in a grand, purpose-built edifice, often next to a large river; the council of beings may be a 'chamber' created by a clearing in a forest. But both refer to gathering in *place,* and this dimension of place is hard to capture with the actor-network metaphor alone, for that metaphor easily describes a technological actor-network of myriad actors who never gather in one place. As Donald MacKenzie argues, carbon emissions trading is one such actor-network.[85] It is, however, governed by the law of price and functions by anonymous exchanges, and in many cases fraudulent trades in 'saved' production emissions which hide still growing consumption emissions and fossil fuel extraction. The large scale of this actor-network gives the largest-scale actors — and especially fossil fuel companies and the banks which provide capital for their activities — undue power relative to the weak, and in particular the victims of climate change. The asymmetry and absence of face-to-face deliberation in such networks hardly gives them the status of a parliament. But parliaments in the time of late capitalism are increasingly giving up democratic powers to 'arms-length', privatised corporate agencies in order to prevent democratic scrutiny of executive power and of the gradual leaching of sovereign

84. John Seed, Joanna Macy, et al., *Thinking Like a Mountain: Towards a Council of All Beings* (Philadelphia: New Society Publishers, 1988).

85. Donald MacKenzie, 'Making Things the Same: Gases, Emissions Rights and the Politics of Carbon Markets', *Accounting, Organizations and Society* 34 (2009): 440-55.

power from nation-states to corporations. This passing of governing power from accountable sovereigns to unaccountable economic corporations has been a feature of the economic corporation from its inception. The extent of the contemporary devolution of sovereignty onto corporations is undermining participative democracy and even the legitimacy of lawful rule. As MacKenzie notes, actor-networks by their very nature obscure questions of agency: the agency of actors is hidden in the network.[86] The networks are like laboratory 'black-boxes', such as computer-calibrated measuring devices, where the calibrations are beyond the control of those who use them. A growing number of financial transactions are performed at high speed by such computerised devices. It is difficult to attribute moral responsibility for harms visited by such 'black-box'–type networks on particular people and places. There is therefore a danger that technologically enabled actor-networks, far from challenging the neoliberal privatisation of sovereign power, the thinning out of democratic practice, and the growing marketisation of civic life and even personal services, will actually underwrite these tendencies. But the reason for this turn of events is ultimately theological and technological; it cannot be explained within the confines of Enlightenment rationalism and the resultant demystification of politics and the spiritualisation of God.

The Christian Cosmopolis of Earth and Heaven

The actor-networks with the longest history in human communities of place are those associated with religious communities, and in Europe with the Christian Church. The places of worship remain the architectural foci of most towns and cities in Europe, and these foci were re-created by Christians on other continents, including the Americas and numerous cities in Asia and Africa. Like parliaments, church buildings are often located in geographically marked places, such as a bend in a river, on a hill, or on an ocean shoreline. Most such buildings are topped by a cross, which is the visible sign of the story of the crucified and risen God who is 'crossed out' in the humanism of the Enlightenment. The first purpose-built Christian churches were built around the tombs of great saints of the locality. Typically the eucharistic table, or altar, was placed immediately on top of the

86. Donald MacKenzie, *Material Markets: How Economic Agents Are Constructed* (Oxford: Oxford University Press, 2008), 34.

saint's tomb, and so it was marked by the saint's corporeal presence beyond death. This feature of Christian churches is not, however, unique. Archaeological evidence indicates that the earliest built environments began as places of burial. The remains of the dead were placed reverentially in the ground; the burial place was marked by a cairn or pile of stones, and around such cairns humans first created the stone-built rooms of the living.[87] In burying the dead, as Robert Harrison argues, human beings have literally *humanised* the earth; human flesh in time becomes humus, as bodies that were living souls are returned to the soil. What is unique, however, in the Christian story of this process is that the Christians believed that the dead are still present as a 'cloud of witnesses' with Christians during their worship rituals and feasts. In the early Church, the actor-networks which constituted the communities of the saints were networks across time as well as space and occasioned the idea of a 'communion of saints' which exceeds the borders both of death and of life as well as the borders of nature and of culture. The resurrection of Christ and his visitation of Hades meant that Sheol, the waiting place of the dead, was abolished, and the dead passed directly into heaven after the coming of Christ. Christian worship was a gathering of the souls of the living and the dead. Thus the oldest surviving built environments of Christian worship — the catacombs on the edge of the ancient city of Rome — are places of burial.

The paradigmatic meeting of Christians from New Testament times was for the sharing of a meal. The meal was accompanied, as were analogous Roman symposia, by readings from classic texts, which for the Christians included the Jewish scriptures, the Gospel records of the life and teachings of Christ, and the letters of the apostles. These readings were interspersed with songs and prayers to God and prayers of blessing over the food, especially over the bread and wine which symbolised the presence of Christ. The earliest account of such a meeting is that of Saint Paul in 1 Corinthians 9. In that account Paul upbraids the church in Corinth for allowing social divisions between rich and poor, free citizens and slaves, to mark the meal and cause dissension. Noting that some brought luxury foods such as meat to the meal while others did not have enough to eat, Paul tells the Corinthian Christians that they should first satisfy their hunger at home, and when they gather together they should all eat the same food as a sign of their equality and fellowship in Christ. He goes on to say that the distinctive

87. Robert Pogue Harrison, *The Dominion of the Dead* (Chicago: University of Chicago Press, 2010).

feature of the Christian polity is that the normal hierarchy of strong and weak, wealthy and poor, is reversed in the body of Christ, so that the weak are given greater honour than the strong. The virtue which sustains this Messianic and anti-hierarchical ethos is that of love or *caritas*.

For early Christians, the eucharist, and eucharistic gatherings, created and performed the social and spiritual order of the kingdom in which Christ as Messiah, and not the emperor, is king, and which Christ inaugurates in his life and ministry and through his death and resurrection. Christian leaders, bishops, modelled their leadership and ministry in the early Church after the Messianic servanthood of Christ, who is also the 'good shepherd'. In early Christian art Christ is depicted as a shepherd rather than as the crucified figure familiar from medieval art, and the apostles are often depicted as sheep; this is why bishops still carry crooks to this day. The churches practised care for the sick; almsgiving; distribution of food and money to the poor, orphans, widows, the elderly, and the unemployed; and visited and fed prisoners. In Roman cities most people were hungry and poor; the majority were slaves owned by the free citizens or in the service of municipal and imperial authorities as labourers and legionnaires. The Church was a community where such people were treated with greater honour than they were by Rome. By such distinctive social practices the Christians modelled a contrast society, and they enacted their belief that the Church was, as Irenaeus of Lyon and other early theologians taught, a Paradise on earth, an alternative reality to that of Rome.

This contrast society was not only a social but a cosmopolitical community which joined Christians and Christ with the Creator of heaven and earth. Hence Christ was worshipped not only as the Good Shepherd but also as *Pantocrator*, who rules over earth and heaven, time and eternity. After the conversion of Constantine, the Christians built churches in large numbers, and the oldest which survive in Italy and Turkey reveal the Christian sacred cosmology in mosaic and stone.[88] Mosaic decoration of the roofs of the first basilicas depict stars and clouds as the heavenly sphere; the supporting arches, and the apse, of these tall buildings depict Paradise, where Christ, as Pantocrator, rules in the company of the apostles and the departed saints, and looks down on the living community of worshippers who gather on the

88. I have only lately begun visiting examples of these early sacred structures myself; their depiction of paradise was first opened up to me in Rita Nakashima Brock and Rebecca Parker, *Saving Paradise: How Christianity Traded Love of This World for Crucifixion and Empire* (Boston: Beacon Press, 2008).

floor of the church.[89] The greatest of these structures that remain is Hagia Sophia in Istanbul. Most of the mosaics are obscured by plaster used to cover the images when Hagia Sophia was converted to a mosque, but the stars of the heavens and the heavenly beasts of Ezekiel's vision can now be clearly seen by contemporary visitors. So too can depictions of Christ and the saints and of vines, fruits, and trees on the upper arches of the structure. As Brock and Parker comment,

> In this three-tiered universe, paradise had both a 'here' and a 'not here' quality. Christians taught that paradise had always been here on earth. Sin had once closed its portals, but Jesus Christ had reopened them for the living. While Christians could taste, see, and feel the traces of it in ordinary life, they arrived most fully in paradise in community worship.[90]

The community worship of the early Christians, then, sustained a sacred cosmopolis which empowered the Christians to challenge the conventional human divisions and hierarchies of the ancient world through works of love while also giving them a vision of the natural world and the heavens which challenged the humanocentrism of classical and Roman thought. For the early Christians, salvation was an ecological as well as a political and spiritual reality in which the earth and all its creatures, as well as human society, were being redeemed through the worship and the witness of the saints and under the kingly rule of Christ.

The text which more than any other set the tone for the Christian vision of Paradise was the Book of Revelation. In the last chapters of Revelation John of Patmos draws on Ezekiel's vision and imagines a heavenly city from which the river of life flows out, giving life to the trees along its banks. Just as Christians believed that the Resurrection of Christ made their entry to Paradise possible again, since Christ had defeated sin and death, they also believed that Paradise was a real physical place. Medieval maps, such as the Mappa Mundi, depict Paradise in the Far East. But the Church was also an allegory of Paradise in every place where the community of Christians gathered. Augustine describes Christian community as an allegory of Paradise:

> No one, then, denies that Paradise may signify the life of the blessed; its four rivers, the four virtues, prudence, fortitude, temperance, and justice;

89. Brock and Parker, *Saving Paradise,* 86.
90. Brock and Parker, *Saving Paradise,* 88.

> its trees, all useful knowledge; its fruits, the customs of the godly; its tree of life, wisdom herself, the mother of all good.

And he continues,

> Paradise is the Church, as it is called in the Canticles [the Song of Songs]; the four rivers of Paradise are the four gospels; the fruit trees the saints, and the fruit their works; the tree of life is the holy of holies, Christ; the tree of the knowledge of good and evil, the will's free choice. For if man despise the will of God, he can only destroy himself; and so he learns the difference between consecrating himself to the common good and revelling in his own.[91]

The restoration of Paradise on earth was for more than a thousand years understood to be the work of the Christian Church. And this was a material and not only a spiritual work, just as Paradise was said to be a real place and not only a mythic or other-worldly reality. It was this same belief in Paradise on earth which empowered and drove the monks to create gardens and herbariums and to develop crafts and workshop techniques and technologies which were labour saving while also increasing the fertility and productiveness of the earth. But as we have seen, the Christian belief in progress toward Paradise takes a particular turn in the Baconian project of progress in the control and domination of life and natural forces of which climate change is the denouement.

The growing domination of science over the material earth led the most influential of modern moral philosophers — Immanuel Kant — to argue for a relocation of the quest for moral progress from the life of the embodied community of Christians on earth to the life of the mind. For philosophers after Kant the moral life is, in Peter Singer's words, 'no part of the structure of the universe'.[92] But this seminal split between the worlds of nature and culture tears apart the fabric of metaphysics and morality so that when faced with the conundrum of climate change, which clearly includes both nature and culture, the moral categories for responding to it in modern moral philosophy are hardly up to the task. Climate change is the hybrid par excellence that pulls down the dividing walls between nature

91. Augustine, *City of God*, 13.21; see also Brock and Parker, *Saving Paradise*, 105.

92. Peter Singer, *How Are We to Live? Ethics in an Age of Self-Interest* (New York: Prometheus Books, 1995), 137.

and culture, science and ethics, theoretical and practical reason, natural and human history, that Kant and his heirs construct. But if it does this, it also challenges the intellectual foundations of cosmopolitan international relations. This is why the cosmopolitan political economic 'solution' to climate change is to construct an actor-network — a carbon emissions market — on the analogy of a market in financial derivatives which is morally blind, driven by self-interested trades rather than moral relationships, and situated in computerised and digital exchanges rather than in practices that care for the earth.

For Kant, the nations are first and foremost gatherings of rational persons who use law to enable them to contract together to live in relative accord with one another. But this Enlightenment account misses the extent to which the nations have deep cultural and natural roots — cultural roots, as I argue in the following chapters, in the Christian idea of the communion of saints as it took root in the histories and communities of particular peoples; natural roots in that each nation is a gathering of souls in geographically bounded *places* in a divine creation. In the chapters that follow I present a communitarian alternative to cosmopolitanism. I argue that the Christian idea and practice of the Church as an embodied, moral, and spiritual community of friendship and love is the root of the first nations in Christendom. The histories of the nations, like the history of the Church, are chains of memory: the best times are remembered as those of divine approval while the dark times as those of judgment. Understanding the story of the nation as a product of history and providence, culture *and* nature, trains citizens in those moral and political virtues which enable them not only to live together peaceably but also to perform works of love which bring Paradise *on earth* closer through mutual submission and mutual action for the common earthly good under the kingly rule of Christ.

6. The *Nomos* of the Earth and Governing the Anthropocene

Climate scientist Dave Reay argues that 'global warming begins at home', by which he means that consumption of fossil fuels consistent with stabilising human climate impacts is primarily a function of household demand for carbon intensive goods and services, including cars, clothing, and computers; domestic appliances such as refrigerators, cookers, and washing machines; electronic entertainment devices; meat; recreational travel; and space cooling, lighting, and heating.[1] On this account householders — and not primarily *nations* as indicated by the UNFCCC — have a duty to limit their consumption of these goods and services in order to conserve an atmosphere conducive to the well-being of future individuals.

There are at least three claims here: a rational and scientifically defensible response to the threat of climate change for the well-being of future generations involves mitigating reductions in the human use of fossil fuels; collective consumption of fossil fuels is primarily the sum of the behaviours and decisions of householders; the key locus of decision-making about use of energy and fossil fuels is private households. But the claim that global warming begins at home is a romantic argument predicated on human society structured around the *oikos* or household and before the rise of the modern nation-state and modern political economy, with their sustaining electricity power grids and energising networks of coal trains and trucks and oil and gas pipelines.

In the city-states of Classical Greece the decisions of householders determined the local supply of goods and services. The same held true in the medieval city-states of Florence, Siena, and Venice, where the modern global market has its historic roots. But as Hannah Arendt argues, modern

1. Dave Reay, *Global Warming Begins at Home* (London: Macmillan, 2005).

political economy has turned from an economy of households into an economy *as* household.[2] Discernment about consumption activities in a household in a modern city where energy is supplied through a global energy market will have no impact on the geophysical causes of climate change, since the global market in fossil fuels ensures that as long as they are extracted they will be burned somewhere. However, the focus on household consumption risks obscuring the geophysical causes of climate change in the extraction of fossil fuels and limestone from the crust of the earth and the destruction of the earth's forests. As we have seen, if the economic corporations and nations which together claim ownership of the earth's subterranean store of carbon continue to extract it, then forgone consumption in virtuous households will merely allow others to consume the energy the householder forgoes at a lower cost elsewhere, given the operation of the laws of supply and demand in the global energy market, as Jevons first realised. Absent decisions by the nations in whose terrain they are to restrict licensing of the remaining fossil fuel reserves to the capacity of the earth to absorb emissions without destabilising the planet, changes in household energy consumption will make no difference to the climate.

Such changes, just because they are not effective in addressing the causes of climate change, may, however, make a crucial difference locally. Householders who have sufficient wealth to choose their sources of fuel may procure renewable power in the form of local biomass, solar, and wind, and are therefore able to reduce their own fossil fuel consumption.[3] In so doing householders contribute to a cleaner local atmosphere and environment, even if their forgone use of coal or oil makes the resource cheaper for someone else. In choosing renewable fuels, such householders also help to establish an alternative and more ethical form of society which is less dependent on fossil fuels, and hence less dependent on destabilising the atmosphere. But the majority of lower income households are locked in to an energy supply model over which they have little control because of

2. Hannah Arendt, *The Human Condition* (Chicago: University of Chicago Press, 1958).

3. Rachel Howell argues that UK householders make decisions concerning the consumption of approximately two thirds of UK greenhouse gas emissions, although this is an average figure and does not address the vast inequalities between rich and poor households in CO_2 consumption; Rachel Howell, 'Lights, Camera . . . Action? Altered Attitudes and Behaviour in Response to the Climate Change Film *Age of Stupid*', *Global Environmental Change* 21 (2011): 177-87. By contrast Korbetis, Reay, et al. find that the majority of UK emissions are generated by the top 20 percent of richest households: Malamo Korbetis, David S. Reay, et al., 'New Directions: Rich in CO_2', *Atmospheric Environment* 40 (2006): 3219-20.

lack of wealth and political influence. In the UK over five million households are defined as 'fuel poor', which means that they spend more than 10 percent of their household income on cooking, lighting, entertainment, heating, and mechanical power in their homes; the fuel poor also mostly live in rented accommodations and are therefore unable to improve the energy efficiency of their homes. Such measures as would enable reduced household consumption of fossil fuels are beyond their control. In these circumstances, if carbon or energy pricing is used as the means to persuade householders to use less energy, this can have perverse consequences. In the UK a large number of low income households are already forced to choose between eating and staying warm. Only if carbon pricing and energy taxes are combined with correlative reductions in payroll taxes, as they are in Germany, or if energy bills take into account household income as they do in Norway and Sweden, could this approach be considered just or sustainable.[4] Alternatively, carbon pricing should be combined with a nationwide effort to increase the energy efficiency of buildings. Materials derived from space travel are now available that can significantly reduce the heating or cooling requirements of older houses, but the thinnest, and therefore most suitable to retrofitting, are very expensive. It is possible to combine building insulation and energy supply in the contract arrangements of public utility companies so they are paid more when they deliver *less* energy. But presently utility supply is organised in the UK and elsewhere on the basis of competitive supply, with no incentives for corporations to work with households and businesses to reduce energy use. Without such measures, however, carbon pricing will merely promote popular resistance to climate change science and policy responses.

The focus on household energy choices neglects the role of collective infrastructural technologies in shaping behaviours and outcomes. In Melbourne, a very fine tram service displaces a number of oil-driven cars from the city centre, but it runs on brown coal. Those who choose to use the tram are acting in ways consistent with reducing household consumption of fossil fuels. But when the tram is fossil fuelled, the intrinsic good of the change in behaviour is undermined. Unless those *nations* who possess significant deposits of coal and other fossil fuels use the near certainty that these de-

4. Tooraj Jamasb and Helena Meier, 'Household Energy Expenditure and Income Groups: Evidence from Great Britain', Electricity Policy Research Group, Department of Economics, University of Cambridge, 2010, and at http://www.eprg.group.cam.ac.uk/wp-content/uploads/2010/02/JamasbMeierCombined-EPRG10031.pdf (accessed 7 February 2013).

posits, if burned, will significantly harm future generations to legally proscribe their extraction, even a large-scale cultural move by individuals to forgo climate heating behaviours — of which there is as yet no evidence — will not change global consumption levels, since the market will price extracted fossil fuels to make their forgone use by some marginally cheaper for other uses.[5]

As we have seen, the Jevons paradox explains why demand management of energy consumption does not work in a global political economy where energy prices reflect global supply and demand. As defined from Adam Smith onwards, an *oikonomie* is not a household but a global trading economy where goods and services are exchanged primarily on the basis of market prices determined principally by the collective and unplanned interactions of the decisions of millions of businesses and households. Smith's metaphor of market exchange is essentially a mechanistic metaphor. Just as classical physics since Newton modelled the cosmos as a machine governed by immutable laws, so modern economists after Smith and Locke model the role of money and markets as a machine governed by immutable laws, laws that work best when human beings trust that providence resides in these laws and that therefore human beings do better to trust the laws — and in particular the 'laws of supply and demand' — than to trust in the efforts of governing bodies in cities or nations to regulate pricing mechanisms.[6]

The earliest legal codes of ancient Mesopotamia and Israel specified the defining purpose of just rule as being to prevent the strong from oppressing the weak and to uphold the cause of the poor; when rulers rule in such a way that the poor receive justice and the children of the needy are saved from want, then 'the mountains shall bring peace to the people' and 'showers shall water the earth' (Psalm 72:3, 6). In Christian ethics, the idea that the core purpose of rule is not only to prevent crime and defend the realm but also to promote justice for all, and hence the common and *cosmic* good, has a long history.[7] However, under the conditions of modern politi-

5. The problem of good intentions leading to bad outcomes in complex technological societies is discussed in Robert Spaemann, *Happiness and Benevolence,* trans. Jeremiah Alberg (Edinburgh: T&T Clark, 2000), and Alasdair MacIntyre, 'Social Structures and Their Threats to Moral Agency', in MacIntyre, *Ethics and Politics: Selected Essays,* vol. 2 (Cambridge: Cambridge University Press, 2006), 186-204.

6. See further Michael S. Northcott, *A Moral Climate: The Ethics of Global Warming* (London: Darton, Longman and Todd, 2007), 70-71.

7. For an influential essay on the common good see Jacques Maritain, *The Person and the Common Good,* trans. J. Fitzgerald (London: Geoffrey Bles, 1948); for an insightful treatment

cal economy the idea of collective action for the common good, which was more dominant in political theory before the rise of Cartesianism, is problematised. Adam Smith is again a key innovator here. He argues that when people trade and exchange they do so first and foremost from the motive of self-interest and the interest of the household, and not in response to a perceived divine conception of justice. As Albert Hirschmann argues, 'the main influence of *The Wealth of Nations* was to establish a powerful *economic* justification for the untrammelled pursuit of self-interest', whereas earlier advocates of the idea of self-interest, such as Montesquieu, still saw it as restrained by recognition of a larger political purpose.[8] Smith, of course, penned his own moral philosophy after the *Wealth of Nations,* but in so doing he gave effect to an influential division of labour between the economy and ethics which extended the nature-culture divide from science and ethics to social science and ethics. Thus, for Max Weber bureaucratic management of the economy and society is no longer intrinsically a morally guided project; the purpose instead is efficient management of competing economic and political interests.[9] Hence politics as a collective activity, requiring and depending upon the participation of the people in civic society and local communities of place, is gradually subverted by the rise throughout the last century of economic management, money, and markets as the dominant instruments of governance in modern nation-states. The political, as Michel Foucault argues, increasingly becomes a domain of sovereign power in which the aggrandising modern state deploys technologies of surveillance and proliferates laws and regulations to control a pliant populace.[10] In this context, it is unsurprising that some argue that climate change also needs to be addressed by a global project of state-driven surveillance, and rationing, of individuals' use of energy.

In response to the rise of economic management as the dominant political form of governmentality, many contemporary political theorists and

---

of commonwealth see Michael Hardt and Antonio Negri, *Commonwealth* (Cambridge, MA: Belknap Press, 2009).

8. Albert O. Hirschmann, *The Passions and the Interests: Political Arguments for Capitalism Before Its Triumph* (Princeton, NJ: Princeton University Press, 1977).

9. Max Weber, *The Theory of Social and Economic Organization,* ed. Talcott Parsons (New York: Simon and Schuster, 1947); see also the critique of the perverse outcomes of the collective pursuit of the goal of efficiency in Jacques Ellul, *The Technological Society,* trans. John Wilkinson (New York: Free Press, 1964).

10. See especially the collected essays in Michel Foucault, *Essential Works of Michel Foucault, 1954-1984,* vol. 3: *Power,* ed. James D. Faubion (New York: New Press, 1994).

philosophers hold that the essential political task is to create the conditions in which individuals pursue individually chosen and rival conceptions of the good.[11] Neoliberal theorists go further and argue that the only proper way in which rival conceptions of the good can be arbitrated is in the marketplace itself. If the state imposes regulations and laws which favour some conceptions of the good over others — such as, say, the preference for forms of energy that do not burden the atmosphere with fossilised carbon, or for food that is not dependent on cruelty to animals or migrant workers — it is argued that the state overreaches its powers and undermines personal freedom. However, though this libertarian argument is regularly invoked to resist government support for climate change–related interventions in energy markets, or efforts to reduce carbon emissions, in reality these markets conform in no way to the idealised model of supply and demand of the economist. Instead, energy supply and demand are regulated by nations who in most cases are the principal owners, and hence permission givers, of fossil fuel extraction; even where private companies claim ownership of oil, gas, or coal fields, they can extract the resource only under government license. Markets, and hence prices, in electricity, oil, coal, and gas are determined by government sales of licenses for fossil fuel extraction, and interact with government subsidies of fossil fuel extraction and government provision and/or licensing of energy infrastructure, such as power grids, roads, and airports. In the United States alone, trillions of dollars annually pass from the taxpayer in one form or another to kinds of expenditure which subsidise the extraction or use of fossil fuels.[12] The G20 nations agreed in 1999 to end fossil fuel subsidies in light of climate change, but no progress at all has been made on that declared shared goal.

Kant argues, as discussed above, that when individuals act as though each of their actions could be universalised without harm they act in such a way as to regulate their own behaviours so as to produce a sum of goods through which they may enhance their own welfare and, cumulatively, the welfare of nations and future generations.[13] The relation of the individual to

11. John Rawls's *A Theory of Justice* (Cambridge, MA: Harvard University Press, 1971) is the single most influential twentieth-century work in which the view of politics as the mediation of the pursuit of incommensurable private goods, rather than of the common good, is advocated; Adam Smith's *Wealth of Nations* achieved even more influence in making the same case.

12. Amory Lovins, *Reinventing Fire: Bold Business Solutions for the New Energy Era* (White River Junction, VT: Chelsea Green Publishers, 2011), ch. 1.

13. Immanuel Kant, 'Idea for a Universal History with a Cosmopolitan Purpose', in *Kant:*

the collective on this account is crucially determined by the sovereign individual as the seat of rational agency, whose actions, when joined with those of other rational individuals, produce law-governed polities. For Kant, it seems obvious that individuals properly educated arrive at individual judgements which do not sharply differ and hence that the universalisability of moral maxims will be sufficient to generate shared respect for the moral law. The unfolding of European history since the Enlightenment has hardly confirmed Kant's account. Nonetheless, influential political theorists such as John Rawls continue to hold that it is possible to describe a framework for collective action and judgement on the basis of an account of autonomous reasoning agents, although Rawls differs from Kant on the degree to which individual apprehensions of the moral law may differ. Others, such as Martha Nussbaum, argue that, with certain revisions, it is possible to extend the essentially Kantian political liberalism of Rawls to problems of international collective judgement such as that posed by climate change.[14] Analogously, Edward Page and Steven Gardiner argue that it is possible to extend the Kantian frame of political liberalism toward the intergenerational as well as collective action problem of climate change.[15]

In the midst of a global financial crisis at time of writing of five years' duration, as well as a much longer-run global ecological crisis, many citizens and corporations, and many of the governments they elect or influence, refuse to acknowledge a shared duty to address global ecological problems, and more especially climate change, in ways that may involve an element of sacrifice either of personal consumption or of national gross domestic product or corporate profits. This was evident in the United States' federal government submission to the second Rio Earth Summit in 2012, in which it insisted that any reference to limiting consumption or to ending fossil fuel subsidies be removed from the agreed text of the summit.[16] This prob-

---

Political Writings, ed. H. S. Reiss, trans. H. B. Nisbet (Cambridge: Cambridge University Press, 1991), 41-53, 44.

14. For a critical discussion of Nussbaum's attempt to internationalize Rawls in Martha Nussbaum, *Frontiers of Justice: Disability, Nationality, Species Membership* (Cambridge, MA: Belknap Press, 2006), see Northcott, *A Moral Climate,* 170-77.

15. Edward Page, *Climate Change, Justice and Future Generations* (Cheltenham: Edward Elgar, 2006); and Stephen Gardiner, *A Perfect Moral Storm: The Ethical Tragedy of Climate Change* (Oxford: Oxford University Press, 2011).

16. The agreed text calls for 'fundamental changes in the way societies consume and produce', but earlier references to biospheric limits on consumption and the ending of fossil fuel subsidies were excluded; Jeff Tollefsen and Natasha Gilbert, 'Rio Report Card: The World

lem is not confined to the United States, though it is particularly evident there. It is characteristic of ecological politics in large and high-emitting developing countries, including Brazil, China, and India, and even some European countries which, faced with the problems of the global financial crisis, resile on collective efforts to reduce greenhouse gas emissions.

Climate change seems, therefore, to pose a distinctive problem in the relationship between liberalism and democracy. Liberals of both left and right affirm that individual citizens and corporations ought, where possible, to have equal opportunities to express competing conceptions of a good life in their consumption and production behaviours. Ecological limits to fossil fuel emissions suggest that such behaviours need collective restraint in relation to the use of fossil fuels. But through corporate manipulation of the democratic process in the marketing of climate denialism and the corporate sponsorship of anti-regulatory 'movements' such as the United States' 'Tea Party', citizens are persuaded to express preferences for political actions which continue to subsidise fossil fuels, limit increases in fossil fuel prices, and hence favour cheap energy for present generations over the long-run interests of future generations in a stable climate.

Climate change challenges modern accounts and practices of the political in a radical way in part because, as we have seen, these accounts neglect that the political is enclosed within and hence interacts with the natural on terms the political does not control. For Bacon, the Elizabethan state is an imperial entity made powerful by the extent to which it dominates nature. But this same conception is also the origin of the modern history of empire, of the domination of a few persons over a very large number of others. For Hobbes, the state is made powerful by its control not over nature but over persons; the state holds the power conferred on it by individuals who contract together to support it as the guarantor of their rights to enjoy their own part of nature — private property — without the interference of others. For Kant, the political is the extension into the national and ultimately the international sphere of the rule of law, which is the form of authority freely consented to by rational individuals. In none of these three is the political circumscribed, and certainly not constituted, by the natural or by a divine Creator: both God and nature are 'crossed out'.

Contrast these modern accounts with the Hebrew idea of a covenantal polity. The nation of ancient Israel is constituted by a covenant which in-

Has Failed to Deliver on Many of the Promises It Made 20 Years Ago at the Earth Summit in Brazil', *Nature* 486 (2012): 20-23.

cludes three parties: God, humanity, and the land. The failure of humanity to honour *either* of the other parties is said to undermine the stability of the covenantal community or polity. It results ultimately in exile from the land and destruction of the seat of both political and spiritual authority — the walled city of Jerusalem and Solomon's Temple. But as Butterfield comments, the ancient Hebrews,

> in their moment of tragedy, turned their very helplessness into one of the half-dozen creative moments in world history. In particular the period . . . associated with the Jewish Exile provides us with a remarkable example of the way in which the human spirit can ride disaster, and wring victory out of the very extremity of defeat.[17]

In the Diaspora, the Hebrew prophets extend the divine election of Israel to all peoples. And in so doing they reinvent Jewish Messianism, which is the root not only of Christian belief in Jesus as the Christ, but also of apocalyptic eschatology and hence of the modern idea of progress.

Climate change represents a cosmopolitical crisis analogous to the ecological as well as political crisis which was the exile of Israel from Palestine, but on an earth-spanning scale: if humanity fails to repent and to abandon its false gods, the exile this time will be from the earth, and not only from a much-contested land to the east of the Mediterranean. Failure to acknowledge a duty to repent in the face of the climate crisis is accompanied by, and quite possibly connected with, the refusal of the nations to acknowledge that they dwell in territories on lands and in the midst of oceans where the Creator God first called all life into being, which community of being over time breathed and shaped a stable climate on which the first agriculturalists, and now the productive powers of the nations, still depend. But like the exile, climate change also represents a messianic moment. The large-scale extraction and use of fossil fuels have promoted gross human inequities within and between nations, and they have promoted large-scale habitat destruction and species loss. The urgent need for a radical turning from fossil fuel dependence therefore represents an apocalyptic moment which is potentially as important to the nations as was the exile to Israel. A radical and rapid transition from fossil fuels to other sources of energy will require the nations increasingly to remap their productive powers onto the capacities of their own ecosystems and terrains to sustain them. At the same

17. Herbert Butterfield, *Christianity and History* (London: Fontana Books, 1949), 100-101.

time it will require radical attention to reducing gross inequality, for only if the fruits of the earth are more equitably shared will an economy in which energy is much more expensive — as it will be if sufficient fossil fuels are put beyond use to prevent dangerous climate change — is not to be one that foreshortens the lives of the asset and income poor.

The suggestion that there is an intrinsic conflict between liberalism and democracy was first made by the controversial jurist and political theorist Carl Schmitt. This suggestion is also present in the critique of liberalism of the neo-Aristotelian philosopher Alasdair MacIntyre. I argue in what follows that the repairs to political economy and liberalism they propose involve the spatial, social, *and* spiritual resituating of the *nations,* and hence of the citizens, corporations, cities, and villages which constitute them. Such an approach ought to foster narratives, practices, rituals, and virtues, and, yes, some democratically agreed regulations, which encourage and enable individuals, corporations, and nations to sacrificially restrain their pursuit of goods which are incommensurable with the common good of generations, nations, and species on a finite planet.

Carl Schmitt and the Long Crisis of Global Capitalism

The covenant in ancient Israel has a cosmopolitical feature which is brought to light by a reading of Carl Schmitt's *Concept of the Political.* This is that the call of Israel and the promise of the gift of good land are underwritten by a strong contrast between themselves and other peoples: the Hebrews come to see themselves as a particular *people,* the *habiru,* gathered from dispersal and slavery in Egypt and tracing a shared genealogy to Abraham and Noah. Their entry into the land involves not only their sense of a shared religio-ethnic identity but also the exclusion, the 'ban', of those — Palestinians, Moabites, Edomites — who preceded them in the land and who do not share their genealogy or their God. So the political in ancient Israel is not simply God, people, and the land, but God, a *chosen* and *particular* people, and the land; and it requires that some are not of the community of this people and are excluded from her benefits. The land belonged to peoples whose sexual, dietary, and other 'manners' the God of Israel is said to abhor: 'You shall inherit their land, and I will give it unto you to possess it, a land that flows with milk and honey: I am the LORD your God, who has separated you from other people' (Leviticus 20:24). And the land, because it depends on the promise of God, will sustain the people of God, and not 'spew them out', only so long as they practise the statutes and judgements of God and do not follow the gods and the customs of the peoples that formerly dwelled there (Leviticus 20:22).

For Schmitt, the concept of the political is the sense that a people in a particular nation are a *people,* and that there are others who are not 'our' people, and who are therefore not friends but 'enemies'. In definitive writings for the Nazi era, he uses this concept of the political to justify the claim of the German *volk* to the exclusive use and defence of the 'fatherland' from the war reparations of the Allies and from the Jews. Schmitt is therefore rightly criticised for underwriting the ideology and the evil acts of the Third Reich. Many of his works were translated into English only in the last twenty years, and some still maintain that he cannot legitimately be studied given his association in particular with the Jewish question.[18] That his works are now being translated is in part because of growing recognition of the perspicacity of Schmitt's definition of the political as the friend-enemy distinction for understanding the political conundrum raised by the construction by the nations in the twentieth century of an increasingly borderless global economy.

For Schmitt, the definitive role of the friend-enemy distinction in the concept of the political means that the defining judgement of the ruler as to who enjoys and who is excluded from the protection of rule within a bordered territory rests upon decisions about the exception and 'borderline' cases. Such cases include decisions at borders about who is friend and who is enemy, and decisions about when to deprive citizens of liberty internally in defence of the institution of private property or of the integrity of the state. For Schmitt, these borderline or exceptional decisions are the hidden basis of the political.[19] But this decisionist foundation of the state is invisible in modern liberal accounts of political sovereignty because they lack a geographic or spatial dimension.

Schmitt's account of the state of exception shines a light on the use of the state's claimed monopoly on violence in the decision about who is friend or enemy, and in the arbitration of claims concerning access to land, natural resources, and private property more broadly. In contrast to Hegel, who had maintained a distinction between state and civil society,

18. See further Ellen Kennedy, 'Carl Schmitt and the Frankfurt School', *Telos* 71 (1987): 37-66.

19. Carl Schmitt, *The Concept of the Political,* trans. George Schwab (Chicago: University of Chicago Press, 2007), 4-9.

Schmitt argues that in the modern state this distinction evolves toward 'the democratic identity of State and society'.[20] Since the state is the arbiter of both in facing decisions about who is friend and enemy, who has the full rights of citizens, and who has property rights, the modern state has taken to itself the social solidarities and ancestral and customary patterns of land use and productive and exchange relations which existed prior to the nation-state. But since claims to property rights and claims to legitimate entry on the border rest more upon economic than political considerations, the monopolisation of these matters by the state, even though democratic, turns the practice of the *political* into the seemingly neutral procedure of economic competition.[21] The priority of sovereign decision about the exception in the modern definition of rule and a ruled terrain — the nation-state — is revealed most sharply by crises, such as that occasioned by external war, or by climate change when it imperils the ability of subsistence farmers to grow enough food for their families. Such crises are 'states of exception' which reveal the true character of the political as the friend-enemy distinction.

Climate change is the definitive border-infringing crisis; it represents a critical threat to the *borders* of states. Greenhouse gas emissions mixed in the atmosphere are intrinsically borderless, although some nations are more responsible for these emissions than others. Extremes of weather and ocean-level rise are leading to growing migration from the nations most severely affected across national borders, which is seen as a 'security' threat to those nations whose productive capacities are less affected by climate change. The militarisation of borders, the growing confinement of migrants to immigration camps or virtual prisons, and military interventions in climate-threatened regions against 'insurgents' — including presently Afghanistan, Mali, Pakistan, and Yemen — thus reveal the political response of nation-states to climate change more clearly than their claimed reductions in greenhouse gas emissions. Judgements about climate migrants reveal climate politics as the politics of the exception, while social scientific and phil-

20. Schmitt, *The Concept of the Political*, 4-24.

21. There is an affinity between Schmitt and René Girard on this point: for Girard modern politics cannot resolve the human tendencies to scapegoating and violence because it rests upon a refusal to admit that these tendencies are a fundamental feature of the human condition apart from grace, and therefore that the redemption of society from violence requires more than lawful rule and procedural power. See further René Girard with Pierpaolo Antonello and João Cezar de Castro Rocha, *Evolution and Conversion: Dialogues on the Origin of Culture* (London: Continuum, 2007).

osophical commentary on climate politics focuses instead on discussions in conference chambers about national greenhouse gas emissions.[22]

Schmitt argues that cosmopolitan and liberal political theorists fail to see the importance of the friend-enemy distinction because of their definition of the human person as *originally* individual, and as possessing as individuals the dignity and rights of personhood, and they lack a sufficient account of the originary significance of land and landed community, and hence of *nomos,* or what in English we call 'the law of the land'. In the liberal perspective, politics is a set of practices necessary for individuals to live together and to secure their person and property, and it therefore rests on a social contract. Politics is, in other words, a cultural and not a *natural* construction. But individuals cannot live apart from a particular territory, or without access to land for dwelling and the product of the land in the forms of clothing, cultural artefacts, energy, food, shelter, and water. Hence the Enlightenment and cosmopolitan claims that the liberal state is founded on equal regard for persons and at the same time defends the rights of persons to choose their own goods are contradictory.[23]

That the modern state is constituted by monopoly powers over the determination of the friend-enemy distinction and property rights is underlined by the fact that it chooses the terms of peace between competing associations — such as workers' unions and commercial businesses — for access to biophysical goods essential to human life. Hence, against the English associationists G. D. H. Cole and Harold Laski, Schmitt argues that the state is unlike other associations because it alone claims the power to declare war:

> The state as the decisive political entity possesses an enormous power: the possibility of waging war and thereby publicly disposing of the lives of men. The *jus belli* contains such a disposition. It implies a double possibility: the right to demand from its own members the readiness to die and unhesitatingly to kill enemies. The endeavour of a normal state con-

22. That Schmitt argued for the use of camps and other 'emergency' devices in Nazi Germany as a response to the dual crises for the German economy and people of the Versailles war reparations and the banking collapse of 1929 does not gainsay the perspicuity of the lens of the friend-enemy distinction and the politics of the exception, in understanding what is already unfolding in the twenty-first century in relations between climate change–threatened nations and the nations that are primarily responsible for climate change; Schmitt, *The Concept of the Political,* 18-22.

23. Schmitt, *The Concept of the Political,* 22-29, and see also 40-41.

> sists above all in assuring total peace within the state and its territory. To create tranquility, security, and order and thereby establish the normal situation is the prerequisite for legal norms to be valid.[24]

The blindness of political liberalism to the importance of *jus belli* friend-enemy distinctions and territorial borders in the constitution of the state reflects a failure to recognise a providential purpose in the separation of humanity into distinct peoples, language groups, and territories. In Genesis, enmity arises first between persons in the original murder of Abel by Cain, and then on a larger scale between peoples; the denouement of large-scale violence is the rise of an imperial power which intends to coerce the people into submission and conflict avoidance by means of the hubristic project of a 'Tower of Babel' in Genesis 11.[25] *After* Babel, God separates the peoples into distinct groups as a response to the collapse of an original universal culture into large-scale violence and autarchic empire. The importance of territorial borders, as well as language, customs, and religious practices, in distinguishing between peoples is subsequently underwritten in Exodus, in the separation of the *habiru* from the Egyptians. Israel is presented in the Torah as the divine repair of the post-Babel separation and of the imperial tendency to dominion. Postexilic prophets later universalise the calling of Israel into a larger divine intent to draw all peoples into a common religion, while retaining their languages and customs. Arguably, then, in Christian history Pentecost is the first realisation of the prophetic vision. But the Church after Pentecost does not press forward a universal culture. Instead, the Church permits people-group churches, which manifest cultural, customary, and linguistic differences, as in the original dual development of Jewish and Gentile churches, and the subsequent development of national churches and Christian nations.[26]

For Schmitt, the increasingly global nature of economic transactions in the late nineteenth and twentieth centuries undermines the theological as well as political validity of the ecological and cultural-linguistic borders and boundaries between nation-states, and reinvents the dangerous universalism of Babel and the related tendencies to total war and empire. In this new situation relations between states are increasingly determined by economic exchanges across borders where trade is a subterfuge for war. Hence judgements

24. Schmitt, *The Concept of the Political,* 46.

25. Schmitt, *The Concept of the Political,* 32ff.

26. See further Adrian Hastings, *The Construction of Nationhood: Ethnicity, Religion, and Nationalism* (Cambridge: Cambridge University Press, 1997).

about exceptions, who or what may *not* come across a border, and when a border infraction is to be resisted become the determinative political acts. Political judgement in a global economy is also no longer so much concerned with whether a person or a group is a friend or enemy. Instead, it degrades into agonistic decisions about power and property, or, in the context of climate change, migration and pollution. Hence, 'were a world state to embrace the entire globe and humanity, then it would be no political entity and could only be loosely called a state.'[27] For Schmitt, neither technical nor economic association can sustain the political and legitimate sovereign rule — whether of monarchy, republic, or democracy — 'in the absence of an adversary':[28]

> The acute question to pose is upon whom will fall the frightening power implied in a world-embracing economic and technical organization. This question can by no means be dismissed in the belief that everything would function automatically, that things would administer themselves, and that a government by people over people would be superfluous because human beings would then be absolutely free. For what would they be free? This can be answered by optimistic or pessimistic conjectures, all of which finally lead to an anthropological confession of faith.[29]

A borderless global economy subverts the political and dissolves the political into the economic. Conflicts will not cease in such a new order. There will still be friends and foreigners, but the dissolution of the political into the economic subverts the category 'enemy', and hence destroys juridical grounds for defence of the nation-state against external threats. Hence in place of lawful war will be 'executions, sanctions, punitive expeditions, pacifications, protection treaties, international police':

> The adversary is no longer called an enemy but a disturber of peace and is therefore designated to be an outlaw of humanity. A war waged to protect or expand economic power must, with the aid of propaganda, turn into a crusade and into the last war of humanity. This is implicit in the polarity of ethics and economics, a polarity astonishingly systematic and consistent.[30]

27. Schmitt, *The Concept of the Political,* 51 and 57.
28. Schmitt, *The Concept of the Political,* 58.
29. Schmitt, *The Concept of the Political,* 57-58.
30. Schmitt, *The Concept of the Political,* 79.

Schmitt's analysis is extraordinarily percipient when read alongside the mobilisation of the discourse of a 'war on terror' to justify military interventions in climate threatened regions such as the Sahel and Central Asia. It also explains the attempt to resolve geophysical limits to greenhouse gas emissions through the automatic economic mechanism of carbon emissions trading. The liberal division between ethics and economics requires that there be no genuine political and participative response to the conflict between the quest for economic growth and the ethical and political duty to conserve and nurture life and to sustain the conditions for human dwelling and livelihood. Behind Schmitt's critique of liberalism in *The Concept of the Political* lies his earlier *Political Theology,* in which he argues that sovereignty arises from the 'state of exception', since only sovereign power may decide when a state is at war or peace, and hence whether normal legal or extralegal power is in place at any particular time: 'in the exception the power of real life breaks through the crust of a mechanism that has become torpid by repetition.'[31]

Schmitt's political theology grounds the legitimacy of lawful governance over biological and personal life in the original law-like ordering of created life by the divine Creator, and in the more particular ordering, by law and the gift of land, of the people of God in Israel. He sets this ecological and theological account of the divine origins of rule against modern social contract theory in which the state is described as a suprapersonal authority with which persons contract because when they live together in particular terrains their rights and property are at risk from harm by others.[32] At the same time, by recovering the ecological and religious roots of the state, Schmitt resists the nature-culture divide, Cartesian dualism, and the secularisation of the political. Hence he recovers a meaning for the nation-state as a *place* where decisions are made about who and what has a right to dwell in a particular territory, and hence the duty of the state to underwrite the stability of human dwelling and its bordered and intrinsically limited or non-global character. On this account, the contemporary descent of democratic governance into market governance in the era of technological globalisation is the fruit of Cartesian mechanistic metaphysics, which advances suprapersonal political procedures as more law-like, rational, and hence 'good' over volitional personhood expressed collectively in communities and households dwelling together in particular terrains and delib-

31. Carl Schmitt, *Political Theology,* trans. George Schwab (Cambridge, MA: MIT Press, 1985), 15.

32. Schmitt, *Political Theology,* 47-48.

erating politically over shared goods and defence from external threats.[33] Hence the political descends from the realm of moral agency in, and personal deliberation over, the sharing of the biological substrate of human life into merely technological and sociological procedures and regulations:

> Today nothing is more modern than the onslaught against the political. American financiers, industrial technicians, Marxist socialists, and anarchic-syndicalist revolutionaries unite in demanding that the biased rule of politics over unbiased economic management be done away with. There must no longer be political problems, only organizational-technical and economic-sociological tasks. The kind of economic-technical thinking that prevails today is no longer capable of perceiving a political idea. The modern state seems to have actually become what Max Weber predicted: a huge industrial plant.[34]

While recognising the anti-political tendencies of Cartesian metaphysics in conceptions of sovereignty from the seventeenth to the twentieth centuries, Schmitt also resists the turn to Romanticism of many of his forebears. Romantics wish to turn things into subjects, and so prevent the decline into impersonalism and mechanism; but their heroic individualism only underwrites the gradual disintegration of the political in the face of atomistic mechanism:

> It is only in an individualistically disintegrated society that the aesthetically productive subject could shift the intellectual center into itself, only in a bourgeois world that isolates the individual in the domain of the intellectual, makes the individual its own point of reference, and imposes upon it the entire burden that otherwise was hierarchically distributed among different functions in a social order. In this society, it is left to the private individual to be his own priest. But not only that. Because of the central significance and consistency of the religious, it is also left to him to be his own poet, his own philosopher, his own king, and his own master builder in the cathedral of his personality.[35]

33. John of Damascus, the 'last of the fathers' of the Church, describes volitional capacity as the essence of personhood in the first Summa of the Christian faith in East or West: St. John Damascene, *The Fountain Head of Knowledge* 22.

34. Schmitt, *Political Theology,* 65.

35. Carl Schmitt, *Political Romanticism,* trans. Guy Oakes (Cambridge, MA: MIT Press, 1986), 20.

For Schmitt the Romantic reaction to economism, and the evolving depersonalisation of the political, is essentially emotivist. This emotivism is revealed as a 'lyrical paraphrase of experience' and a turn to the aesthetic, which should not be confused with real 'political energy':[36]

> An emotion that does not transcend the limits of the subjective cannot be the foundation of a community. The intoxication of sociability is not a basis of a lasting association. Irony and intrigue are not points of social crystallization; and no societal order can be established on the basis of the need, not to be alone, but rather to be suspended in the dynamic of an animated conversation. This is because no society can discover an order without a concept of what is normal and what is right.[37]

The decline of the political for Schmitt is from two sources. The first is the declining influence of the mythos of Catholic Christianity, which is the religious root of European political order. Romanticism does not reverse this tendency because it does not remythologise the political but rather puts the feelings and sensibilities of the private individual in place of the founding myth of Christianity and the *nomos* of Christian political theology. The second is the nature-culture, science-ethics divides bequeathed to modernity by the Copernican and Cartesian revolutions:

> Modern philosophy is governed by a schism between thought and being, inwardness and reality, mind and nature, subject and object, that was not eliminated even by Kant's transcendental solution. Kant's solution did not restore the reality of the external world to the thinking mind. That is because for Kant, the objectivity of thought lies in the consideration that thought moves in objectively valid forms. The essence of empirical reality, the thing in itself, is not a possible object of comprehension at all.[38]

The schism between earth and mind, nature and culture is not repaired by the liberal ideal of a neutral sphere of parliamentary debate and democratic governance. Instead, the technical governance of the external world takes hold over parliamentary deliberation over justice, and the romantic *sen-*

36. Schmitt, *Political Romanticism*, 159-60.
37. Schmitt, *Political Romanticism*, 161.
38. Schmitt, *Political Romanticism*, 52.

sibility for justice cannot sustain its recovery in the 'real' world of private property and economic power.

For Schmitt, the medieval mystical synthesis of the external and internal worlds holds the key to why the Christian mythos originally produced a European order with tendencies that took up the Roman *pax romana* and bracketed and restrained war within and between nations:

> Only God is infinite possibility and, at the same time, each concrete reality. He united in himself *posse* and *esse,* what could be and what is, as the suspension of all the oppositions of the infinite and the finite, motion and rest, possibility and reality. As the curious form of words of Nicholas of Cusa has it, he is the *Possest:* the unified identity of possibility and actuality. That is a mystical resolution, not romanticism. Here, too, the romantic attitude is that of the subject that does not commit itself. What the medieval mystic had found in God, the romantic subject attempted to take upon itself, but without giving up the possibility of assigning to the two new demiurges, humanity and history, the problem of such a unification.[39]

Schmitt's critiques of liberalism and romanticism and his analysis of the post-political condition has come to the ascendant, despite his Nazist sympathies, since the 1980s. This is because his scepticism about the capacity of a neutral sphere of parliamentary party politics to sustain sovereignty in the absence of a unitive metaphysics or a spatial *and* spiritual mythos, as provided by the friend-enemy distinction of the Christian empire, has proven remarkably predictive of the current global political era. His politico-theological insights are also instructive in relation to the global emergency of climate change. This is because, alone among twentieth-century political theorists, Schmitt locates the political in the interstices between air, earth, and sea, and in the agential role of *earthly* forces in the formation of the borders and laws of the nations, and hence of the political.

Nomos in the Earth, Sea, and Sky

In *The* Nomos *of the Earth* Schmitt moves beyond his critique of liberalism toward a more constructive project in which he brings together his account

39. Schmitt, *Political Romanticism,* 67.

of the mythological and religious origins of the European idea of nation and the new geospatial ordering of the world that emerged in the post-Columbian era, of which economic globalisation is the nadir. For Schmitt, the ground of the state is a physical *ground* — literally a space in which people have evolved a culture of artefacts, customs, and propertied holdings. A people can function politically only in a shared space, when a people has defined others who are not *their* people and hence has designated others as outsiders. Liberal democratic theorists fail to account for the extent to which land and land appropriation, or *expropriation,* lie at the origin of sovereign rule both ancient and modern and hence of the sovereign decision to distinguish between friend and enemy. The earth in mythical language is therefore the 'mother of the law' because 'the earth is bound to law in three ways. She contains laws within herself, as a reward for labor; she manifests law upon herself as fixed boundaries, and she sustains law above herself, as a public sign of order.'[40] The origins of lawful governance and nationhood therefore still lie, as they did definitively in ancient Israel, in the boundaries of agrarian worked fields, hunter-gatherer forests, and the delineation of tribal and then village, region, and national land borders: 'all property and legal order has land as its precondition and is derived from the original acquisition of the earth's soil.'[41] Here Schmitt is in accord with Kant when he argues that 'first acquisition of a thing can be only acquisition of land' and land appropriation; hence the 'law of mine and thine', which goes back to the Hebrew patriarchal narratives, is the origin of true law, rather than the positivistic law of modern nation-states.[42]

As we have seen, in the post-Renaissance and Baconian 'Age of Discovery' a new relation to the earth as a whole emerged, and this relation included for the first time seas as well as land in its purview. This resulted in a new European '*nomos* of the earth' in which law was founded not on land alone but on land and *sea.* Schmitt speculates that, through technology and plane travel, the *air* would also become part of this new *nomos,* as people 'transform their planet into a combination of produce warehouse and aircraft carrier', something which has now in effect happened with global manufacturing supply chains and the growth of air freight.[43] Whereas pre-

40. Carl Schmitt, *The* Nomos *of the Earth in the International Law of the* Jus Publicum Europium, trans. G. L. Ulmen (New York: Telos Press, 2003), 42.

41. Schmitt, *The* Nomos *of the Earth,* 48; see also Karl Marx, *The Grundrisse,* Notebook IV.

42. Schmitt, *The* Nomos *of the Earth,* 48.

43. Schmitt, *The* Nomos *of the Earth,* 49.

vious empires such as Egypt and Rome considered themselves to be the centre of the *world,* in competition with a barbarian or Mongol outside, the borderless character of modern economic world order, because it includes lands and seas throughout the whole planet, is a novel imperial phenomenon in world history.

The nations prior to the age of discovery had incommensurate law codes, but in medieval Catholic Europe there emerged an 'encompassing unity of international law' which was called a *res publica Christiana.* This expressed for the first time in history the idea of a law-governed unity of peoples across borders. It also translated the Christian idea of peace as a normative general concept into 'one oriented concretely to the peace of empire, the territorial ruler, of the church, of the city, of the castle, of the marketplace, of the local juridical assembly.'[44] For Schmitt, the Christian empire has a related eschatological purpose as the *katechon,* which is 'the historical power to *restrain* the appearance of the Antichrist', as attributed to Rome by Saint Paul in 2 Thessalonians.

> I do not believe that any historical concept other than *katechon* would have been possible for the original Christian faith. The belief that a restrainer holds back the end of the world provides the only bridge between the notion of an eschatological paralysis of all human events and a tremendous historical monolith like that of the Christian empire of the Germanic kings.[45]

The idea of the *katechon* has grown in Christian political theology as an enduring answer to the problem of the long delay of the return of the King. Given the many references in the words of Christ and in the writings of the early Christians to the imminent eschaton and the return of Christ as heavenly Judge, later theologians had to explain the long delay of the eschaton and the endurance of an aeon between the first and second comings of Christ. The idea of a Christian or Holy Roman Empire, which emerged after the conversion of Constantine in the Latin West and Byzantium, as *katechon,* or the restraining force in history, hence underwrote the sense of an enduring stability of Christian peace beyond the early Christian era, and its preservation 'against the overwhelming power of evil'.[46]

44. Schmitt, *The* Nomos *of the Earth,* 59.
45. Schmitt, *The* Nomos *of the Earth,* 60.
46. Schmitt, *The* Nomos *of the Earth,* 60. Julia Hell misses the Christian origin of *kate-*

The idea of a spiritual empire which united distinct peoples and their monarchs had significant implications for the relation of politics and territoriality, and hence for the medieval and early modern origin of international law. Within Europe the *res publica Christiana* had a spatial and spiritual unity of diverse powers and loci of authority which was not a simple split between spiritual and temporal power, as it is often portrayed. Instead, there were 'diverse orders' of power and authority which were not centralised in one person but lay between the pope and the various Christian monarchs claiming the title of emperor. Wars within Europe were rights disputes between rulers who existed within a common realm, rather than friend-enemy conflicts. The areas outside of Europe, however, were defined as a heathen realm into which European rulers were given papal mandates of missions or crusades to extend Christian rule. *Extra-European* conflict did not therefore undermine the intrinsic law-governed spatial and spiritual unity of Christendom, and for some was not even governed by the laws of war which were supposed to restrain war within Christendom.

For Schmitt, conceptions of rule and power from the late Middle Ages lacked the restraining eschatological conception of *katechon* and declined toward the notion of Caesarist empire. Medieval political theorists therefore proposed a unity of authority and power in the monarch (thus the late medieval invention of the 'divine right' of kings) and thence the new and absolutist claims to power of the early modern state.

The common realm with diverse loci of political and spiritual authority and power took two dominant forms once the medieval synthesis began to break down.[47] In northern European countries after the Reformation, mon-

---

chon when she ascribes it to Schmitt's post-imperial intent to justify modern European empires even as they were crumbling after the Second World War: Julia Hell, '*Katechon:* Carl Schmitt's imperial theology and the Ruins of the Future', *The Germanic Review: Literature, Culture, Theory* (2009): 285-326. By contrast, Meyer, Schetter, and Prinz argue as I do that *katechon* and the recognition of the spatial in Schmitt's thought both have origins in Catholic political theology: G. Meyer, C. Schetter, and J. Prinz, 'Spatial Contestation? The Theological Foundations of Carl Schmitt's Spatial Thought', *Geoforum* 43 (2012): 687-96.

47. Schmitt argues, with John Neville Figgis, that the medieval synthesis of a body politic which included temporal and spiritual, lay and clerical power, broke down under the influence in the thirteenth century of the Aristotelian idea of a *societas perfectae* and promoted the idea that Church and State are *severally* societas: Schmitt, *The* Nomos *of the Earth,* 61. As John Neville Figgis has it, 'in the Middle Ages Church and State in the sense of two competing societies did not exist', and hence '*ecclesia* and *republica* are more often than not convertible terms in medieval literature'; Figgis, 'Respublica Christiana', *Transactions of the Royal Historical Society,* Third Series 5 (1911): 63-88.

archs claimed spiritual as well as temporal power within their domains. This helps to explain in the English case why the state turned so viciously against the commons and customary use rights of peasant farmers and villages after the Tudor revolution; the modern alienation between land and labour had its origins precisely in the demise of the diverse political orders of the medieval Church which restrained absolute claims to sovereignty over land. In the French and American cases, after their respective revolutions, the governing powers took a different path: in both countries the unitary claim to sovereignty by the organs of the state within the terrain of the state was pursued by laicising the state and disregarding the customary, organic, and religious roots of political association in prerevolutionary history.[48]

For Schmitt, these developments bequeath to the twentieth century's dominant Old and New World powers a 'state-centred international law' in which '*discovery* and *occupation*' become the dominant metaphors for the state's outside. This is evidenced in the Bonapartist descent into total war after the French Revolution, which in turn provoked the rise of Prussian and then German nationalism and imperialism. But this imperial turn is also strangely a-spatial or *u-topic,* since, despite the spatial origins of the modern claim to state sovereignty, it involves a 'conclusive fundamental separation of (political) order and (geographical) orientation' that is the *unique* characteristic of the modern era.[49] This is the decisive root of the illegitimate and violent extraterritorial claims and counterclaims of Britain, France, Germany, and later the United States, which led to the total wars of the nineteenth and then twentieth centuries.

The root of the contemporary collapse of the political is therefore the breakdown of the relation between land and law that had been so important to the ecclesial body politic of the medieval synthesis, and earlier to the Hebrews, but which is absent from most Classical and modern conceptions of law. Modern political theory in effect excludes land as its source and fosters instead a bureaucratic idea of legality arising from a centralised state which is self-referential and lacks the foundational *nomos* of the ordering of fields and pastures, and hence of a *terroir* or terrain shaped by generations of a people.[50]

48. Schmitt, *The* Nomos *of the Earth,* 63.

49. Schmitt, *The* Nomos *of the Earth,* 66 (words in parentheses mine).

50. On the potential of *terroir* in the food economy as means to resistance to economic globalisation see F. Fort and H. Remaud, 'Le processus de mondialisation dans la valorisation des produits agroalimentaires à travers le concept de terroir: contrainte ou opportunité?' Supagro-Montpellier Working Paper at http://afm.cirad.fr/documents/5_Agro_industries/Syal/FR/REMAUD_FORT.pdf (accessed 29 August 2012).

Before modernity, *nomos* referred less to law than to a bounded *space* of political cohabitation; it therefore carries a range of meanings prior to 'law', including dwelling place, district, pasturage, forest, grove, or woods. Rule which is derived from *nomos* is a kind of rule that is oriented to both space and time, geography and the generations, nature and culture, and is therefore ecologically situated in deep time, both human and more than human. Modern law, by contrast, is positivist, apersonal, and denatured; its authority does not derive from an inherited claim to a pasture, nor from the personhood of spiritual and temporal powers and authorities. This results in a politically unmediated and positivist, and geographically displaced, conception of law and sovereignty which Schmitt characterises as *nomomarchy*.

The Breakdown of the Spatiality of the Christian *Res Publica* and the Rise of the New World

The spread of a despatialised idea of sovereign rule to the New World presented Europe with a novel set of problems, of which climate change is just the most recent.[51] The division between Europe and its outside had been a spatial as well as a spiritual division of Christian and heathen. But as the global authority of the pope declined after the Reformation, infractions into new territories beyond Europe began to involve sets of relations that were no longer governed by the treaties and friendships that, even as the medieval synthesis broke down, continued to govern relations between the nations within Europe. Hence for Schmitt the roots of the modern borderless global economy, and of modern global wars, are to be found in the breakup of the Christian empire. In the course of that breakup after the Reformation, amity lines, meridians, and other devices emerged from the pre-Reformation era to limit, if not end, conflict in the Southern Hemisphere and the New World, both between colonial powers and between colonists and colonised.

However, when it suited them the Spanish, the English, and other colonial powers would claim that the new terrains were not only heathen but lawless, and therefore beyond the reach of European moral or legal norms; hence, having abolished slavery within their terrains, they reinvented slavery in the New World. This move was resisted by theologians and jurists

51. Schmitt, *The* Nomos *of the Earth*, 83.

such as Francisco de Victoria and Bartolomé de Las Casas, and by a papal bull. However, in the domain of the sea few were prepared to argue that conflict could be restrained by land-based conventions, or 'law of the land', as the English call it.[52] The sea was 'free', according to Grotius, and on the model of the law of the 'free sea' emerged modern law and regulations governing 'free trade' and economic competition between global market actors, including states as well as purely economic corporations. The new conception of sovereign rule beyond Europe is, for Schmitt, the root of normless economistic conflict and of technologically enhanced total war in the twentieth century.

The problem of the legality and morality of New World land acquisition was dealt with critically by Francisco de Victoria in the framework of just war arguments.[53] For Victoria the conversion of the heathen was a just cause, and the Columban mission to the New World was underwritten by the 'papal missionary mandate'.[54] There were therefore just wars between the conquistadores and barbarians, but this did not mean that barbarians were to be treated as subhuman or possessed as slaves. On the contrary, and as Schmitt underlines, if they were true *enemies* in war then relations with them in war were subject to just law rules, and hence analogous to relations between Christian peoples at war. Therefore *as* enemies they were to be accorded rights *equal* to those of Christians.[55] For Victoria, just cause in relations between Europe's inside and outside — between Christians and heathen, friends and enemies — is determined by the authority of the Church.

By contrast, the modern juridical conception of war sets up all nations as equally sovereign in declaring war. Relations between peoples in boundaried territories have become subject to the positivist juristic conception of the state precisely as moral and theological arguments have been gradually erased from juridical and political arguments. This produces a conception of war governed not by just cause but by *justus hostis* (just enemy), which degenerates to the current terminology of 'rogue' or 'failed' states.[56] Under the influence of non-theological and statist conceptions, war is increasingly seen as punitive, with the result that the enemy is no longer treated as a

52. Schmitt, *The* Nomos *of the Earth,* 98.

53. Franciscus de Victoria, *De Indes and De Jure Belli Reflectiones,* trans. John P. Bate (London: John Wiley and Sons, 1964).

54. Schmitt, *The* Nomos *of the Earth,* 111.

55. Schmitt, *The* Nomos *of the Earth,* 124.

56. Wouter G. Werner, 'From *Justus hostis* to Rogue State: The Concept of the Enemy in International Legal Thinking', *International Journal for the Semiotics of Law* 17 (2004): 155-68.

just *enemy,* but as a criminal whose lands may be legitimately destroyed or looted, while war itself is increasingly subject to assertions of *force majeure* (a superior power). In this situation, moral discrimination over proportionate means, of the kind mandated by just war arguments, is abandoned.

This analysis provides a powerful account of why traditional theological approaches to just war are now so often neglected, and in particular in the United States of America, which is also the key leading developed nation and the single most influential political entity in resisting international efforts to limit damage to the climate.[57] From the perspective of Schmitt's account, this is no coincidence. The 'discovery' of the New World leads to a new extra-European legal synthesis in which international law is increasingly framed in an extra-historical manner that lacks the territorial but also the personalist and moral aspects of international law present in the medieval synthesis. This produces the Grotian and then Kantian frame of the 'law state', or liberal state, and its corollary of liberal individualism; in this situation corporate personality resides legitimately only in the state, and only by state concession in so-called 'civil society' entities, such as religious associations, trade unions, and joint stock-holding corporations.[58] The descent of political theory into statist theory is a central corollary 'of the new spatial order emerging with the European land-appropriation of the New World'; this produces a situation in which legal and juridical concepts grounded in the historical and spatial realities of Europe no longer apply, and hence political concepts are adopted that 'lack any spatial sense'.[59]

If the opening of the New World underwrote a growing post-spatial turn in European political thought, the *way* to the New World, the oceans, even more fully underwrote this turn, above all through the island of England. It was also in England that the word and idea of *utopia* was coined by Thomas More as a way of conceiving progress toward a new world that was not oriented by or toward history and tradition or the old world. Schmitt identifies England's resultant seminal role in shaping modern world order as the seminal mediator between the old land-based political order of Europe and the new global — because maritime — order of the emergent industrial and economic order of the nineteenth and twentieth centuries. From a buccaneering island of the 'free sea' over which England claimed su-

57. See further Cass Sunstein, 'The World vs. the United States and China? The Complex Climate Change Incentives of the Leading Greenhouse Gas Emitters', *UCLA Law Review* 55 (2008): 1675-1700.

58. Figgis, 'Respublica Christiana', 62-63.

59. Schmitt, *The* Nomos *of the Earth,* 135.

premacy for three hundred years, and from a Tudor's imagining of a utopic new world, emerged the new post-territorial, industrial, technological, and trading orders which have so signally shaped earth history since the eighteenth century and therefore the historic roots of climate change.

As we have seen, it was not, however, an Englishman but a Dutchman who first fully elaborated the law of the free sea. The English in their ships, by dint of their maritime supremacy, expected to be shown honour by all other passing ships. They also preyed at will on the ships of other nations for booty. Grotius, in arguing against their claimed dominion of the oceans, suggests that the oceans could not be subject to any overarching claim to sovereignty and that instead they were free, and ships ought therefore to be permitted freely to ply the oceans without honouring the English or being subjected to their piratical raids. Grotius's argument with the English, and for a 'free sea' untrammelled by claims of spatial dominion or property rights, is for Schmitt a key element in the dissolution of a '*jus publicum Europaeum*'. But far more influential were the claims of settlers of a New World who wished to kick over the traces of the Old, as More had first done in literature in his *Utopia,* and hence enabled the emergence of a maritime-based 'free, global economy ignoring all territorial borders' in which 'the old *nomos* of the earth determined by Europe dissolved'.[60] Not even the maritime English were as powerful in support of this dissolution as the emergent New World power of the United States of America, although the fact that the global meridian still passes through Greenwich indicates the extent to which England came close to being *the* mediating realm, the *katechon;* and for a time in the nineteenth and early twentieth century, England did actually sustain the balancing global equilibrium between the Old World and the New.[61]

For Schmitt, the twentieth-century rise of the United States as the dominant world power makes it inevitable that the idea of a League of Nations as a stateless means of governing relations between states was first proposed by the United States and Woodrow Wilson. However, the League was never formally ratified by the United States, which instead pursued the Monroe Doctrine of hemispheric influence. The Monroe Doctrine was affirmed in

60. Schmitt, *The* Nomos *of the Earth,* 226.

61. Schmitt, *The* Nomos *of the Earth,* 238; Schmitt here is remarkably close to William Blake in his account of the mediating role of Albion (which for Blake includes Scotland and Wales) in the emergent history of the industrial and politico-economic world order. This mediating role is also recognised in Tolkien's title 'Middle Earth' for the realm of the Shire and other regions that ultimately restrain Mordor and Sauron in *Lord of the Rings.*

the Charter of the League of Nations, and so led to the formal abandonment by the Europeans of their own historic and spatial conception of international order.[62]

The United States' absence from the League of Nations, but the League's recognition of U.S. hemispheric influence and the U.S. conception of global order, meant that the League of Nations underwrote the formal separation of economics and politics in international order which U.S. influence as described in the Monroe Doctrine represented in the Western Hemisphere:

> The most important character of this influence was that it was based on free trade, i.e., on trade free of the state, on an equally free market as the constitutional standard of international law, and on ignoring political territorial borders by utilizing devices such as the 'open door' and 'most favored nation'. Thus, in the sense of the separation of politics and economics, official absence meant only *political* absence, while unofficial presence meant an extraordinarily effective — economic — presence and, if need be, also political control. Until today, the separation of politics and economics is considered by many French, English, and American theoreticians to be the last word in human progress, and to be the criterion of the modern state and civilisation. But in reality, given the primacy of the economic motive, it only intensified the disorder occasioned by the unsolved problem of the spatial order of the earth.[63]

In this analysis the cause of the nightmarish collapse of the emergent spatial order of the earth represented by the Second World War was fundamentally economic. Whereas Bonapartism and the First World War represented the dying gasps of the old spatial division of the *jus publicum Europaeum,* the roots of the Second World War lay in the spatial chaos of the League of Nations, which led to the Versailles Treaty — never ratified by the United States — at the centre of which were crippling war reparations imposed on Germany and which, together with the Great Depression, were for Schmitt the economic root of the state of exception which the Nazi regime declared.[64]

Schmitt here is entering a special pleading against recognising the Nazi phenomenon, and, even worse, the Holocaust, as related to specifically German cultural or territorial phenomena. But nonetheless he is right

62. Schmitt, *The* Nomos *of the Earth,* 254.
63. Schmitt, *The* Nomos *of the Earth,* 256.
64. Schmitt, *The* Nomos *of the Earth,* 267.

in acknowledging that the approach taken up in the Versailles Treaty was misleading in ascribing the origin of the First World War exclusively to Germany and in ignoring the complicity of other parties in the roots of war. The ascription of blame exclusively to Germany also carried with it the implication that only one side in this war had fought a just war, in the old terms of just war, while the other was treated as a criminal nation requiring punishment. This shift toward the outlawing of war was formalised in the 1924 Geneva Protocol, which is the origin of the treaty form invoked by the Kyoto Protocol of 1992 in relation to climate change. Neither the English nor the Americans ever signed the Geneva Protocol, and it did not become international law. Instead, the subsequent Kellogg-Briand Pact of 1928 expressed a formal concept of the illegality of wars of aggression. In the emergence of this pact the United States took its definitive place in the new global order, an order resting upon the economic presence, but political *absence,* of the United States from Geneva, the headquarters of the League of Nations. The Kellogg-Briand Pact also underwrote the Monroe Doctrine of American hemispheric influence, European isolation, and the American separation of economic from political — and hence territorial — power. In this way the U.S. took up the role first declared for it by Hegel as the new Europe, the first fully enlightened nation, which had left behind the spiritual rootedness of Europe in the history of Christendom and had constructed instead a new ahistorical, apolitical, and supra-spatial *ratio* of international order.[65] Schmitt, however, believes that this supra-spatial conception of order was dangerously misleading, and he argues that only when the United States takes up the mantle of responsibility for maintaining a world order with borders which are rooted in geospatial, and hence cultural and religious, history will the New World at last reach a stable rapprochement with the Old.

War through the Air

Schmitt's account of the historic and spatial origins of law and the political provides valuable resources for understanding the political and theological

65. Schmitt cites many political historians in support of this view, but the most evocative citation is the last sentence from Rotteck's *Allgemeinen Geschichte vom Anfung der historischen Kenntnis bis auf unsere Zeiden:* 'Europe will view the holy fire, which until now it preserved from beyond the Atlantic Ocean'; Schmitt, *The* Nomos *of the Earth,* 291, n. 15.

challenges of climate change. The planet is now scientifically surveyed and legally claimed in the mostly settled borders of the modern nations. But it may only be governed in a way that restrains or brackets war, and hence does not descend into uncontrolled violence, by an ongoing determination to situate *nomos* in the history of the spatial divisions of the earth, and not merely in positivist legal norms and the political no-man's-land of borderless trade. Such a determination, we might say, involves discrimination between imperious acts of resource appropriation — including appropriation of atmospheric carbon sinks — and just and legitimate acts of peoples in laying claim to their own boundaried spaces, deriving a decent and dignified living therein, and fending off external threats, in particular those of predatory global corporations that subvert the viability of local, regional, and terrain-based economies with cut-priced imports based on the coerced labour of other peoples, the destruction of natural resources, and the extraction and burning of fossil fuels and forests in other domains.

For Schmitt there is only one way of ensuring that such discrimination is sustained, and this is through the emergence of a new restraining power against the forces which threaten the collapse of international order, a new *katechon* which can take the place that for Schmitt the Holy Roman Empire once occupied in Christendom. It is this view which led Schmitt to support the rise of the Third Reich against what he saw as the chaos unleashed on the German people by the unjust Versailles Treaty in combination with the global financial collapse following the Wall Street crash of 1929. With the definitive end of the European determination of geospatial history after the collapse of the Third Reich at the end of the Second World War, Schmitt was compelled to reconfigure his Hobbesian view of history, and in *The* Nomos *of the Earth* he argues that the United States is the only power that can henceforth restrain the chaos and impose order on a post-European world order threatened by communism as well as by ecological limit problems.

Schmitt's political analysis as applied to the geopolitics of climate change would play out as follows. The United States is historically both the greatest polluter of the atmosphere with greenhouse gas emissions and, on Schmitt's account, the one nation whose influence on the globe is capable of generating a coherent and collective response to the problem of global climate change. This is because the United States, far more than Europe, is accustomed to treating the whole world — East and West, North and South — as a borderless realm which atmospherically it more clearly is than when it is viewed from the perspective of peopled lands. Whereas Europe remains caught between a spiritual and spatial determination of sovereignty and the extra-legal or *exceptional* space of economic power, the United States is 'the only political entity capable of resolving the crisis of global order' since it treats the whole earth as an economic sphere of its influence.[66] If Schmitt is right, then this is even more true for climate change.

Schmitt's analysis helps explain the failures of the nations to hold back and restrain the coming borderless crisis of climate change. The Monroe Doctrine and the related separation of the political and the economic imply an imperial separation between the domination of space and the rule of law. Schmitt sees the first manifestation of this separation during and after the Second World War, with the movement of war located on land and sea to war located increasingly, and often primarily, in the air. The distinctive feature of air war is that it involves a deeper spatial separation between combatants and potential civilians caught up in conflict than land or sea war:

> Independent air war dissolved the connection between the force applying power and the population in question to a greater degree than exists in a sea war blockade. With air bombardment, the lack of relation between military personnel in the air and the earth below, as well as with inhabitants thereon, is absolute.[67]

This new type of war 'demonstrates the absolute disorientation and, therewith, the purely destructive character of modern air war.'[68] It thus serves technologically to underwrite the emergent evolution of the European bracketing of war from the Christian theory of just war toward the conception of war as criminal and requiring punitive response. This helps explain the extent to which the United States since the end of the Second World War has become the dominant air power, far more even than the dominant sea power. It also explains the rise of undeclared aerial warfare as an instrument of American foreign policy, most brutally in the 'secret' carpet bombings of Cambodia and Laos in the Southeast Asian conflict, and more recently in the extensive aerial destruction of civilian as well as military infrastructure in the second Iraq war. This new type of warfare also anticipates the rise in the twenty-first century of aerial drone warfare in which combatants are distanced not just vertically from the attacked population but also horizontally: many drones are controlled by combatants, or

66. G. L. Ulmen, 'Translator's Remarks', in Schmitt, *The* Nomos *of the Earth,* back cover.
67. Schmitt, *The* Nomos *of the Earth,* 320.
68. Schmitt, *The* Nomos *of the Earth,* 320.

nonmilitary agents such as CIA employees, located on different continents from the place of their deployment. The drones are also deployed in nations where no war has been declared.

The absence of natural embodied relations — of spatial *correspondence* — between combatants and populations in drone warfare is a powerful analogy to the spatially and temporally distanced damage that greenhouse gas emissions do to global terrains far distant from the source of emissions. The United States invokes its 'right' to attack 'terrorists' on the soils of other nations regardless of the European-originated rules of war on the basis of a claimed 'state of exception': targeted terrorists are said to be beyond the rules of civility and polity and therefore not deserving of treatment conventionally owed between combatants in the bracketed wars of Christendom, or twentieth-century attempts to recast these conventions in 'laws of war'. Similarly, the United States reserves its right to continue to pollute the atmosphere, and hence destabilise the terrains, of other states through an exceptional per capita loading of greenhouse gas emissions because the United States takes the view that its economic interests are superior to the United Nations' quest, through the UNFCCC, for a global bracketing of greenhouse gas emissions. That the United States is the leader both in drone warfare and in historic greenhouse gas emissions underlines the problem that the dissolution of the old European spatial and spiritual model of international order presents for the resolution of the crises of the present century.

The United States has both the cultural energy and the economic and technological power to develop on its own soil a politico-economic society that is no longer based on fossil fuels.[69] But far from developing such a society, the United States has, since the end of the presidency of Jimmy Carter, who first announced the need for a post-oil economy, significantly increased its dependence on fossil fuels. Carter believed the United States could become energy independent through energy conservation and so reduce its ambiguous dependence on Arab dictatorships for oil. Indeed, if the vehicle efficiency standards and fiscal incentives for renewable power that Carter put in place in his presidency had endured under his successors, the United States would have had no need of foreign oil in the last two decades. The United States is presently pursuing Carter's vision of energy independence, but not by conserving energy, nor through renewable en-

69. For an insightful and realistic account of what a post–fossil fuel North American economy looks like see Lovins, *Reinventing Fire*.

ergy, although renewable energy investment is growing as a source of power in some states. Instead, it has turned to unconventional gas and oil from shale, sand, and sedimentary rock to achieve energy. These technologies are highly controversial and environmentally damaging. New horizontal drilling techniques, combined with hydraulic fracturing of rock substrata, force methane out of myriad small holes in the rock and up to surface wellheads. Studies indicate that as much methane is emitted into the water table, and into the atmosphere, as is captured by the wellheads.[70] While 'fracked' gas has reduced the price of natural gas in the United States substantially, and even lowered the price of coal because of reduced demand from power stations, it is doing this at the cost of the toxification of groundwater and of greenhouse gas emissions on some estimates equivalent to or above those of coal extraction. Sand oil extraction is similarly a highly climate-polluting process, as discussed above.

Jimmy Carter presented his vision of energy independence through energy conservation and renewable power as a moral project which would require his fellow citizens to be more careful in their use of energy so that they might become less dependent on foreign wars, and more respectful of the environment their children and grandchildren would inherit.[71] But the idea of the use of energy as a test of national character was rejected by his successors in the White House, and by the United States Senate and Congress.

The failure of the United States to lead in the political resolution of the threatening chaos of extreme climate change does not gainsay the central role of the United States in the delineation of the post–Second World War international architecture, including the United Nations, the World Bank, and latterly the World Trade Organization. The United States took the lead in the establishment of all three institutions, and this role is matched in the role of U.S. scientists and the military in the development of a new global architecture of weather prediction and climate governance. Under the umbrella of the newly established United Nations, the World Meteorological Organisation was set up in 1947; its purpose was to create and maintain

70. Stephen G. Osborn, Avner Vengosh, et al., 'Methane Contamination of Drinking Water Accompanying Gas-well Drilling and Hydraulic Fracturing', *Proceedings of the National Academy of Sciences* 108 (2011): 8172-76; and Robert W. Howarth, Renee Santoro, et al., 'Methane and the Greenhouse-gas Footprint of Natural Gas from Shale Formations', *Climate Change* 106 (2011): 679-90.

71. On the remarkable combination of moral vision and spiritual leadership that characterised Jimmy Carter's presidency see Randall Balmer, *God in the White House: A History* (New York: Harper One, 2009).

a new global system of weather prediction which could be utilised in air travel, weather forecasting, and climate modelling. As Paul Edwards argues, the WMO in effect represents the 'vast machine' first imagined by John Ruskin as the means by which a world 'Meteorological Society' would develop to enable scientific weather recording and forecasting.[72] He shows how the 'vast machine' drew together nineteenth-century theories of atmospheric circulation and twentieth-century attempts to map the variables of air pressure, clouds, insolation, precipitation, and wind into mathematical and then computerised models of climate which are designed to have predictive power.

Until the end of the Second World War, the principal means of weather prediction remained historic records. Forecasters working for the Allies would draw on real-time weather records and compare these with historic records of weather in similar states and attempt to make accurate forecasts. The method was quite successful and gave the Allies a particular advantage in the D-Day Normandy landings, when the forecasters predicted a window in a run of strong storms which would enable a surprise landing on the Normandy beaches.[73] After the war, meteorologists redoubled their efforts to gather and standardise global weather data from transponders and weather stations on airplanes and ships as well as on land. The intention was to feed this now-global set of data into giant calculations which would enable the prediction of weather based on the laws of physics rather than on historic records.

The prototype of the first large-scale computing machine was developed in the Manhattan Project to compute the myriad physical equations involved in the making of the atom bomb. The United States not only provided the key hardware for this new global project of weather prediction. It saw the WMO as a project that might lead in due course to weather *control* and give the U.S. and NATO the upper hand in the Cold War. Hence U.S. government lawyers instructed the WMO's predecessor body, the International Meteorological Organisation, to include only 'sovereign states' in the WMO, as recognised by the newly established UN. This excluded scientists from nations occupied after the war, including the divided Germany and countries behind the Iron Curtain, as well as the People's Republic of China, and colonised states such as India and Malaya.

72. Paul N. Edwards, *A Vast Machine: Computer Models, Climate Data, and the Politics of Global Warming* (Cambridge, MA: MIT Press, 2010), 192-202.

73. Edwards, *A Vast Machine,* 116.

The scientific relation of war and weather first came to the fore in the use of chlorine and other gases as atmospheric weapons in the trenches of the First World War.[74] It was important to have good meteorological information before using such weapons to ensure the wind conditions would take the gases toward enemy lines, and not blow over the troops firing them. This association between weather prediction and war is, of course, much older. Even the Old Testament records instances of battle plans in which weather conditions — such as mist or low cloud — play crucial roles. And it explains why in Germany, and then in the United States, Britain, and France, meteorology research institutes were set up as departments of the military.

The interconnection of meteorology with war is also indicated in the growing scientific and military quest for means to control the weather and not merely to predict it to aid battle plans. In the Cold War the United States used cloud seeding technology in an effort to wash out the Ho Chi Minh trail in Cambodia and disrupt the supply lines of the Vietcong. Cold War intervention in the atmosphere reached not only into the clouds but even into the stratosphere with the Soviet Union's atmospheric testing of the atom bomb, closely followed by its successful completion of the first manned orbital space flight. The 1957 *Sputnik* launch triggered a new front in the Cold War which the United States was determined to win. The Soviet acquisition of the atom bomb triggered a terrifying new arms race.

The space race and the threat of atomic war triggered a new interest in the atmosphere of the earth, both as a source of potential threat and as a new frontier for the human species. The ambiguity of this frontier is indicated in a speech by John F. Kennedy:

> Can a nation organized and governed such as ours endure? That is the real question. Have we the nerve and the will? Can we carry through in an age where we will witness not only new breakthroughs in weapons of destruction, but also a race for mastery of the sky and the rain, the ocean and the tides, the far side of space, and the inside of men's minds? That is the question of the New Frontier.[75]

For the superpower architects of the nuclear arms race and the space race, the *katechon* is the controlling metaphor which describes the partnership of

74. Peter Sloterdijk, *Terror from the Air* (Cambridge, MA: MIT Press, 2009).

75. John F. Kennedy, 'The New Frontier', Democratic National Convention Nomination Acceptance Address, 15 July 1960, Memorial Coliseum, Los Angeles.

government and science. The restraint of evil in the enemy — the communist or capitalist other — through mastery of the skies is the guiding logic of both races. With the threats of a global war poisoning the atmosphere and of weapons in space, the atmosphere and the stratosphere are rendered more culturally influential and acquire new significance for scientists. Science, and especially atmospheric and atomic science, rocket propulsion, and guidance systems, acquire critical roles in resisting the enemy.

The beginnings of climate modelling of the kind that is central to climate science are also to be found in the Cold War. Scientists in North America were commissioned to estimate the effects of atmospheric fallout from an international exchange of nuclear weapons using the emergent computer technologies of the era. The computerised Global Circulation Models on which climate modelling now depends were developed both to aid in the global guidance of thermonuclear weapons and to answer questions about the atmospheric dispersal of radiation fallout and other atmospheric effects, including ozone depletion and the dispersal of soot from nuclear fires, an effect which Paul Crutzen would evocatively call 'nuclear winter'.[76] The prospect of global fallout from an exchange of nuclear weapons combined with the use of cloud seeding technology in warfare led in the 1970s to growing cultural concern with human modifications of the atmosphere. This concern was given added cultural weight by the publication in 1972 of the Club of Rome report *Limits to Growth*. Donnella Meadows and her team of natural and social scientists used the new computer simulation models to treat human economic and ecological activities as a set of systems within a finite 'earth system'. They predicted that exponential growth in human numbers and in economic activity and natural resource consumption would eventually hit planetary limits, causing an eventual civilisational crisis from depleted commodities such as copper, oil, and phosphates.[77]

These related uses of climate and earth simulation models, combined with the oil crisis and the development by the Carter Administration of a large-scale synthetic fuels program after the OPEC oil crisis, led in 1978-79 to the first scientific assessment of the possible climate effects of industrial emissions of CO_2 into the atmosphere. Teams of American scientists at NASA, NOAA, MIT, and the RAND Corporation drew together a run of Global Circulation Model predictions and estimated the likely effects of a

76. Paul J. Crutzen, Ian E. Galbally, et al., 'Atmospheric Effects from Post-nuclear Fires', *Climatic Change* 6 (1984): 323-64.

77. Edwards, *A Vast Machine*, 357-72.

doubling of atmospheric CO_2 at an increase somewhere between two and three-and-a-half degrees Celsius. This first attempt at scientific modelling of the greenhouse effect led to the World Climate Conference in Geneva in 1979, which set up a global scientific climate assessment program sponsored by the WMO, which in 1988 became the Intergovernmental Panel on Climate Change under the auspices of the UNFCCC.[78]

The United States is not only the progenitor of atomic weapons, climate simulation models, cloud seeding technologies, and drone warfare; it is also the principal sponsor to this day of Global Circulation Models and hence of climate science. Of the six large-scale Global Circulation Models currently in operation, four are located in United States research institutions. It is a double irony, then, that the United States is the most responsible for historic greenhouse gas emissions, the first nation to explode atomic bombs and to use cloud seeding technologies in warfare, and the principal sponsor of climate science, while at the same time the political organs of the United States deny that climate science has any implications for U.S. political and economic policy.

From Cold War *Katechon* to Warm War Refusenik?

For Schmitt, the United States was the only global power after the Second World War capable of donning the mantle of the *katechon* and restraining the tensions arising from the post-Christendom borderless condition; and, on the dominant reading of the history of the Cold War, he was right. It was the power and persistence of the United States, its economic as well as scientific and military prowess, which ultimately led to the collapse of the Soviet empire and the end of the division of Europe into East and West.

The twenty-first-century threat of a destabilising global division and war is not a Cold War but a Warm War. With the milestone of 400 parts per million of atmospheric CO_2, which the planet passed in 2013, international conflict from planetary heating, depletion of groundwater, desertification, and large-scale migration look increasingly likely before the middle of the century. In the 'Warm War' the divisions will be less ideological — communist and capitalist — than economic, since the developing nations in the tropics and subtropical regions will experience the greatest damage to their food-growing and water-harvesting capacities. The evidence that the

78. Edwards, *A Vast Machine,* 372-77.

United States might also play the role of *katechon* in relation to this new 'Warm War' is actually quite extensive. Not only is it American scientists who first modelled climate sensitivity to CO_2 and constructed the first computerised Global Circulation Models. It is also American scientists and social scientists who constructed the first social scientific model of the earth system which suggested that there are earth system limits to the growth of industrial civilisation. Furthermore, the scientific and technological infrastructure which the United States created out of the global nuclear standoff and the numerous proxy battles of the Cold War, from submariner conflict and the wars in Southeast Asia to the space race, now provides much of the scientific and technical infrastructure, from satellite, atmospheric, ground, and ocean data collection to Global Circulation Models and research institutes, which undergird the scientific work of the IPCCC and the UNFCCC. Key data sets charting climate change, such as Charles Keeling's pioneering decision to begin monitoring atmospheric CO_2 levels at Mauna Loa in Hawaii in the 1950s, and key papers making the case for climate change, such as Michael Mann's infamous hockey stick graph and James Hansen's paleoclimate modelling, emanate from the United States' extensive and very well resourced scientific establishment.

At the root of the refusal of many in the United States to accept the consensus view of the scientific community that industrial-scale greenhouse gas emissions are capable of destabilising the earth's climate is the North American narrative of freedom, conventionally known as political liberalism or, in its more extreme forms, economic neoliberalism. As we have seen, there is a deep contradiction in the concept of the political in liberal democratic theory which arises from the failure to acknowledge that modern concepts of citizenship, political freedom, and democratic participation are interconnected with the power of the state to decide who is and who is not to be accorded political recognition. Nation-states are therefore conditioned by territorial borders, and the political as a realm of deliberation and judgement only works within borders. When human migrants, natural resources, traded goods, and monetary wealth cross borders without restraint, the political is therefore always at risk. Decisions about such movements indicate that the determination of who or what belongs to a political community and who does not is the hidden decisionary basis on which the political really is founded. Hence the liberal claim that sovereign power in modernity derives authority from the free consent of citizens in a social contract obscures the true nature of the political, and it hides the agential role of the earth and the history of human-nature relations in particular terrains, in *producing* nation-states.

That Schmitt's political theory is a foundational challenge to liberalism is indicated in the extent to which freedom has been the overarching political concept in modern European thought since at least Voltaire. Enlightenment accounts of freedom have, however, consistently treated freedom as a primarily political concept. Hence, at least since David Hume, the determination of freedom and other guiding moral and political concepts have been delineated in a realm of human culture in which earth is assumed as a stable background for their realisation. Philosophers and historians since the Enlightenment have assumed that humans, and not gods or nature, are the authors of human history. Natural history becomes a separate and assumed fixed background to human history. The essence of being 'enlightened' is the claim that humans, and not the gods or natural forces, are at last free to determine the outcome of their own history.

In the Anthropocene this new power to determine the outcome of human history is turning into an unsought-for power to determine the outcome of *natural* history. Enlightenment philosophers and historians had carefully constructed a boundary between human history and natural history, setting aside premodern religious accounts of the intrinsic relation of the two as magical and superstitious.[79] But human history and natural history are converging again, and this convergence threatens the core Enlightenment belief in the expansive power of humans to free themselves from the determination of fate by the will of the gods or of natural forces.

The American vision of Enlightenment liberty is in some ways even more at odds with the unsought postmodern convergence of human and natural history than that of Europe. The settlement of the New World was a frontier project in which the inability of some to make their own history in the Old World of Europe became a goad for migration and settlement of new lands on an expanding frontier where monarchic power and aristocratic landholdings were no longer limits to individual ambition. Hence American ideas of liberty, property, and rights take more individualistic and atomistic directions than those of many Europeans. At the same time the domination of nature in the 'virgin' wild lands of the North American continent is more complete because the makeover of the land for coal, gold, mineral, oil, and timber extraction and agricultural production is beset with fewer cultural obstacles than in the heavily settled and more endur-

79. Dipesh Chakrabarty, 'The Climate of History: Four Theses,' *Critical Inquiry* 35 (2009): 197-222.

ingly culturally marked lands of Europe.[80] This is not to deny that for Native Americans the landscape of North America was deeply and culturally significant. But in most cases neither native customary usages nor sacred taboos were recognised or honoured by the colonial settlers. The frontier mentality therefore marks North American liberalism more strongly than colonial expansion marked European liberalism, since the expansion was within the borders of the new territory; and where borders were encountered that the colonists did not like, they simply fought to move them, as in the long southward move of the Mexican border. When the settlers reached the immovable Californian coast, expansion did not stop there. Hawaii, Guam, the Philippines, and most recently outer space all represented new frontiers where settlers could pursue their expansive dream of liberty through the subjugation of new territory.

The Anthropocene is a different kind of 'new world', different even from the expansive limit of the Pacific Coast of the American West. If a highly energy-dependent form of liberty is to endure on the new frontier of the Anthropocene, this will require new kinds of work than those which frontiers normally involve. There is no expansion offered by the prospect of this new 'new world', although a tiny number of humans might find it possible to set up a technological Noah's Ark on Mars or another planet.[81] If the work required to be 'free' on this new frontier is no longer territorial expansion and settlement, then it will require a different kind of struggle, an *interior* struggle with greed and the lust for power, if humanity is to learn to find freedom and yet live peaceably on this new frontier.

In the 'new world' of the Anthropocene there are basically two options. There could be a war over the atmosphere, through which the most powerful attempt to steer geopolitics, under the guise of a *katechon,* toward neglect of planetary limits and ongoing growth in the consumption of limited natural resources, leading to a large-scale collapse of food-growing areas and water sources, after which the strongest assert their will over the weakest. In a way this war is already begun, and hence some use the rhetoric of war to describe the failing negotiations of the UNFCCC.[82] Alternatively,

80. See further George Grant, *Technology and Empire: Perspectives on North America* (Toronto: House of Anansia Press, 1969).

81. I draw here on comments made by Bruno Latour in an interview for the documentary film *Anthropocene Observatory I* by Armin Linke and Anselm Franke and shown at the Haus der Kulturen Der Welt, Berlin, visited on 29 April 2013, and excerpted at http://anthropocene.territorialagency.com (accessed 14 May 2013).

82. Jeremy Leggett, *The Carbon War: Global Warming and the End of the Oil Era* (Lon-

there could be a new post-liberal politics in which the threat of exile from the formerly stable climate of the Holocene provokes a willingness within the nations to find a new political covenant which is not one of a fearful Leviathan or *katechon* restraining the contest for the earth's last habitable lands and potable water; instead, a covenantal politics is together embraced by the poor and the powerful in which the rich and powerful accept limits on their will to accumulate power and wealth, and there is a rebalancing of asymmetric relationships between rich and poor and between humans and other agencies in the earth's habitats and climate system.[83]

Arguably, this covenantal ideal is advanced by the UNFCCC. But the framing of the UNFCCC is too close to the liberal idea of politics as a social contract of originally free individuals who contract together for certain purposes, and in particular to prevent individuals or nations from harming each other. Hence the abiding goal of the UNFCCC is not the promotion of a new covenantal ideal of climate and cosmic justice, but the identification of a 'factish' target — territorial emissions — which does nothing to prevent the powerful from continuing to extract fossil fuels and market them in ways which advance the wealth accumulation of the few and the global harm of dangerous climate change.

The failure of the UNFCCC is a powerful instance of the contradictions of democratic political liberalism as identified by Schmitt. The Conferences of the Parties act as gatherings of a small number of representatives — bureaucrats and politicians — of national governments. The *earth* has no representative, nor its future citizens, nor the other creatures threatened by climate change: there is no earth-centred perspective, only national governments defending their rights and the property rights of their wealthiest citizens and corporations against the desires of other peoples — and especially the small island states and the states of North and Sub-Saharan Africa, which are already experiencing politically destabilising levels of climatic change.[84]

The CoP representatives of the United States government have been among the most assertive of the Parties to the UNFCCC in insisting that a

---

don: Routledge Chapman and Hall, 2001); and Stuart Sim, *The Carbon Footprint Wars: What Might Happen If We Retreat from Globalization?* (Edinburgh: Edinburgh University Press, 2009).

83. I refer again to Jürgen Moltmann's seminal essay 'Covenant or Leviathan? Political Theology for Modern Times', *Scottish Journal of Theology* 47 (1994): 19-42.

84. Bruno Latour, *Politics of Nature: How to Bring the Sciences into Democracy,* trans. Catherine Porter (Cambridge, MA: Harvard University Press, 2004).

global treaty to 'fix' climate change must not prevent the United States and other fossil fuel owners from continuing the central strategic and geopolitical goal of ensuring the continued economic growth and prosperity of its corporations and wealthiest citizens, and their rights to harvest fossil fuels and pollute the atmosphere without limit in pursuit of that goal. The single most powerful driver of fossil fuel extraction — the quest for economic growth — is said to be non-negotiable by the most powerful parties to the UNFCCC and the Kyoto Protocol. Contrast this with the Montreal Protocol. Once Paul Crutzen and others had established with a reasonable degree of certainty that chlorinated fluorocarbon atoms from refrigerants and industrial propellants were damaging the atmosphere, the nations agreed to a rapid phase-out of those chemicals. But twenty-five years after the founding of the UNFCCC, the IPCCC has yet to articulate in any of its assessments a plan to reduce the governmental and corporate *extraction* of fossil fuels to within earth system limits. Treaty commitments have focused instead on the downstream emissions of national, household, and corporate users of the fuels.

Another feature of Schmitt's analysis also helps to explain the failure of the UNFCCC, and this is the growth of the borderless and legally, if not physically, 'off-shore' economy, in which corporations and wealthy individuals use accounting devices to evade national regulations and taxes. The world's greatest imperial power before the Second World War, Britain, at the end of empire turned to the remnant of its imperial possessions — islands in the Caribbean and off the coast of mainland Europe — to grow an offshore economy, linked to the City of London, to draw in a pool of monetary wealth to replace its former on-shore imperial wealth.[85] Half of global personal and corporate wealth is now held in offshore domains.

The growth of the offshore economy reflects the growing neoliberal belief that wealth should be 'free' from constraint. But why do wealthy liberals think like this? Dipesh Chakrabarty argues that freedom has been the dominant theme of political liberalism for the last two centuries, and most of the freedoms the rich claim above other planetary inhabitants are dependent on fossil fuels.[86] If, as the Anthropocene now reveals, even wealthy industrial humans and corporations are *earthbound,* then there are not only ecological limits to growth. There are also political limits to the wealth of earthlings.[87] If

85. See further Nicholas Shaxson, *Treasure Islands: Tax Havens and the Men Who Stole the World* (London: Bodley Head, 2011).

86. Chakrabarty, 'The Climate of History'.

87. Latour used the terms 'earthbound' and 'earthlings' in his 2013 Edinburgh Gifford Lectures, 'Facing Gaia: A New Enquiry concerning Natural Religion', February 2013, available

a minority of people — a global 10 percent or around seven hundred million people — claim the right to continue to drive, fly, heat or cool their homes and workplaces, eat meat, and so on without regard for the growing limits their climate pollution places on the prospects for deriving a secure livelihood from the earth of billions of less favoured people, and if the rich refuse to acknowledge that there are ecological limits to their excessive consumption, then the Schmittian claim that the borderless condition destroys the political, and liberalism's failure to acknowledge that the role of borders renders it an incoherent political philosophy, are both confirmed. All that will stand in the way of an all-out war later in the century will indeed be an imperious *katechon,* and again the likely candidate will be the United States, even although it undermines the other *katechon* — the UNFCCC — for whose scientific architecture it is nonetheless still largely responsible.

Liberal Politics Versus Climate Politics

That politics involves physics, and not just models based on physics, since polities are situated in a geophysical cosmos, runs counter to the key innovation of the first liberal philosophers of the Scottish Enlightenment, David Hume and Adam Smith. For Hume and Smith, societies governed by reason rather than religion are directed toward the collective good by the freedom given to individuals and businesses to pursue their own 'interests'. What makes interest expression coalesce into a politics is not a religious or naturalistic norm but social *utility.* Beneficence or charity for Hume is not an intrinsic good but is valuable because those who show beneficence acquire honour and respect among their fellows, and therefore through charitable works advance their *own* interests.[88] The French moralist Claude Helvétius similarly argues that 'as the physical world is ruled by the laws of movement so is the moral universe ruled by laws of interest.'[89] This is a genuine innovation in moral and political philosophy which suggests that self-interest is the guiding motivation of men and women, and that a society that organises its

at http://www.ed.ac.uk/schools-departments/humanities-soc-sci/news-events/lectures/gifford-lectures/series-2012-2013/lecture-four (accessed 13 March 2013).

88. David Hume, 'Of Beneficence', section II of *An Enquiry Concerning the Principles of Morals* (London: A. Millar, 1751), 11-32.

89. Helvétius, *L'Esprit,* cited in Albert O. Hirschman, *The Passions and the Interests: Political Arguments for Capitalism Before Its Triumph* (Princeton, NJ: Princeton University Press, 1977), 43.

affairs accordingly is the most likely to produce social harmony and an increase in well-being and wealth. The purpose of political governance is therefore not to restrain avarice, greed, or wealth accumulation. Such traditional moral ideals are condemned by Hume as 'monkish virtues' which 'neither advance a Man's Fortune in the World nor render him a more valuable Member of Society'.[90] Instead, the purpose of government is to enable individuals to maximise their interests by pursuing their interests. In this seminal narrative of liberal capitalism, individuals achieve flourishing only when they are free to maximise their interests by maximising their preferences for goods which they themselves individually choose. Growth in *distribution* of opportunities to maximise individual preferences therefore becomes the overarching goal of liberal capitalist politics, even if the outcome is that the goods preferentially chosen — such as car journeys, meat-based meals, fashionable clothes, large bank balances held in tax havens — decline in value as city streets, animals, sweatshop workers, or economically depressed nations groan under the burden of reproducing these in sufficient quantity to continue to maximise the individual preferences of the 'free'.

In contrast to the interest theory of the human good of liberal capitalism, climate science is premised on an idea of an intrinsic good — an equable climate — which is sustained by complex agential networks between humans and other creatures mediated in the habitats of soils, ocean, and atmospheres. On this account relational interconnections and mutual dependencies of life forms in the earth system together construct a substantive good — a stable climate — which is intrinsic to human flourishing and whose maintenance requires negotiation between the parties, who have no other earth to move to. This is because this geophysical good is not immutable; it is not like law in the Newtonian sense. The atmosphere is changing beyond all historic parameters because of human interventions. That it continues to change, despite clear scientific forecasting, is because modern political economy is premised on the interest theory of human well-being, which neglects the reality that to live together in bordered territories, and a finite earth, the rich and powerful need to restrain their interests for the sake of the good of those other persons and creatures with whom they share the earth and whose co-productions of food, water, and other essentials cannot ultimately be supplied by money accumulated in offshore bank accounts!

Most social scientific analyses of climate change resist the conclusion that liberal capitalism and a stable climate are inimical. Instead, social sci-

90. Hume, *An Enquiry Concerning the Principles of Morals,* 174.

entists look for explanations for the lack of public response to the science in other fields of social science, among which psychology has become particularly prominent.[91] One study of Swiss attitudes to climate change found a lack of preparedness to acknowledge personal responsibility for climate mitigation, and a related tendency to blame governments for failure to act, or for a tendency toward pragmatic denial of the implications of climate science. But the premise of the study is that individual attitudes and consumption practices are where change most needs to occur, even although, as we have seen, the block is not primarily in public attitudes but in political economy, and especially the extractive designs of the statist and private corporate owners of fossil fuels and forests.[92]

Another psychological study found that beliefs about the capacity of *individuals* to remedy climate change is a good predictor of individual preparedness to conserve energy in response to climate science.[93] But paradoxically, belief in the power of individuals and their choices to make and shape societies for good is also a powerful predictor of climate change denial. The social group most likely to disbelieve climate science are 'conservatives' who believe that individuals are most able to find flourishing in a free market economy which permits a maximal array of choices and individual preference satisfaction and is characterised by minimal lawful regulation by government agencies.[94] The problem with this turn to psychology is that it neglects the deep psychological tenor of liberal capitalist societies as indicated in the interest account of political economy, a tenor sustained by the constant commercial bombardment of individuals in liberal capitalist societies with messages which connect material possessions, shopping, and status goods with the good life and social honour.

The contemporary political quest for a shared response to climate change is mired in public disagreement because liberal capitalist democracies are mired in disagreement about the ends and goals of human life.

91. Anthony Leiserowitz, 'Weather, Climate, and (Especially) Society', *American Meteorological Society* 4 (2012): 1-4.

92. S. Stoll-Leemann, Tim O'Riordan, and Carlo C. Jaeger, 'The Psychology of Denial concerning Climate Mitigation Measures: Evidence from Swiss Focus Groups', *Global Environmental Change* 11 (2001): 107-17.

93. Jon A. Krosnick, Allyson L. Holbrook, Laura Lowe, et al., 'The Origins and Consequences of Democratic Citizens' Policy Agendas: A Study of Popular Concern about Global Warming', *Climatic Change* 77 (2006): 7-43.

94. Aaron M. McCright and Riley E. Dunlap, 'Defeating Kyoto: The Conservative Movement's Impact on U.S. Climate Change Policy', *Social Problems* 50 (2003): 348-73.

The core disagreement that stymies climate change politics arises from the claim that the global economy has 'borders' or ecological limits; this claim is resisted by many because the role of borders in nation-state formation, as described by Schmitt, is neglected by political theorists, and it is subverted by the borderless nature of a global economy.

Climate scientists, and even climate activists and politicians, believe that in making the claim that there are ecological limits to the economy they are making a claim grounded in natural science. Thus in the United States, Vermont Senator Bernie Sanders argues that climate change is about 'physics not politics', while climate activists protesting a proposed new coal-fired power station in Kent, England, hold placards saying 'we stand by peer reviewed science'.[95] But of course the claim that there are ecological limits to human consumption of fossil fuels is a political claim and by no means only a claim about natural science. If there are physical limits to consumption, these become knowable and real only when they are acknowledged in the political and economic sphere. That there are geophysical limits to human activity, and that these restrain individual liberties, runs counter to the increasingly influential neoliberal view that the good life is realised by maximal preference satisfaction in economic production and material consumption. It also suggests that the moral life and political community are subject not only to human judgement and intuition but also to the non-human or more-than-human earth. The political and personal implications of climate science therefore challenge the nature-culture divide which is foundational to moral and political liberalism. To put this another way, moral judgements about climate change not only arise from the subjectivity of human beliefs but also are relative to another source of authority — what Christians call natural law — which is set into the structure of the cosmos.

The Agential Role of Nature in the Laws of Peoples

The most influential proponent of natural law morality in modern philosophy is the British philosopher Alasdair MacIntyre. MacIntyre argues that modern liberal societies are more riven by major moral disagreement than premodern societies because they lack a shared understanding, and shared

95. Bernie Sanders, 'The Planetary Crisis of Climate Change Is Not Political . . . It's Physics', *Common Dreams,* 16 February 2013, at http://www.commondreams.org/view/2013/02/16-11 (accessed 23 February 2013).

narratives, of the end or ends of human life. They also lack acknowledged moral authorities, or an authority structure in which deference is given to those more skilled in moral judgement.[96] Hence moral judgements tend to be based not on rational argument about the truth of certain ends, or shared deliberation about ways of achieving these, but on feelings which dispose individuals to choose particular moral preferences. MacIntyre characterises the resultant moral condition as emotivist individualism.[97]

'Emotivism' is the word used to describe an approach to moral judgement based on feeling, as first laid out by Hume.[98] In eighteenth-century Calvinist Scotland, biblical revelation was the dominant cultural moral authority. In his *Enquiry concerning the Principles of Morals,* Hume's guiding aim is the rational derivation of morals apart from the Bible. The 'research question' this project involves is 'whether they [morals] be derived from Reason, or from Sentiment; whether we attain the knowledge of them by a chain of argument and induction, or by an immediate feeling and finer internal sense.'[99] If, as had been generally believed until Hume, moral qualities are objectively real and are reflected in the structure of the external world, then they are perceptible by rational minds, which is close to the view that Kant adopted. Against this, Hume argues that moral qualities are subjective, that they are not qualities which are evident from the external world but instead reflect the emotions or sentiments of individuals and groups of individuals.

Hume uses the analogy of beauty in support of his view. Beauty, he argues, is a 'matter of taste', and such taste is the outcome of cultural training; therefore ideas of beauty, like ideas of good and bad, vary in different cultures and different individuals. Against the rationalists, Hume therefore held that 'reason alone can never be a motive to any action of the will', and that 'reason is and ought to only be the slave of the passions'.[100] Hume's ad-

96. On the role of deference in promoting group learning and cooperation see the empirical study of Natalie Henrich and Joseph Henrich, *Why Humans Cooperate: A Cultural and Evolutionary Explanation* (Oxford: Oxford University Press, 2007).

97. Alasdair MacIntyre, *After Virtue: A Study in Moral Theory* (London: Duckworth, 1981), 6-35.

98. MacIntyre, *After Virtue,* 6-35. See also A. C. MacIntyre, 'Hume on "Is" and "Ought"', *The Philosophical Review* 68 (1959): 451-68; and A. C. MacIntyre, ed., *Hume's Ethical Writings: Selections* (London: Collier Books, 1965).

99. David Hume, *Enquiry concerning the Principles of Morals* (1777; Edinburgh: James Murray, 1912), 4.

100. David Hume, *A Treatise of Human Nature,* vol. 2 (Sioux Falls, SD: Nuvision Publishers, 2008), 296.

vocacy of passions over rationality and of inner sensibilities over external facts sets up crucial features in post-Enlightenment accounts of morality, including the current vogue for moral psychology. However, as MacIntyre argues, Hume's emphasis on emotions over the external world, and on passion over reason, is reductionist and relativistic. It misses too much of what in most cultures, as well as in Classical and Christian moral philosophy, really constitutes moral character, moral reasoning, and the virtues.

For MacIntyre, 'morality is not what it once was' because the attempt to ground morality in emotions, rationality, or utility neglects the transcendent roots of morality in human and natural history and, ultimately, in the divine origination and destiny of life on earth. In the twentieth century this produced the doctrine, most clearly enunciated by G. E. Moore, 'that all evaluative judgements and more specifically all moral judgements are nothing but expressions of preference, expressions of attitude or feeling, insofar as they are moral or evaluative in character.'[101] Thus the statement 'fossil fuel consumption, beyond the capacity of the earth to absorb emissions without changing the climate, is wrong' is for emotivists a statement of preference; the individual who makes it prefers a stable climate. But it is also, for the climate scientist and the climate activist, a statement of fact about human-nature relationships as well as a value judgement. This hybrid character is a feature of many if not most moral judgements, since human beings are not only minds but also bodies who dwell in a world which is physically and therefore objectively real.

Modern moral philosophers resist the claim that there is any significant difference between contemporary and premodern morality, or that moral disagreements are graver now than in the past, because they also neglect the historical character of morality. It is a matter of indifference that Plato was a Greek, Hume a Scotsman, and Kant a Prussian.[102] It therefore becomes impossible to compare contemporary moral disagreements with former eras, since all versions of morality are considered as reflecting equivalent private preferences about what is good or evil. There are therefore no eminently rational grounds on which modern moral disagreements, including disagreements about climate change, may be resolved. The net effect has been to push morality from the public to the private sphere, so that increasingly people no longer offer reasoned grounds for their moral judgements but instead make claims such as 'I *feel* it is right (or wrong) to do x (or y)'.

101. MacIntyre, *After Virtue,* 11-12.
102. MacIntyre, *After Virtue,* 11.

The individuation of ethics is a principal cause of the loss of a shared understanding of common or public goods which reasonable people accept as worthy of assent and collective validation. The extent of contemporary moral disagreement is illustrated in the growing violence of arguments about abortion in the United States, and by the violent rhetoric and death threats that climate denialists have used toward climate scientists.[103] The loss of a transcendent ground for moral judgements in contemporary societies, and the related demise of collective agreement about common goods, results in the rise of what Max Weber called 'bureaucratic rationality', according to which the dominant goal of modern corporations and governments is economic efficiency in the use of scarce resources, and in which the dominant social instruments are those of managerial power and market coercion. The rise of managerial authority based on cost-benefit analysis involves the growth of forms of organisational culture in private companies and public agencies in which individual behaviours, decisions, and feelings are coordinated and pressed into conformity with the ultimate bureaucratic end of 'efficient' cost-benefit allocation of scarce resources.[104] This represents the growth of a decisionist tendency in public morality, first identified by Nietzsche, according to which those with the more powerful will, and the resources to back their will, impose their choices on those with less power and resources.[105] As MacIntyre argues, Nietzsche and Weber provide 'the key theoretical articulations of contemporary social order; but what they delineate so clearly are the large-scale and dominant features of the modern social landscape'.[106] Emotivism *plus* decisionism produce a tendency toward psychological control as means to underwrite the authority of managerial power. Such psychological control is particularly prominent in managerial techniques associated with personnel or 'human resources' management, but it is also prominent in the realm of marketing and advertising. As Vance Packard first observed, the turn to Freudian desire management in the 1950s by the advertising industry served the corporate managerial good of growth in profits once the consumer society had spread the basic necessities of a modern and comfortable material existence — refrigerators, washing machines, radios,

103. MacIntyre, *After Virtue,* 21.

104. MacIntyre, *After Virtue,* 25; the superiority of economic efficiency above other ends or values in twentieth-century societies is a central analytic concept of Jacques Ellul in *The Technological Society,* although MacIntyre does not engage Ellul.

105. MacIntyre, *After Virtue,* 108.

106. MacIntyre, *After Virtue,* 108.

bicycles and/or cars, air conditioning, heating — to the great majority of Western households.[107]

MacIntyre's analysis explains why cost-benefit analysis has become the über moral discourse of modern societies, and it is no surprise that such analysis is front and centre of most attempts to judge the conflicting moral claims of present and future persons, or of American automobile drivers and Bangladeshi fishers, in relation to anthropogenic climate change. Despite the emergence of the Anthropocene, most individuals still believe that moral judgements are ultimately expressions of personal preference. And they are encouraged in this belief by the ubiquity of advertising and commercial media as well as by the plots of modern novels, modern moral philosophy, and political theory.[108] The imaginary freely choosing agent of the modern 'democratised self' is an idealised self stripped of narrative and natural history, of personal roles such as those of child, parent, citizen, and worker, and hence of substantial biological, geographical, historical, and spiritual identity.[109]

MacIntyre's critique of the democratized, emotivist, liberal self is the analogy in philosophical liberalism to Schmitt's account of the dehistoricised, despatialised, and despiritualised nation in political liberalism and international relations. Both discern a conflict between the liberal account of the autonomous individual, and the familial, geographical, historical, national, and religious contexts of individuals gathered into political collectives; and both engage Nietzsche in their critique. Both argue that the post-Enlightenment demise of historical, geophysical, and spiritual narratives of the ends and identities of persons, cities, and nations is crucial to the demise of public agreement about these. Absent these ecological, historical, and spiritual moorings, the societal goal of economism (Schmitt) or economic efficiency (MacIntyre) becomes the *deus ex machina,* the 'fictitious believed-in-reality' which substitutes for God and geography in modern political judgement.[110] Consequently, claims to leadership and

107. Vance Packard, *The Hidden Persuaders* (Harmondsworth: Penguin, 1957).

108. MacIntyre traces to Cartesianism the split between cost-benefit analysis as the guiding organisational logic of modern liberal societies and emotivism as the dominant liberal account of how individuals make moral decisions; *After Virtue,* 32-33.

109. MacIntyre, *After Virtue,* 32-33.

110. MacIntyre, *After Virtue,* 76. For MacIntyre it is Hume who first identifies the inability of a secular enlightened public to reach agreement about moral ends or values on rational grounds, and hence who proposes to ground morality instead in public and individual sentiment; MacIntyre, *After Virtue,* 49-50. For an analogous argument see Charles Taylor, *A Secular Age* (Cambridge, MA: Belknap Press, 2007), especially 463-69.

power in modern societies tend to rest not upon ethnic or communitarian traditions or narratives but instead upon a 'moral fiction' or 'state of exception'.[111] Paradoxically, then, post-Enlightenment claims to political power and expressions of sovereignty increasingly rest on a suspension of reason and of reasoned judgement and involve an arbitrary imposition by some on others of a decisionist and emotivist mode of moral and political judgement.[112]

In the climate change debate it is not hard to see a powerful exemplification of this analysis. Scientists claim expertise over the interpretation of present-day and historical climate and weather patterns and atmospheric and oceanographic chemistry. They judge that increased CO_2 from industrial activities in the atmosphere and oceans correlates to global temperature rise and increased water vapour and hence to a less stable climate prone to more weather extremes. Scientists pass on the results of their scientific expertise to corporate and political managers to decide on the economically most efficient way of resolving the 'values' problem exposed by the 'factual' data. Many corporate managers take the view that it is in the best interests of profit- and rent-seeking corporations to lobby for and marketise an alternative version of the facts to those presented by climate scientists. Many political leaders and managers take the view that in order to hold on to power it is best not to alienate corporate fossil fuel interests by attempting to reduce subsidies to fossil fuels, thus raising the price of fossil fuel energy to consumers and hence reducing demand. The 'experts' are frustrated that their expertise does not command authority and produce changes in behaviour sufficient to mitigate climate change. Both scientists and politicians, and many activists and environmentalists, then blame individual 'consumers' or 'voters' as being unprepared to countenance sacrifices for the good of a stable climate, and hence as being the reason why agreed paths of action may not be decided upon.

The division of labour that MacIntyre's analysis exposes between the public rationality of economism and the private irrationality of desire is exposed in this set of events. The reason public action on restraining fossil fuel extraction and deforestation cannot be agreed on is the godlike supremacy of the moral *rationality* of economic efficiency, but the reason ascribed to

111. For MacIntyre such claims are typically exercised by artists, scientific experts, and bureaucratic managers; MacIntyre, *After Virtue*, 73.

112. MacIntyre, *After Virtue*, 73-79; C. S. Lewis argued the same in his *The Abolition of Man* (Oxford: Oxford University Press, 1943).

the public for inaction is *irrational* individual preferences, consumer desires, and democratic choices.

The failure of the nations to act for the global common good of a stable climate, and for the interests of future generations, is a powerful instance of the failure of political judgement conceived in economistic terms as a means of deliberation on costs and benefits where, in the absence of an overarching biological and/or religious narrative of human origins and destiny, no larger agreement about moral or political ends can be assumed. But what repair does this analysis suggest for what Mike Hulme calls the 'wicked' moral disagreement of climate change?[113] Whereas Schmitt offers only a modified Hobbesian account of international relations, in which the responsibility for global order rests with the most powerful nation, MacIntyre develops a thicker account of the kinds of repair to moral practices and moral theory necessary to enable individuals to act together for agreed or common goods. At the heart of his repair is a repristination of Aristotelian and Thomist virtue ethics, and an account of how habits, practices, and rituals in moral communities educate and form moral agents who are capable of directing their desires and interests toward shared goods. This analysis is combined with a growing engagement in MacIntyre's most recent work with insights gleaned from moral psychology and animal ethology which underwrite the natural law character of morality.

Virtues for Living in the Anthropocene

Most modern moral philosophers, when they do not resort to emotivism, tend to resort to various accounts of law-governed or rule-governed behaviour in order to explain the nature of moral judgement. Thus Rawls holds that what the ancients called virtues are sentiments which dispose individuals to act, or not to act, according to prior conceptions of right.[114] But in the Classical schema, the virtues describe the essence of what it is to be a good person, or to possess a moral character, and they are taught through stories, not through law or philosophical accounts of the good or the right. The self of the heroic sagas of Greece, Iceland, or Ireland is not an idealised autonomous agent who takes on roles and performs duties as

113. Michael Hulme, *Why People Disagree about Climate Change* (Cambridge: Cambridge University Press, 2009).

114. John Rawls, *A Theory of Justice,* 192 and 436, cited in MacIntyre, *After Virtue,* 119.

she chooses, or as guided by rational responses to law. Instead, the self is born into a household, a set of ethnic relations, and social roles in which individuals know what is due from them, and what is due to their kin; hence in Icelandic the word *skyldr* ties 'ought' and 'kin to'.[115] This is because the heroic self lacks the capacity of the modern self to detach herself from any particular social role or duty since 'in heroic society there is no "outside" except that of the stranger'.[116]

Belief in the specificity of certain moral excellences or virtues — such as courage, justice, patience, and temperance — and their role in guiding individuals and groups through harms and dangers gives historical depth and narrative unity to the self in Classical and Christian philosophy as well as in many European sagas. The virtues in the Aristotelian and Christian traditions have a narrative, relational, and teleological shape which enables a fit between the actions of an individual and the common good of the community. They therefore facilitate the realisation of the goods for which persons are made, and in particular the this-worldly end of *eudaimonia* or fulfilment and the other-worldly end of the contemplative or beatific vision of the soul. The emphasis of the virtues tradition on fulfilment, happiness, or well-being, and on the ultimate end of human being as the perfection of the soul, gives to this moral tradition a distinctive shape. This shape is lacking in modern moral philosophy, which tends to reduce morality either to inner states or sentiments which are not clearly connected with history, biology, the social, and the spiritual or to rules which force individuals to behave in ways they would often not prefer. By contrast, Classical and Christian virtue ethics is a process of education of intention and sentiment so that individuals choose the right action in particular situations because their training in the virtues has disposed them toward the right and the good.

Ethics on this account is a communitarian project, an '*education sentimentale*', which trains individuals in the performance of practices in which those things which are truly admirable, excellent, good, and beautiful are honoured, which are goods not relative to cultural or historical context.[117]

115. MacIntyre, *After Virtue*, 122.

116. MacIntyre, *After Virtue*, 126; the similarities with Schmitt's account of the centrality of the friend-enemy distinction are marked, though not noted by MacIntyre.

117. While some cultures or contexts may not acknowledge that the good, the true, and the beautiful are part of the given structure of human being, and of the biophysical context of being more broadly, MacIntyre maintains, with a number of natural law thinkers, that there is an underlying shape to the human good that exceeds cultural disagreement. See further

Education and practice in the virtues, and the goods which are internal to their practice, provide a *habitus* in which individuals are enabled to 'act from inclination formed by the cultivation of the virtues', and do 'what is virtuous *because* it is virtuous':[118]

> An Aristotelian theory of the virtues does therefore presuppose a crucial distinction between what any particular individual at any particular time takes to be good for him and what is really good for him as a man. It is for the sake of achieving this latter good that we practice the virtues and we do so by making choices about means to achieve that end. . . . Such choices demand judgement and the exercise of the virtues requires therefore a capacity to judge and to do the right thing in the right place at the right time in the right way. The exercise of such judgement is not a routinizable application of rules.[119]

But if ethics is not reducible to sentiment or rules, how is the life of the virtues socially sustained? MacIntyre argues that in a community committed to a common enterprise, such as the building of an art gallery, hospital, or school, there need to be two kinds of evaluative practice. The community 'would need to value — to praise as excellences — those qualities of mind and character which would contribute to the realisation of their common good or goods.' It would also need 'to identify certain types of action as the doing or the production of harm of such an order that they destroy the bonds of community' and so make the completion of the project impossible.[120] Such actions would need to be proscribed by law. Hence the purpose of law is not to promote good actions but to provide a basis for communal judgement, and hence prevention, of those actions which do not promote the good. Justice, on this account, 'is the virtue of rewarding desert and of repairing failures in rewarding desert within an already constituted community: friendship is required for that initial constitution.'[121]

---

my discussion of natural law in Michael S. Northcott, *The Environment and Christian Ethics* (Cambridge: Cambridge University Press, 1996). This is also the position of Elizabeth Anscombe in 'Modern Moral Philosophy', *Philosophy* 33 (1958): 1-19, the essay which may be said to have inaugurated the modern revival of virtue ethics.

118. MacIntyre, *After Virtue,* 149.

119. MacIntyre, *After Virtue,* 151.

120. MacIntyre, *After Virtue,* 151.

121. MacIntyre, *After Virtue,* 156.

For Aristotle, the *polis* describes the political community which determines the bounds of friendship and provides the education in the virtues which form individuals to discern and perform the good. But such an account of moral community, while it may describe some educational, philanthropic, or religious associations in modern societies, does not conform with the modern conception of political community. This is not only because modern conceptions of polity do not rely on slavery and the exclusion of women and children as moral agents. It is also because of the failure of modern political theory, as Schmitt also notes, to account for friendship and for geography.

Friendship 'involves affection' in a set of relations 'defined in terms of a common allegiance to and a common pursuit of goods'.[122] In Aristotle's *polis* there is no conflict between affections for one's friends and for the larger political community, because the city-state is a relatively small, boundaried entity. But in modern moral and political theories friends are chosen because they are *liked,* while the political entity and geographical region in which a person happens to be born or to reside is more often seen as incidental to the expression of a person's preferences and the formation of their affections. Thus, for moderns friendship describes a *chosen* good which seems to provide affinity between Aristotle's account and modern liberal morality. But absent a recognition of the role of geography in friendship and in the formation of political communities, the affinity is an illusion. With the aid of fossil fuel–powered mobility and computers, many modern individuals do increasingly experience friendships and even political affiliations in ways which exceed communities of place or regions. This technological borderlessness underwrites the conceptual borderlessness of modern liberal individualism. Consequently, individuals in particular polities or nation-states increasingly lack strong affiliations with their neighbours in these polities, often sharing stronger affiliations with people on the other side of the world. Hence, nations increasingly comprise individuals who lack a sense of inhabiting a shared narrative of what is the good for persons in the nation-state, and hence of the common good in the face of the climate emergency.

Critics of MacIntyre's communitarian philosophy argue that his account of the Classical virtues neglects the extent to which Classical societies were intolerant of difference and sustained hierarchies that are contrary to modern ideas of human rights and equality. But MacIntyre acknowledges important limitations to Aristotle's virtue ethics, including his metaphysical

122. MacIntyre, *After Virtue,* 157.

biology and his acceptance of slavery and the class and gender structure of Athenian society, and hence the structure of the *polis.*[123] MacIntyre identifies important modifications to Classical and heroic virtue ethics in the Christian medieval synthesis, and in particular the practice of forgiveness and the theological virtue of *caritas,* or love — revisions to Classical morality which are, of course, rooted in the moral teachings of the New Testament and the Church fathers of East and West.[124]

For MacIntyre, ethics in most times and places is not about an 'either/or' in relation to the existing political order. While at times this choice is real, and resulted in martyrdom in the pre-Constantinian era, in modernity, as in medieval society, Christian ethics is not about 'how to die as a martyr but how to relate to the forms of daily life that the Christian has to learn.'[125] For MacIntyre, the New Testament account of virtue is at least as much Stoic as it is Aristotelian. In particular, the New Testament comes close to a cosmic ethic when it proposes that

> the standard to which a rightly acting will must conform is that of the law which is embodied in nature itself, of the cosmic order. Virtue is thus conformity to cosmic law both in internal disposition and in external act. That law is one and the same for all rational beings; it has nothing to do with local particularity or circumstance. The good man is a citizen of the universe.[126]

While the law in the New Testament primarily stands for the Jewish law, nonetheless MacIntyre suggests that the New Testament account of the cos-

123. What MacIntyre does not acknowledge in *After Virtue* is the extent to which Aristotelian virtue was an armed virtue. The defining virtue of courage was to be learned in warfare, in defense of the city-state, and only young men who had been soldiers were able to become free citizens capable of moral judgment and political participation in the affairs of the city. Cf. John Milbank, *Theology and Social Theory,* 2nd ed. (Oxford: Blackwell, 2006), 333-34, and Charles Pinches and Stanley Hauerwas, *Christians among the Virtues: Theological Conversations with Ancient and Modern Ethics* (Notre Dame, IN: University of Notre Dame Press, 1997).

124. On the roots of theological virtue in the New Testament see, among others, Daniel Harrington and James F. Keenan, *Jesus and Virtue Ethics: Building Bridges between New Testament Studies and Moral Theology* (Oxford: Sheed and Ward, 2002); and Vigen Guroian, 'The Shape of Orthodox Ethics', in Guroian, *Incarnate Love: Essays in Orthodox Ethics,* 2nd ed. (Notre Dame, IN: University of Notre Dame Press, 2002), 13-27.

125. MacIntyre, *After Virtue,* 167.

126. MacIntyre, *After Virtue,* 169; and see further my discussion of natural law in *Environment and Christian Ethics.*

mic origins of law is close to that of the Stoics. The apostles affirmed that the covenant between the Creator and humanity first revealed in the covenant with Noah, Abraham, and Moses is a cosmic covenant between the Creator of the earth and 'all the kindreds of the earth' (Acts 3:25), and hence that the failure of humanity to honour the covenant has cosmic, and not only human, implications (Romans 8:22).

The New Testament transforms the moral import of the law in a way analogous to the Stoics when the apostolic writers affirm with the Hebrew prophets that the new covenant and the new law are not like the old: 'This is the covenant I will make with them in those days, saith the Lord, I will put my laws into their hearts, and in their minds I will write them; and their sins and iniquities I will remember no more' (Hebrews 10:16). For the New Testament writers, the coming of Christ and the gift of the Spirit inaugurate a new moral beginning for humanity *and* for creation. The first moment in this new beginning is repentance in response to Christ's transforming message of the forgiveness of sins and the defeat of sin and death on the Cross and in the Resurrection of Christ. These events redeem the intergenerational curse of sin and death resulting from the fall. The second moment is the gift of the Holy Spirit, who, after Pentecost, and through the sacramental life of the Church, indwells the hearts and minds of Christians to turn their wills to perform the just requirements of the law, so that Christian ethics is not in essence about conformity to law but about living in freedom to love what is truly desirable.

The unfolding of this Spirit-inspired ethic of forgiveness and love in Christian history has momentous consequences in European history and beyond, as both Schmitt and MacIntyre acknowledge. For Schmitt, the possibility of the nation as a spiritual community in a geospatial territory where war is bracketed arises first in human history in the medieval Christian synthesis. For MacIntyre, the medieval practices of penance and forgiveness, which extended to the requirement that kings do public penance, are historically distinctive fruits of Christian forgiveness and *caritas* in history and represent a profound revision of heroic and Classical virtue.[127] Hence the Christian ethic of love does not simply add another virtue to the Classical list. Rather, 'its inclusion alters the conception of the good for man in a radical way; for the community in which the good is achieved has to be one of reconciliation. It is thus a community with a history of a

127. MacIntyre, *After Virtue*, 173.

particular kind.'[128] The centrality of divine reconciliation and love sets the ultimate end of union between God and the soul in a different light to that of contemplation in Aristotle. Worship of God in this life as well as in the next becomes the orienting educational and moral practice of the virtuous community, and it indicates that the ultimate end of the soul is intimately linked with the resurrection of the human body and the reconciliation of all things, all divinities, persons, and creatures.

The Christian body-soul conception of regeneration is not something Aristotle would have appreciated any more than he would have been able to accept, had he been alive at the time, that Jesus Christ had risen from the dead.[129] The medieval synthesis is therefore also more profoundly *ecological* than Aristotle's virtue ethics because it situates the moral life of the Christian in the history of divine creation and redemption, and more especially in the fulfilment of that history in the life, death, and Resurrection of Jesus Christ and the gift of the Spirit. Its conception of history is not, therefore, the modern progressive conception, but instead one that is as much rooted in the past as it promotes hope in the future.[130] Christian care for the future and the outcome of history — even the outcome of the present climate crisis — rests not upon present choices or decisions but upon the new direction toward which history is pointed in the forgiveness of sins, the defeat of sin and death, and the revelation of the depths of divine love displayed in the Cross and Resurrection of Jesus Christ. This is not to say that Christians do not also share the heroic and Classical conception of life as a journey

128. MacIntyre, *After Virtue*, 175.

129. The centrality of the resurrection to Christian ethics is the guiding theme of Oliver O'Donovan's *Resurrection and Moral Order: An Outline of Evangelical Ethics* (Leicester: Intervarsity Press, 1984); it was also an important idea for Ludwig Wittgenstein, as the following indicates: 'What inclines even me to believe in Christ's Resurrection? It is as though I play with the thought: — If he did not rise from the dead, then he decomposed in the grave like any other man. He is dead and decomposed. In that case he is a teacher like any other and can no longer help; and once more we are orphaned and alone. So we have to content ourselves with wisdom and speculation. We are in a sort of hell where we can do nothing but dream, roofed in, as it were, and cut off from heaven. But if I am to be REALLY saved, — what I need is certainty — not wisdom, dreams or speculation — and this certainty is faith. And faith is faith in what is needed by my heart, my soul, not my speculative intelligence. For it is my soul with its passions, as it were with its flesh and blood, that has to be saved, not my abstract mind'. In Ludwig Wittgenstein, *Culture and Value*, ed. G. E. M. Anscombe and G. H. von Wright, rev. ed. (Oxford: Wiley-Blackwell, 1998), 38. For my own account of the relationship between resurrection and Christian ethics see my *Environment and Christian Ethics*, ch. 5.

130. MacIntyre, *After Virtue*, 176.

toward a goal, or of the life of the virtues as a quest in the midst of trials and evil. But it is to say that for Christians tragedy is not ultimate, for the ultimate end of all things is already revealed in Christ and in the communion of the saints who precede Christians in the quest for the life of love that Christ opened up.

At this point it may be argued that the trajectory of this book has taken such a theological turn that it seems to leave those who do not share the Christian faith out of the picture. Given the multicultural and multinational challenge of climate change, such a turn would hardly seem helpful. If the nations must all be Christ-followers before they can learn to respect the climate, then no solution to the problem will be found. MacIntyre's moral and political philosophy has attracted many philosophers and political theorists who are not practicing Christians and who have also adopted MacIntyre's turn to virtue ethics.[131] Nonetheless, if the claims made here for the distinctiveness of the Christian medieval synthesis are to be sustained in relation to the discussion of climate change, then it is necessary for some evidence to be presented that this synthesis has been operative in relation to environmental goods in a distinctive way in the past, and, even better, that there is biological and historical evidence of the truth of this synthesis beyond the Christian tradition.

Overcoming the Nature-Culture Divide Through Ecological Ethics

One of MacIntyre's earliest essays is on the 'is-ought' question first identified by David Hume. According to Hume, it is necessary after the rise of modern science to detach language about what ought to be the case in the human moral life from language about what is the case in the biophysical order. The split between science and ethics, nature and culture, long preceded Hume, as I argue above; but Hume brought things to a significant point, and MacIntyre challenges this split front and centre in his account of the virtues. For MacIntyre, the human good is objectively real, and statements about it therefore have a factual as well as a moral meaning, and biological as well as conceptual roots. But he does nonetheless acknowledge the inadequacy of Aristotle's metaphysical biology. In *Dependent Ra-*

131. See further Ronald Sandler and Philip Cafaro, eds., *Environmental Virtue Ethics* (Lanham, MD: Rowman and Littlefield, 2005); and my discussion of environmental virtues in *Environment and Christian Ethics.*

tional Animals he offers an account of virtue ethics which helpfully situates it in modern biological science and animal ethology, while also elucidating more fully both the distinctively Christian features of his account of virtue ethics and the importance of virtue ethics for the sustaining of practices which serve the human, and hence planetary, common good. I have given an essay-length exposition of this book elsewhere, but, in brief, MacIntyre makes three principal claims in this book.[132]

The first claim is that the human moral experience is *embodied* in a way that modern moral philosophy does not allow, and that this embodiment is particularly manifest in relations of dependence and nurture that humans, like other animals, experience at birth, in childhood, and, often, in the last months or years of life; the experience of dependence unsettles the narrow modern liberal conception of the individual moral reasoner choosing her own goods and deciding autonomously on paths of action to realise them. Second, and related to the first claim, the nurturing and group behaviours of other animals reveal that practical reasoning and the virtues do exist, albeit in rudimentary form, in other than human biological communities and have therefore a biological reality which exceeds the cultural and historical contexts in which virtue theory was developed. But because human beings, because of their uniquely volitional and personal character, have a *greater* capacity than other animals to fail to achieve their good, they therefore stand in need of *more* education and ritual legitimation of the moral life than do other animals. Third, the experience of dependence radically revises the Classical account of heroic and armed virtue, for it suggests that the experience of dependence is truer to the human condition and to the biological condition than the experience of autonomy.

MacIntyre's account of the centrality of nurture to the moral capacities of individuals and groups to express the virtues reflects the perspective the feminist philosopher Carol Gilligan calls the 'ethics of care' and the importance she and other moral psychologists place on good parenting as a foundation and context for the development of morally sensitive individuals capable of expressing the virtues.[133] Moral psychologist Darcia Narvaez offers a complementary perspective that provides more evidence for the reason for the associations between appropriate nurture and the forma-

132. This exposition draws on a fuller account in Michael S. Northcott, 'Do Dolphins Carry the Cross? Biological Moral Realism and Theological Ethics', *New Blackfriars* 84 (2003): 540-53.

133. Alasdair MacIntyre, *Dependent Rational Animals* (Chicago and La Salle, IL: Open Court, 1999), 92.

tion of moral character. Her research suggests that there are three ethical 'orientations' that are common to all individuals in varying degrees: 'security', 'engagement', and 'imagination'. Security concerns individual survival needs. Engagement is the capacity to empathise with others and to engage with their projects as well as one's own. Imagination refers to higher moral and aesthetic functions, including deliberative reasoning, cultural activities such as play, and artistic and intellectual endeavours.[134] Nurture of children plays a critical role in the ability of individuals to develop a personality in which these three ethical orientations are well balanced. In particular, there is a close relationship between physical touch of infants and young children and their later ability to empathise with and care for others.

Narvaez speculates that the growing use of child care facilities and especially nurseries outside the home has resulted in the formation of generations of adults in the United States and elsewhere who are less balanced in the three kinds of ethical orientation because for much of their early lives they are in situations where human touch is denied them.[135] Richard Louv analogously argues that increased urbanisation and reduced embodied interaction with natural beings — other animals, trees, grass, the 'outside' — plays a role in the malformation of contemporary children and young adults, and that this is manifest in the prevalence of conditions such as attention deficit disorder as well as increased tendencies to violence and narcissistic individualism.[136]

Both Narvaez and Louv make the claim that embodied experience of dependence on others, both human and wild, is key to the formation of individuals who can express empathy and care to others, human and wild. This account of the crucial role of dependence and the foundational importance of physical attachment and love in the nurture of children is reminiscent of New Testament analogies between child development and the development of disciples capable of living the 'way' of Christ. Christ on more than one occasion suggests that 'becoming like a child' is an essential trait of a true disciple, and he suggests that those who treat children harshly

134. Darcia Narvaez, 'Triune Ethics: The Neurobiological Roots of Our Multiple Moralities', *New Ideas in Psychology* 26 (2008): 95-119.

135. Darcia Narvaez, 'Triune Ethics Theory and Moral Personality', in D. Narvaez and D. K. Lapsley, eds., *Personality, Identity and Character: Explorations in Moral Psychology* (Cambridge: Cambridge University Press, 2009), 136-58.

136. Louv coins the phrase 'nature deficit disorder' in Richard Louv, *Last Child in the Woods: Saving Our Children from Nature-Deficit Disorder* (Chapel Hill, NC: Algonquin Books, 2005).

commit a great crime: 'whoever shall offend one of these little ones who believe in me, it were better for him that a millstone were hanged about his neck, and that he were drowned in the depth of the sea' (Matthew 18:6).

Christ not only pays surprising and countercultural attention to children in his teaching and example, but he attends frequently to other weak and vulnerable groups, including prostitutes, the disabled, the sick, and the poor. All of these groups were regarded in Roman society and Jewish custom as socially marginal and treated with disrespect. Christ on the other hand treats them with special care. This concern for the weak and the vulnerable sets up a cultural dynamic in the Christian tradition which is sometimes described as the 'moral priority of the weak'. This moral priority is underwritten by Christ's moral teachings, and in particular his insistence that his followers should not return evil for evil and should love their enemies; it is also underwritten by his life story, in which he overcomes evil not by a display of messianic power but by nonviolent goodness and innocent suffering in the way of the Cross. What Saint Paul calls the 'folly of the Cross', and the victory of the Resurrection over death and hell, reveal that incarnate, embodied vulnerability and weakness are the divinely chosen way to overcome evil and redeem creation.

Some will dismiss MacIntyre's efforts, and my own here, to link the Christian moral narrative with contemporary moral psychology as an unnecessary spiritual embellishment of empirically validated moral theory. I would want to claim against this that the Christian narrative and ethics, grounded in the Christ events, offer a way of reading the human moral drama through history, as well as the findings of modern moral psychology, as a tradition in which the priority of love over justice, of nonviolence over violence, of care over power is validated both by Christian history and by modern moral psychology.

If the forgiveness and divine love revealed in Christ represent a new direction in human history, we would expect to see marks of this in Christian history, and more especially in Europe, which has been the most fully Christianised continent until very recently. For Schmitt, bracketed war and the foundation of the nation as a geophysical and intergenerational entity are key achievements of the Christian medieval synthesis in Europe which liberal modernity has failed to appreciate or to sustain adequately with its aspatial conception of the nation, its borderless economy, and its ahistorical conception of the political. For MacIntyre, the key achievements of Christian virtue are in practices, and especially the forgiveness of sins and charity, and their fruits in the medieval synthesis, including the just wage and the just price, the origin of guilds, and charitable institutions such as hospitals and schools.[137] To this list I would add the ending of slavery, which was a unique achievement of medieval Christendom, as compared to previous world civilisations, until its reinvention in the New World.[138] I would also add the unfolding in Europe of the practice of the respect of the commons and the post-feudal distribution of use rights to agricultural land, as discussed above.

MacIntyre's thought, far more than Schmitt's, is strongly influenced by Marx as well as by Aristotle. Marx held that it is not ideas but practices that change the world. For MacIntyre, the moral life is a learned life, and the moral community that is capable of forming moral agents is a community that trains its members in the practices of the good life, provides occasions for moral choices to be made rightly, fosters rituals which enable right choices to become habitual, and sustains laws and procedures that punish and restrain wrongdoing.

On this account, the central duty of moral and political communities in relation to the mitigation of climate change is to train their members in new ritual practices of daily living in which care for the environment and the climate is honoured; these include using less energy, using wherever possible energy that is not derived from fossil fuels, and living more lightly on the earth by eating less meat, consuming less 'stuff', flying and driving less, and turning down the heating or the air conditioning, particularly when they are powered by fossil fuels. This view of the political role of moral communities in relation to climate change is exemplified by the development of movements such as 'Transition Towns' and 'Eco-Congregations', which foster local, place-based community practices such as energy conservation, self-sufficiency, renewable energy generation, local food growing, composting, and other features of low consumption, low carbon lifestyle; I discuss these movements in the last chapter.[139] And as I argue elsewhere, the ecclesial rituals of pilgrimage, sanctuary, and eucharist play a distinctive and shaping role in sustaining such everyday rituals among Christians.[140]

The communitarian direction of this kind of community-based virtue ethics appeals to romantic and organic conservatives such as Wendell

137. On the societal fruits of Christian practices see MacIntyre, *Ethics and Politics: Selected Essays,* vol. 2.

138. This is the central claim of Hilaire Belloc, *The Servile State* (London, 1912).

139. See further Rob Hopkins, *The Transition Handbook: From Oil Dependency to Local Resilience* (Dartington, UK: Green Books, 2008).

140. See further Northcott, *A Moral Climate,* chs. 5, 6, and 7.

Berry and Roger Scruton, for whom it is action at the level of the farm, the family, the firm, the household, the neighbourhood, the village, and the workplace — and not governmental action — that is truly capable of fostering virtues and practices that conserve a habitable world.[141] However, as I have already argued, if action on fossil fuel reduction is confined to the household or the local community and is not accompanied by more concerted global action to address the continued extraction of fossil fuels and the corporate media marketing of fossil fuel–intensive goods and lifestyles, such local exemplary action will not prevent the continued global growth in greenhouse gas emissions and the threat of planetary heating that will see millions, and eventually billions, of people driven from their lands and homes by drought, extreme heat, sea level rise, and violent storms. Nation-states, and especially the United States, Canada, Australia, and Saudi Arabia — to name just four of the more egregious — have been reluctant to educate and train their citizens and corporations in practices and lifestyles that use fewer fossil fuels, or to restrain economic corporations from extracting ever-larger quantities of climate-polluting fossil fuels from their territories. Instead, as I have shown, they have turned to economistic mechanisms, such as the creation of ineffective markets in carbon emissions, while continuing to subsidise fossil fuel–based growth and fossil fuel extraction and consumption by investing in more carbon-intensive infrastructure, by continuing significant tax subsidies for fossil fuel extraction investment, and by continuing to engage in foreign wars for oil.

MacIntyre's account of the descent of modern moral codes into cost-benefit analysis and managerial coercion explains why modern nation-states are unwilling to train their corporations as well as their citizens in virtues such as temperance and justice which are needful for ecological moderation if people and species are not to be harmed or even extinguished by climate change. The modern state is incapable of inculcating the virtues in its citizens, as the Classical polis or medieval guilds did, because it is founded on an illusion: that there are no biospheric constraints in the determination of those goods of bodies and souls toward which human life *ought* to be directed other than economic efficiency, and no sacrifices the state can ask of its corporations or citizens for these common goods other than their preparedness where necessary to defend the security of the state.

141. Wendell Berry, *The Unsettling of America* (San Francisco: Sierra Club Books, 1978); and Roger Scruton, *Green Philosophy: How to Think Seriously about the Planet* (London: Atlantic Books, 2012).

The core business of the state is therefore to promote economic efficiency as the means to expand the choices of citizens and corporations and hence maximal growth and maximal consumption. In virtue ethics, by contrast, the core work of the polis is to foster *participation* in deliberation on the nature and ends of the good life, and in practices that serve the good. But in the modern era the 'centralised large-scale nation-state' and the 'large-scale market economy' are too riven by conflicts of interest and grave inequalities of wealth and power to be capable of fostering such shared deliberation and effective decision-making.[142] Hence, for MacIntyre such deliberation is only conceivable in 'small-scale local communities' such as cooperative farms, fishing crews, households, local churches, neighbourhoods, schools, small towns, and villages.[143]

Between the Scylla of the *Katechon* and the Charybdis of the Small Platoon

MacIntyre's anti-state rhetoric leads many political theorists to dismiss him as a communitarian conservative who is uninterested in the larger moral repair of the political and the public realm. However, MacIntyre's political writings are more influenced by Marx than by Burke, and some characterise him not as a conservative but as a 'revolutionary Aristotelian.'[144] MacIntyre reminds his readers that the modern state and global market economy not only are poor at fostering the virtues but also are founded, as Marx also argues, on historic acts of land theft and coercive accumulation of use rights to natural resources from peasants and indigenous peoples; the resultant structural injustices that large-scale institutions sustain ought therefore to be resisted by people who know their history and what charity and justice require.[145] In this perspective, the medieval practices of just wage and just price, where guilds set the terms of trade, would represent radical challenges to contemporary capitalism, and it is therefore no coincidence that neoliberal capitalists seek so strenuously to undermine the vestiges of guild socialism in the trade union movement.[146] Analogously, modern corporate

142. MacIntyre, *Ethics and Politics: Selected Essays*, 2:39.

143. MacIntyre, *Ethics and Politics: Selected Essays*, 2:39.

144. Paul Blackledge and Kelvin Knight, eds., *Virtue and Politics: Alasdair MacIntyre's Revolutionary Aristotelianism* (Notre Dame, IN: University of Notre Dame Press, 2011).

145. MacIntyre, *Ethics and Politics: Selected Essays*, 2:146.

146. MacIntyre, *Ethics and Politics: Selected Essays*, 2:147.

outsourcing of jobs to non-unionised labour in weakly regulated and non-unionised domains has reduced many citizens to the status of consumer, or to forms of work that are in one way or another merely consumptive. This undermines the possibility of responsible engagement in co-agential world maintenance through the practices of good skilled work in craftsmanship, on the land, in the professions, and in the measured use of technologies, which for MacIntyre exemplify ways in which individuals learn the virtues.[147]

There is another reason why the large-scale state and economic market foster ethical corruption and duplicity rather than virtue, and this is the extreme mobility of many modern citizens, and the related tendency to compartmentalisation and fragmentation of human roles. Place-based communities enable the formation of moral virtues and shared practices toward their realisation because these communities have geographic boundaries and a sense of belonging — of who is in and who is out, and hence of friend and enemy in Schmitt's terms. The unrelenting mobility of people and goods promoted by the modern borderless economy undermines the integrity of small-scale economies and small-scale political deliberation. Consequently, cheating is normalised in a way that it is not in small-scale communities.[148] At the same time people in the modern era belong to multiple communities, and in these communities they behave according to different and often competing logics or mores. MacIntyre cites a study he was engaged in of electric power company executives, who, when they worked for the power company, expressed a preference for maximal power consumption and hence for the maximum profitability of the company, whereas as householders and citizens they expressed a contrary preference for energy conservation and price and other incentives that minimised consumption.[149] This compartmentalisation of social roles promotes an 'ethics of deception' and structural injustice in which people in the corporate sphere or the political sphere will regularly do things, such as lying, that they believe would be wrong in their close personal relationships.[150]

MacIntyre's virtue ethics promotes a local rather than a global conception of how to overcome the moral failings of nation-states and corpora-

147. MacIntyre, *Ethics and Politics: Selected Essays,* 2:149.

148. This is why for E. F. Schumacher the reorganisation of modern life into small-scale communities is central to the resolution of the ecological crisis; E. F. Schumacher, *Small Is Beautiful: Economics as if People Mattered* (London: Blond and Briggs, 1973).

149. MacIntyre, *Ethics and Politics: Selected Essays,* 2:196.

150. MacIntyre, *Ethics and Politics: Selected Essays,* 2:199.

tions to address the climate change conundrum. But it suffers from the opposite problem of Schmitt's concept of the political. For Schmitt, the failure of men and women to live together peaceably within boundaried terrains is an intrinsic feature of a post-Christendom world in which bracketed war and the concept of the political, operative among friends within national boundaries, are increasingly replaced by sovereign rule and the state of exception. Like Schmitt, MacIntyre sees friendship as an intrinsically political form of association. Like Schmitt, he understands that only within geographically limited places can the bonds of friendship and kinship develop into the kinds of moral and political association that foster shared understandings of the common good. But MacIntyre has no account of the moral or the political beyond such small-scale communities. The Greek city-state is perhaps the nearest he comes to describing a working political community beyond the Burkean small platoon. But where does this leave the nation-state? Is it hopelessly bound to destroy the bonds of community and friendship necessary for the formation of agents capable of eschewing asymmetric lusts for power and wealth and of agreeing to share common wealth in a common realm? In the final chapter I explore the concept of the political as messianic that first emerged in the history of Israel and has remained the driving force behind apocalyptic discourses and literatures of the kind that first emerged in the Hebrew prophets, and have lately appeared in association with the threat of extreme climate change. In the face of the threat of exile from a habitable earth, I argue that there is a messianic alternative to the imperial *katechon,* and that it may yet draw the nations from the Babylon of fossil fuelled economic growth and heedless consumerism to a new covenantal community between creatures, humans, and the heavenly realms.

7. Revolutionary Messianism and the End of Empire

The root of Carl Schmitt's idea of *katechon* is Saint Paul's proposal in 2 Thessalonians 2:6-7 that there is a restraining force set into world order, a *katechon,* which will hold back the apocalyptic revealing of the antichrist until the return of Christ at the end of time. The Thessalonians were so excited by the prospect of the end of history and the return of Christ that they were abandoning normal life and disrupting public order in the city. Saint Paul writes to them that there is a *katechon,* or restraining force in history, which is holding back the full force of iniquity and the antichrist (2:6-7). This withholding force creates an interval in time in which Paul charges the Thessalonians to 'hold fast to the traditions you have received by word or our epistle' and in the love of God revealed in Jesus Christ 'to find consolation and hope through grace' (2:15-16).

The Cult of Catastrophism and the City of God

I am reminded of the Thessalonians' enthusiasm for the imminent end when I read articles and blogs by climate change activists and radical environmentalists, some of whom seem to almost welcome the prospect of an imminent demise of summer ice in the Arctic. They envisage that this will provoke climate change 'tipping points', such as a vast release of subterranean subarctic methane, and then major weather effects which will rapidly affect crop outputs and lead to what some have called 'feral industrial collapse' and then 'near term extinction' of other species and ultimately *Homo industrialis.*[1] For the prophets of collapse, the near-term melting of Arctic

1. The phrase 'near term extinction' is coined by Guy McPherson in his blog *Nature Bats*

ice will undermine the scientific claim that there is no risk of an imminent shift into climate extremes. Melting ice, rapidly rising temperatures, famines, droughts, and heat waves will also expose the lies of the fossil fuel companies about climate science and the foolishness of consumers and politicians in continuing with life as 'normal' when life as normal is soon going to be no longer possible. If for national governments the IPCC is the *katechon,* set up to hold back and resolve the 'scientific' problem of climate change, for climate activists of a certain doom-laden ilk the IPCC is a coterie of scientific false prophets who hold back the *true* revelation of the end time events which the Arctic is bringing on.

Schmitt's theology of the *katechon* is an influential theological version of the familiar twentieth-century idea of a 'balance of power' in international relations. Arnold Toynbee at the end of the Second World War expressed the concern that the war had left the former imperial powers of Britain, the Netherlands, Spain, and Portugal exhausted and that there was a risk of a power vacuum on the world stage.[2] The balance of power idea has a long history, as its roots in Paul's theology might be said to indicate. It is also reflected in the 'crisis theology' of twentieth-century neo-orthodox theologians, including Karl Barth, Reinhold Niebuhr, and Jacques Ellul, for whom the global crises and wars of the twentieth century were rooted in the idolatrous salvific claims of political liberalism, technocentric scientism, and the social scientific attempt to model human sociality on the laws of physics. Against the idols of the modern age, Christian realists advance a view of the political which makes space for personal morality and virtue, while challenging the salvific and quasi-sacral claims of political liberalism, scientism, and the nation-state. There is, however, a pessimistic nihilism to the realist view of history which assumes that, absent an overarching power, men and women will destroy one another, and it is this same pessimism that is manifest among some of the wilder imaginings of climate activists.

Such deep pessimism is no longer confined to climate change activists. As noted above, in May 2013 the Mauna Loa Observatory recorded 400 parts per million of atmospheric CO_2.[3] The statistical 'milestone' prompted sci-

---

Last at http://guymcpherson.com. McPherson is a conservation biologist who retired early from the University of Arizona to establish a permaculture community in New Mexico with family and friends.

2. Nicolas Guilhot, 'American Katechon: When Political Theology Became International Relations Theory', *Constellations* 17 (2010): 224-53.

3. The full record of the Mauna Loa measurement is at the Earth System Research Laboratory website, ftp://ftp.cmdl.noaa.gov/ccg/co2/trends/co2_mm_mlo.txt (accessed 16 May 2013).

entists and government representatives to suggest that the earth is passing a threshold of climate change beyond which the prospects for preventing catastrophe are increasingly remote. Professor Robert Watson, the former chair of the IPCC and a chief scientific adviser to the UK government, said:

> Passing 400ppm of carbon dioxide in the atmosphere is indeed a landmark and the rate of increase is faster than ever and shows no sign of abating due to a lack of political commitment to address the urgent issue of climate change — the world is now most likely committed to an increase in surface temperature of 3C-5C compared to pre-industrial times.[4]

The sense in the scientific community of political failure to slow the rate of growth in global greenhouse gas emissions is leading to growing interest in geoengineering. Hence wealthy corporate leaders such as Richard Branson and Bill Gates, as well as government research agencies, are sponsoring a small but increasingly influential group of climate scientists in conducting geoengineering experiments in the ocean and the atmosphere. For advocates of geoengineering, the threat of civilisational catastrophe must be held back; the scientist and the technician must use their power to take up the role of a cosmic balance of power on the atmospheric border between the earth and the sun. But if, as Reinhold Niebuhr argued, there is risk of imperial overreach in the United States' assuming the mantle of defender of the free world in holding the balance of power against Nazism or communism, there is an even greater risk of overreach in the claim that geoengineering can prevent and hold back climate catastrophe.[5] The climate is not a machine; despite the vast record of historic climate states in the databases of the WMO, weather forecasters can still not predict whether a presently cool spring in northern Europe will turn into another cool and wet summer. It is therefore foolish to imagine that climate scientists could reliably know in

4. Quoted in Damian Carrington, 'Global Carbon Dioxide in Atmosphere Passes Milestone Level', *The Guardian*, 10 May 2013.

5. On United States overreach see Reinhold Niebuhr, *The Children of Light and the Children of Darkness: A Vindication of Democracy and a Critique of Its Traditional Defense* (Chicago: University of Chicago Press, 1944). I am grateful to Dane Scott for his presentation 'A New Heaven and a New Earth: Technological Power, Christian Ethics, and Geoengineering', at the workshop 'Religious and Spiritual Perspectives on Climate Engineering' at the Institute for Advanced Sustainability Studies in Potsdam, 24-26 April 2013, in which he invoked Niebuhrian criticism of imperial overreach in relation to geoengineering.

advance the outcomes of large-scale interventions in the oceans and the atmosphere sufficient to offset warming from greenhouse gas emissions. But that such imaginings are becoming more common indicates the continued lure of the imperial; the temptation to look for an overarching power to stand against an evil tide endures despite all the vaunted hopes of political liberals that we live in more cosmopolitan and egalitarian times.

There is a certain paradox in the association of Christian realism and the *katechon* with Saint Paul. Paul, more than any other New Testament writer, advances a view of history and theology which, if not anti-imperial, is at least what today might be called 'postcolonial'. For Paul, the powers that be were trounced by the crucified Christ; they thought they had killed the Son of God and Lord of Glory, but instead through the Cross and the Resurrection Christ 'led captivity captive', freed the dead souls in Sheol, and set up a new kingdom on earth, a 'new creation' of those called out from among all the peoples to be a new polity, the 'body of Christ' on earth. The distinctive feature of this new polity, as contrasted with the polity of Rome, is that the strong defer to the weak, those with least worldly honour are given most respect, and if there is competition it is in performing works of love and service rather than in the lust for power or status. Paul himself claims some of the messianic lustre of this new community when he says in his second letter to the Corinthians that God works through his weaknesses rather than his strengths, for 'my power is made perfect in weakness' (2 Corinthians 12:9).

The trope of a kingly ruler who changes the course of history and sets up a new form of political sovereignty through the power of innocent suffering is not, of course, original to Paul or even to Christ, but was adopted first by the Hebrew exilic prophet Isaiah. In his *Occidental Eschatology,* Jacob Taubes argues that the Babylonian captivity of Israel provoked a revolutionary new reading of the meaning of time and history which is deeply intertwined with the history of the West and is the root of all apocalyptic literature. The Exile, as the apparent end to the history of Israel, led the Hebrew prophets to the idea that history is not cyclical but is directed toward an end, and therefore that human and divine action in history shapes history before the end, and that the end is new and unlike the beginning because there is no eternal return. Exile is therefore the origin of apocalyptic thought, and for Taubes it is apocalyptic which is the constant source of the revolutionary turn in history *against* empire and from the underside of history.

Against the liberal account of the fulfilment of history in the Enlightened rise of human freedom, Taubes argues that the crises of the twentieth century smashed 'the framework of the modern age' and hence that 'the aeon demarcated by the milestones of Antiquity — Middle Ages — New Age comes to an end.'[6] Taubes traces this tripartite reading of history to Joachim of Fiore. Joachim's Trinitarian division of history describes the first age of the Father as from Eden to the birth of Christ. The second age of the Son is from Christ's Incarnation through the Holy Roman Empire to the High Middle Ages. The third age of the Spirit, the 'new age', Joachim dates from A.D. 1200. This is the age he announced in which would come the collapse of Catholic Christian civilisation and its replacement by groups of spirit-inspired individuals, dedicated not to the continuance of an earthly kingdom of this world but to living in anticipation of a final end of history and the second coming of Christ.

Joachim revived chiliasm in a new and influential 'pursuit of the millennium' after it had been in abeyance for many hundreds of years.[7] The decline of chiliasm goes back to the second and third centuries of the Christian era, when Christian theologians, including most seminally Origen and Augustine, gradually accommodated Christian eschatology to the delay of the *parousia* and to the endurance of the present age, which they redefined as the time of the Church. Hence, for Augustine, the millennial reign of the saints announced in the Apocalypse of John does not refer to the future coming of the Kingdom of Heaven; instead, it refers to the coexistence of the earthly empire — and especially Rome — and the Catholic Church or 'City of God' for a thousand years, from the time of Christ until the second coming. For Augustine, the tension between earth and heaven which is indicated in the petition of the Lord's Prayer 'Thy will be done on earth as it is in heaven' points to the potential alignment within human history between the two cities — earthly and divine. This alignment will come about only when a sufficiently large and influential body of people turn from pagan idols to the worship of the God of heaven and earth, who is revealed in Jesus Christ:

> Justice is to be found where God, the one supreme God, rules an obedient City according to his grace, forbidding sacrifice to any being save himself alone; and where in consequence the soul rules the body in all men who belong to this City and obey God, and the reason faithfully rules the vices in a lawful system of subordination; so that just as the in-

6. Jacob Taubes, *Occidental Eschatology*, trans. David Ratmoko (Stanford, CA: Stanford University Press, 2009), 93.

7. Norman Cohn, *The Pursuit of the Millennium: Revolutionary Millenarians and Mystical Anarchists of the Middle Ages* (London: Secker and Warburg, 1957).

> dividual righteous man lives on the basis of faith which is active in love, the love with which a man loves God as God ought to be loved, and loves his neighbour as himself. But where this justice does not exist, there is no 'association of men united by a common sense of right and by a community of interest'. Therefore there is no commonwealth; for where there is no 'people', there is no 'weal of the people'.[8]

Against the pagan philosopher Scippio, Augustine argues that Rome is not yet a *commonwealth* because its rulers do not pursue justice for all, but only for free-born citizens. The heavenly city is the only true commonwealth, for only there is the true God worshipped, and hence its citizens are able to bring their disordered desires for worldly goods and for power over others into subjection to their love for God and neighbour.[9]

For Augustine, the politics of the heavenly and earthly cities come closer through the cure of souls. The cure of souls is more often described in association with Christian than with Classical thought, but its origin is as much Platonic as Christian, and Augustine is Western Christianity's foremost interpreter of Plato. Training for the soul and mind in the Classical oratorical schools, such as Augustine taught in before his conversion to Christianity, involved reading the great texts and contemplating through mathematics, music, and the movements of the stars the 'Supreme Measure' which through the mind of God orders all things.[10] Through the perfection of the soul in such disciplines, bodily appetites and sensory engagement with the world and with society are reordered and repaired:

> [N]o soul, being perfect, is in need of anything; and, while it takes whatever seems necessary for the body if it is at hand, if it is not at hand, the lack of such things will not crush it. For every wise man is brave and no brave man fears anything. Therefore, the wise man does not fear either bodily death or sufferings for whose repelling or avoiding or deferring he would need those things that he is capable of lacking.[11]

8. Augustine, *The City of God against the Pagans,* ed. R. W. Dyson (Cambridge: Cambridge University Press, 1998), 19.23.

9. George Klosko, *History of Political Thought: An Introduction,* vol. 1: *Ancient and Medieval,* 2nd ed. (Oxford: Oxford University Press, 2012), 232.

10. Paul R. Kolbet, *Augustine and the Cure of Souls: Revising a Classical Ideal* (South Bend, IN: Notre Dame University Press, 2010), 94-98.

11. Augustine, *De beator uita liber unus* 29.78, cited in Kolbet, *Augustine and the Cure of Souls,* 100.

But for Augustine, the training of the soul, once she becomes a Christian, is perfected in the spiritual reading of the divine scriptures and in contemplative worship more than in reading the great texts or in contemplation of numbers, music, and the cosmos. Reflection on creation in the light of divine wisdom revealed in the scriptures and contemplation of the holy mysteries in sacramental worship engage the five bodily senses in seeing the 'works of God' and deepen the understanding of divine Wisdom, toward which Classical rhetoric was also directed. Thus, for Augustine, a deeper and truer divine 'rhetorical economy' is revealed in the incarnate *logos* than in Classical teaching, but there is also continuity: Christianity heals the soul by deepening rather than denying the Classical cure of souls.[12]

For Augustine, the training of desire is essential in order that the soul sees her true end in God, and this training is Augustine's analogue for the Platonic *paidea* of the soul.[13] If the means for such education are spiritual reading and sacramental participation and contemplation, another way of describing these is to invoke, as Augustine does, the love of God. Love, above justice, is the supreme spiritual and political virtue of the Christian as opposed to the Classical tradition. But immediately, and not least for the post-Christian reader, this raises the question, why the love of *God* above all things? Hannah Arendt attempts an answer to this question when she argues that for Augustine, as for Plato, love for God is *therapy* for the soul, which is otherwise distracted by sense impressions and by desires for the things of the world for their own sake.[14] For Augustine, the answer to the distraction of the self is the focus of the soul on the love of God, for 'when I love God' I do not love 'the beauty of bodies, nor the splendor of time, nor the brightness of light', but still 'some kind of light, some voice, some odor' which are closer to my 'inner man' and to the light that 'shines within my soul'.[15] Love for God is the only kind of love that can rescue persons from the being unto death which is otherwise their fate, as the fate of all being in time. Death poses the ultimate challenge to the meaning of human life; for Augustine, only in the love of God as the eternal source and future destiny of the self does the quest for transcendence — for eternal life — beyond

12. Kolbet, *Augustine and the Cure of Souls,* 106-14.

13. Augustine, *On the Morals of the Catholic Church* 17.31, in *Nicene and Post-Nicene Fathers,* vol. 4, ed. Philip Schaff (New York: Cosimo, 2007).

14. Hannah Arendt, *Love and Saint Augustine* (Chicago: University of Chicago Press, 1996).

15. Words in quotation marks are from Augustine, *Confessions,* book 10, as cited by Arendt, *Love and Saint Augustine,* 25.

death find an answer: 'hold fast the love for God, that as God is eternal, so you too may remain in eternity: since such is each as his love'.[16] As Arendt comments,

> Man loves God because God belongs to him as the essence belongs to existence, but precisely for this reason man *is* not. In finding God he finds what he lacks, the very thing he is not: an eternal essence. This eternal manifests itself inwardly — it is the *internum aeternum,* the eternal in so far as it is eternal. And it can be eternal only because it is the 'location' of the human essence. The 'inner man' who is invisible to all mortal eyes is the proper place for the working of the invisible God. The invisible inner man, who is a stranger on earth, belongs to an invisible God.[17]

For Augustine, as for the Classical tradition, human desire cannot be sated by the things of the material world. The soul is prone to insatiable desire, and from such insatiability all forms of addictive and disordered behaviour — including the addiction to consumerism — spring. This insatiability is answered in the Christian tradition by the revelation of God in Christ who is 'the desire of the nations', the Incarnation of the divine source of all being and material form, the *ratio* who is also *nous,* the body who is also spirit, the human who is mortal and yet overcomes death. Christ overcomes death, restores the human image, and so reunites the immortal soul with her maker.

A frequent criticism of the Augustinian tradition, as well as of Platonism, is evoked by Arendt's suggestion that the 'invisible inner man' is a 'stranger on earth' who 'belongs to the invisible God'.[18] If humanity is in exile on earth it may be said that the human abuse of nature is par for the course. What else might be expected of a being who, according to the Platonist tradition, is in an alien environment of temporal, spatial, mortal objects with which persons refuse fully to identify? If the only way to find meaning is to turn away from these things, does this not suggest a deeper alienation between humanity and the cosmos that the Platonist spiritual *paideia,* far from healing, only underlines and deepens?

For Augustine, as for the Hebrew Exilic prophets, it is precisely the sense of exile from heaven that is the source of the great spiritual turning

16. Augustine, *Homilies on the First Epistle of John* II, cited in Arendt, *Love and Saint Augustine,* 26.

17. Arendt, *Love and Saint Augustine,* 26.

18. Arendt, *Love and Saint Augustine,* 26.

and the possibility of the perfectibility of souls, and even of cities, in the human pilgrimage on earth:

> For the city of saints is on high, although it produces citizens here below, in whose persons it is a pilgrim until the time of its kingdom shall come. Then it will call together all those citizens as they rise again in their bodies; and then they will be given the promised kingdom, where they will reign with their Prince, the king eternal, world without end.[19]

The quality that distinguishes the two cities is not temporality or spatiality, for both cities endure through the history of the earth, but love:

> Two cities, then, have been created by two loves: that is, the earthly by love of self extending even to contempt of God, and the heavenly by love of God extending to contempt of self. The one, therefore, glories in itself, the other in the Lord; the one seeks glory from men, the other finds its highest glory in God. . . . In the Earthly City, princes are as much mastered by the lust for mastery as the nations which they subdue are by them; in the Heavenly, all serve one another in charity, rulers by their counsel and subjects by their obedience. The one city loves its own strength as displayed in mighty men; the other says to its God, 'I will love Thee, O Lord, my strength.'[20]

The visible form of the Kingdom, of the City of God, before the end of time is principally, according to Augustine, the Church. As Augustine puts it, 'the Church now begins to reign with Christ among the living and the dead.'[21] And for Augustine the growing influence of the reign of the Church on the kingdoms of the world will result in time in their gradual conversion to the Christian commonweal:

> Let those who say that the doctrine of Christ is incompatible with the State's well-being, give us an army composed of soldiers such as the doctrine of Christ requires them to be; let them give us such subjects, such husbands and wives, such parents and children, such masters and servants, such kings, such judges — in fine, even such tax-payers and tax-gatherers — as the Christian religion has taught that men should be, and

19. Augustine, *City of God,* 15.1.
20. Augustine, *City of God,* 14.28.
21. Augustine, *City of God,* 20.9.

then let them dare to say that it is adverse to the State's well-being; yet rather let them no longer hesitate to confess that this doctrine, if it were obeyed, would be the salvation of every commonwealth.[22]

While Augustine in two key passages in *City of God* condemns empire and imperialism explicitly, he does not consign all commonwealths or nations to the same judgment.[23] Instead, he favours a universal society of small nations, which would be a condition, as John Figgis puts it, 'in which there should be as many States in the world as there are families in a city.'[24] For Augustine, the Church is the true embodiment of heavenly rule on earth and has the potential to be *both* a trans-generational Communion of Saints and a 'universal world-wide polity'. Through the Church the multitudes of the nations are grafted into Christ, 'the odour of whose name fills the world like a field.'[25] Through the Church the harmonious rule of heaven is at last realised on earth in a way that links the living and the dead who are all members of the Church, and who are the Kingdom of God on earth as well as in heaven.[26]

Augustine's politics of the two cities is designed to take the 'eschatological sting' out of Christian chiliasm. John of Patmos's vision of the futurity of the thousand-year reign is being made present, in the extent to which the Church in Augustine's time had become the restraining power, the educator, and lawmaker of Rome, and therapist to Roman souls. Hence, as Taubes puts it, instead of the eschatological expectation of the end of history, Augustine sets the destiny of Christian souls centre stage.[27] It is then entirely consistent with this recentring of Christian eschatology from a 'new heavens and a new earth' to the therapy of the soul, and the Church's penitential and sacramental mediation of that therapy, that Augustine's most influential heir in the Western tradition, Thomas Aquinas, should narrow the domain of grace from the whole material creation to the souls of humans. Beastly, rather than creaturely, theology may begin in the Middle Ages, when it becomes possible for Aquinas to hold that there is no intrinsic reason why a man should not treat a beast cruelly, for it has no intrinsic value; but this theology has its roots in Augustine's realised eschatology of the soul.

22. Augustine, *Ad Marcellinum* 138, cited in John Neville Figgis, *Political Aspects of St Augustine's* City of God (London: Longmans Green, 1921), 57-58.

23. Augustine, *City of God*, 3.10 and 4.3, 15.

24. Figgis, *Political Aspects of St Augustine's* City of God, 59.

25. Augustine, *City of God*, 13.23, 16.37.

26. Augustine, *City of God*, 20.9.

27. Taubes, *Occidental Eschatology*, 80.

From the Ptolemaic Era of the Church to the Copernican Revolution

Augustine is an anti-chiliast, and his account of a potential partnership between the Church and the nations as the form of Christian polity in the thousand years' delay of the return of Christ in many ways describes rather prophetically the emergence, after the fall of Rome, of the Holy Roman Empire. Joachim of Fiore more than any other is responsible for the revival of chiliasm, for he lived at the end of the thousand-year delay. And for Joachim, as later for the Protestant Reformers, the Holy Roman Empire, while it may once have been the Kingdom of God on earth, would no longer be the *katechon* holding back the inevitable end of history, for the time of the reign of Christ from heaven through the mediation of the Church on earth had ended. There would in the second millennium no longer be a correspondence between heavenly and earthly rule. The Christian empire would decline and come under the same divine judgment as the empires that preceded it. In the 'New Age', the age of the Spirit, mystics, seers, and spiritualists would come into their own, living as those who are preparing for the imminence of the end of time.[28]

For Taubes, Joachim is not only the most influential apocalyptic thinker of the second Christian millennium, but he is also the prophet who inaugurates the true Copernican turn in historiography and popular culture, and hence the true herald of the modern era. After Joachim, heaven and earth are no longer in alignment. There is no correspondence between the will of the heavenly beings and the direction of human history and human rule. If there is a heaven, it is no longer knowable, its influence no longer palpably displayed in the unfolding of human history on earth. For Taubes, then, the announcement by astronomers of a lack of physical correspondence between the human sense perception that the earth stands still while the heavenly bodies move around it is preceded by the Joachimite vision that there is no longer a correspondence between human history in the second millennium and divine will.

For Joachim, the end time had begun not only in the turn from the first to the second millennium, but in the plagues and wars of the Middle Ages which followed. The *katechon* was no longer restraining; instead of the Holy Empire, other rulers would come who desired power for its own sake. In this new age of plagues, famines, and wars, rulers of the darkness of

28. The key work is Joachim of Fiore, *Expositio in Apocalipsim (Exposition of the Book of Revelation)* (Frankfurt-am-Main: Minerva, 1964).

this present world, the antichrist himself, would be unleashed, and the true saints, the Spirit-inspired saints who refuse the idolatrous compromises the Church would make with the antichrist and faithfully await the second coming of Christ, must expect to be accused by priests and rulers alike of heresy and to become martyrs.

In Taubes's insightful pairing of Joachim and Copernicus, before Joachim the history of the world turns on the history of the Church on earth, just as before Copernicus and Galileo the heavens revolve around the central orb of the earth. The word 'revolution' therefore originated as a description of the relation of the heavens and the sun as they revolve around the earth. After Copernicus, the word 'revolution' described a radical turn in human affairs. After Joachim, individuals who take charge of their history before the end of time are the true revolutionaries; they are those who truly align the human spirit to the coming end of the age, while the Church and earthly rulers pursue the illusion of their enduring power. For Taubes, then — the real innovator of the Middle Ages, the one who first anticipates the modern age, the detachment of history from the gods of the heavens and the fate of the earth, and hence the scientific revolution, the rise of reason, the age of Absolute Spirit — is not the nominalists Duns Scotus and William of Ockham, but Joachim of Fiore:

> Joachim's theology of history shatters the foundations of medieval theocracy. The slogan *ecclesia spiritualis* destroys the equation that has applied since Augustine and on which medieval theocracy is founded: Church = Kingdom of God. Joachim and the Spirituals proclaim God's Kingdom as the *ecclesia spiritualis,* as a spiritual kingdom. Joachim's theology of history is taken to its conclusion by Thomas Müntzer's theology of revolution.[29]

For Taubes, the history of European political thought in the modern age is rooted in this radical Joachimite genealogy. Hence Hegel's equation of the Kingdom of God with the 'kingdom of the mind', and the left-Hegelian call for a violent revolution to bring in the Kingdom on earth, is the logical outcome of Joachim's philosophy of history. As Taubes puts it, 'a history of European revolution is identical . . . with the history of the loss of Europe's Christian-Catholic essence':

29. Taubes, *Occidental Eschatology,* 86.

> The medieval Church is characterized by the Ptolemaic worldview. The world, as it is, is an image of its archetype, and by elevating its proper nature to a higher plane, the imperfect image of this world approximates to its archetype. The earth, according to Ptolemaic theory, is beneath the heavens, and everything which happens on it is an image of the archetype, a symbol. Ptolemaic man, who still believes the world to be the image of its own archetype, seeks fulfilment in rising toward the ideal. The Church as the body of Christ acts as an accomplice of Christ. The medieval Church is a *charismatic* form of Christianity; at the heart of medieval religion is the Mass, where heaven and earth become one. The history of European revolutions is the history of resistance to the Ptolemaic Kyrios Christos cult of medieval Christianity.

With the collapse of the medieval synthesis, and after Joachim and Copernicus,

> the earth mirrors no heaven, and the reality of this world is gained by *Copernican* man, not by having the world emulate a superior archetype, but by revolutionizing the world in terms of an ideal that lies in the future. The Ptolemaic world is ruled by the Platonic concept of *eros,* which attracts the lower sphere to the upper sphere. The Copernican world is ruled by the *spirit,* which invariably presses ahead. The ethics of Copernican man is an ethics of the future.[30]

It is the gradual collapse of the Ptolemaic world, and its substitution with an idea of space and time oriented to the end of history, and hence the future, that the *potenz* of humanity as the most powerful agency in history comes to the fore in the new science, in mercantilism, in European imperial expansion beyond the Mediterranean region, and in the Enlightenment and the subsequent age of revolutions. And this new age is also the age of the return of Prometheus, who reappears in the 'shadow of Christ'.

The return of Prometheus is grandiosely discussed by Kant in an essay entitled 'The End of all Time', in which Kant argues that reason, though triumphant, also imagines its end in the abyss of reason. Humanity after the Enlightenment of the Age of Reason still has a transcendental eschatology:

> Why does mankind expect the world to end *in the first place?* And even if this is conceded, why does it have to be a terrifying end (for most of

30. Taubes, *Occidental Eschatology,* 88.

> the human race)? . . . The reason for the first assumption seems to be that reason tells man that the continuation of the world only makes sense insofar as the rational beings inhabiting it focus on the ultimate goal of their existence. If this is unattainable then creation seems pointless to them, like a drama lacking a resolution and a reasonable purpose. This perspective arises from the view that the whole nature of the human race is so corrupt that there is no further hope; and therefore the only proper solution, the wisest, most just (according to the majority of people) solution is in fact to bring it to a dreadful end.[31]

As Taubes comments, for Kant and Enlightened humanity, the 'price of freedom is independence from God', and hence 'the fate of the world and that of mankind is tragic.'[32] Friedrich Schelling was the first to point up the Promethean implications of Kant's new humanist chiliasm, which he characterises as follows: 'the world having progressed as far as Zeus, a new possibility arises for the human race that is independent of him and has its origins in a different world order, a possibility which becomes a reality through the prescience of Prometheus.'[33] For Schelling, Kant's philosophy of spirit is intrinsically Promethean because it sets the struggle for the sublime in the Enlightened ego at the centre of the cosmos:

> Prometheus is the thought in which the human race, having once conjured up the whole panoply of the gods from its innermost resources, turns back on itself and becomes aware of itself and its own fate. Prometheus is that principal archetype of humanity which we have called *spirit;* and where previously there was weakness of spirit, he brought reason and self-awareness to the soul. He atones for the whole of humanity, and is in his suffering the pure sublime model of the human ego.[34]

As Taubes comments, 'the Promethean philosophy of Copernican man is the Kantian philosophy of the ego', and it anticipates the secular turn away from God and toward the powers of reason, science, and technology in *this* world as the means for human salvation, and as the determining forces of

31. Immanuel Kant, 'The End of All Things', in *Gesamtausgabe,* 6:397, quoted in Taubes, *Occidental Eschatology,* 140.

32. Taubes, *Occidental Eschatology,* 143.

33. Friedrich Schelling, *Sämmtliche Werke,* 11:483, cited in Taubes, *Occidental Eschatology,* 142.

34. Schelling, *Sämmtliche Werke,* 11:482, cited in Taubes, *Occidental Eschatology,* 142.

the end of history.[35] For Kant, the moral kingdom of ends becomes the invisible but true Church which in the present age of Enlightenment becomes the apocalyptic agency that at last presides 'over the separation of good from evil' which is the 'final outcome once the divine state has been fully established'.[36]

Back to the Future of Revolutionary Messianism

For Schmitt, as we have seen, the collapse of the medieval synthesis and the anomic borderlessness of the modern European political economy together produce the crises of the nineteenth and twentieth century, which can only be resolved by the recovery of the Roman *katechon*. Taubes, however, takes a totally opposite path. For Taubes, there is no going back behind Copernicus or Joachim. Indeed, such a project, far from resisting the Prometheanism of Kant and Hegel, merely reproduces it and therefore fails to resist the tragic chiliasm of the 'final solution' of the Third Reich. Here is the critical parting of the ways between Schmitt and Taubes, the two political theologians who are acting as our guides in the 'new world' of the Anthropocene.

Taubes and Schmitt are at one in their insistence that the modern crisis of the political is at root a theological crisis. Europe, the Mediterranean world, the culture of the West, and its expansion across the globe cannot be understood without understanding the Jewish and Christian roots of European civilisation. For both, the secular and social scientific turn of post-Enlightenment thought is a mistake which hides the true root of the political in the theological. But for Taubes it is not possible to understand, let alone resolve, the crises of the twentieth century without also recognising the seminal role of apocalyptic in shaping Europe's second millennium. Joachimism, not Augustinianism, is the truest political theology precisely because it is revolutionary theology from the underside of history. It is anti-imperial, anti-Egyptian, anti-*katechon* theology; and it is *Jewish,* and exilic, as Taubes explains in his last set of lectures on chapters 8–11 of Paul's Epistle to the Romans, which he introduces with an ironic remark:

> Today I see that a Bible lesson is more important than a lesson on Hegel. A little late. I can only suggest that you take your Bible lessons more se-

35. Taubes, *Occidental Eschatology,* 143.

36. Immanuel Kant, *Gesamtausgabe,* 6:312, cited in Taubes, *Occidental Eschatology,* 148.

> riously than all of philosophy. But I know I won't get anywhere with that, it isn't modern. But then I never wanted to be modern.[37]

The first political assertion of Paul in Romans that Taubes highlights is from the preface: Jesus Christ, who 'descended from David according to the flesh', 'was declared to be the power of God according to the spirit of holiness by the resurrection from the dead' (Romans 1:2-3). Taubes calls these opening words 'a declaration of war against Rome'. Christ, not the emperor, reigns after the Resurrection, and he reigns in Rome, the imperial city. His reign is not only for the Jews but, through Paul's own apostleship, it is 'to bring about the obedience of faith among all the Gentiles'. Taubes argues that 'obedience of faith' here is a 'polemical variant of obedience of laws'; the *nomos* of Rome no longer rules history (and here we already see the argument with Schmitt opening up).[38] This is not a quietist introduction, nor a self-effacing one; it would have been read by Rome's censors. Paul writes to a congregation he did not found and announces that he is its progenitor, and its purpose is to worship Christ and not the emperor in the imperial city.

Taubes then moves to Romans 8, where he notes that Paul's ringing declaration of the superiority of love over law also includes more than a hint of separation: 'I am convinced that neither death, nor life, nor angels, nor rulers, nor powers [this all in some way amounts to the Gnostic doctrine of the Archons], nor things present, nor things to come, nor height, nor depth, nor anything else in all creation, will be able to separate us from the love of God in Christ Jesus our Lord' (Romans 8:38-39). Why does Paul fear being cut off from the love of God? Well, as he goes on to say in Romans 9:1-3, he is prepared to become 'accursed, and cut off from Christ for my brethren, my kindred according to the flesh, who are Israelites' (9:3). Paul wishes that he might be cut off from the love of Christ if that would mean that his people, the Jews, who are *his* people (he is writing to Gentiles who are not his people), could again be reconciled to God, whose Son Christ Jesus they refused as the Messiah. But there is a reason for their rejection: 'through their stumbling salvation has come to the Gentiles' and 'their stumbling means riches for the cosmos' (11:11). Their stumbling means that they have even become 'enemies of the gospel' but, and here Taubes openly puts down the great lawyer Carl Schmitt and his 'racist theozoology', this does not mean

37. Jacob Taubes, *The Political Theology of Paul*, trans. Dana Hollander (Stanford, CA: Stanford University Press, 2004), 5.

38. Taubes, *The Political Theology of Paul*, 14.

they have become enemies of Christians or ultimately of God; rather, they are become 'enemies for your sake; but as regards election they are beloved, as for the sake of their forefathers' (11:28).[39]

Christians are called to love their enemies. That is the high claim of the good they are called to obey. And that good can never be subordinate to any power other than that of God, including the authority of the state, as some wrongly deduce from the words about submission to authority in Romans 13. The true authority is Christ, and the true law is the new command to 'love your neighbour as yourself', for 'love is the fulfilling of the law' (Romans 13:9, 10). Here Taubes finds the original imprint of Christian messianism: 'it is the love not of the Lord, but of the neighbor that is the focus here. No dual commandment, but rather *one* commandment. I regard this as an absolutely revolutionary act.'[40] It is in the love of neighbour that the old order is finally challenged, the time is fulfilled, and salvation comes near: 'It is now the moment for you to wake from sleep. For salvation is nearer to us now than when we became believers. The night is far spent, the day is at hand' (Romans 13:11-12). The call, then, to which Taubes, with Paul, exhorts his hearers is that even though 'the appointed time is grown short' and 'the present form of this world is passing away' (1 Corinthians 7:29, 30), this is not the occasion for a revolutionary challenge to the powers that be, not even 'if tomorrow the whole palaver, the entire swindle were going to be over.'[41] The true apocalyptic way is the refusal to resist the enemy, as Schmitt had proposed, and instead to recognise a friend in the enemy by loving one's neighbour as oneself.

Why is love the ultimate revolutionary act? Because 'love is the admission of my need.' 'The point in Paul is that even in perfection I am not an I, but we are a we. Meaning that need consists in perfection itself. Just as it says in 2 Corinthians: "Your power is made perfect in weakness". *Telos,* perfection, is a notion from mysticism, from the language of the Mysteries, but also from physics. And the punch line is: *en astheneia,* "in weakness." '[42]

For Taubes, the ground of the state in Christendom cannot be fully described without understanding that Jesus Christ, and his revolutionary command to love the neighbour, is the charismatic ground of the idea of the nation-state in modern Europe. For Taubes, Schmitt is right that sover-

39. Taubes, *The Political Theology of Paul,* 52.
40. Taubes, *The Political Theology of Paul,* 53.
41. Taubes, *The Political Theology of Paul,* 54.
42. Taubes, *The Political Theology of Paul,* 56.

eign rule conceals its origin and that 'in the exception the power of real life breaks through.'[43] But the true exception which is concealed is Christ himself, and the love command. It is not so much the determination of who is the enemy that is the ground of a would-be Christian nation but the refusal to treat the stranger as an enemy.

The ultimate purpose of Schmitt's political theology is to shore up the power of the state. But the purpose of Paul's messianism is to recognise that the Messianic Christ is not the continuation of history; Christ shores nothing up. Christ is the *end* of history, not the rebirth of theocracy. Romans is not a text about the endurance but the *destruction* of the Roman Empire. And this is why for Paul all creation 'groans in travail', having been 'subjected to futility' until 'the creature itself shall be delivered from the bondage of corruption into the glorious liberty of the children of God' (Romans 8:21, 22). This is why 'Paul has very peculiar worries about nature. Of course they are not ecological worries. He's never seen a tree in his life.' For Paul, as for Kafka, 'nature appears only as judgment', as an 'eschatological category' which is at risk of decay, just as the political is on the way to destruction.[44]

The *meaning* of nature is not, then, to be found in an immanent Protestant piety, which is the source of the Romantic turn back to nature and of the revival of 'nature religions' as means to resist the Baconian and Kantian reincarnation of Prometheus. Instead, for Taubes the messianic points toward the transcendent meaning of suffering, including the suffering of the natural. But the messianic claim is that the exception is more real than the law, and this is first truly known in the exception who is the crucified Christ. This is what natural science, like political science, denies: 'the question is whether you think the exception is possible — and it is on the exception after all that the whole law of natural science runs aground, because natural science is based on prognosis. . . . The question is: How can one think about suffering in a transcendent-Christian way?'[45]

For Paul, as for Taubes, there is only the way which acknowledges that behind the moral law, and the good will, is a 'profound powerlessness' and a failure to do what the good will even intends. Hence, as Paul believes, 'humanity and the cosmos are guilty', and guilt remains a substantive experience for human beings which only the entanglement of the crucified

43. Carl Schmitt, *Political Theology,* trans. George Schwab (Cambridge, MA: MIT Press, 1985), 22, cited in Taubes, *The Political Theology of Paul,* 65.

44. Taubes, *The Political Theology of Paul,* 73.

45. Taubes, *The Political Theology of Paul,* 85.

Christ in human sin can turn aside.[46] Paul's messianic reading of history reveals that only the sacrifice of Christ can turn away sin, and at the same time that *we* killed him, and not the Jews.[47] We become the enemies of God; and then God calls us his friends. This is why Paul is right and Schmitt is wrong. There is a new law, and it is not a law of the preservation of the state. It is instead the revolutionary law of love that refuses to accept that guilt or corruption or decay or suffering can occasion the need for anyone to hate or resist the stranger, or the outcast, or the migrant, as Taubes puts it in the conclusion to a letter to Armin Mohler:

> law is finally not the first and the last after all, because there are 'even' between man and man relationships that 'exceed', 'transcend' law — love, mercy, forgiveness (not at all sentimentally but in reality). I wouldn't know how to take a single step further in my wretched and often warped life (and indeed I don't know how to take a step further) without holding fast to 'these three', and this leads me again and again — against my 'will' — to Paul.[48]

Taubes, a *Jewish* philosopher, discovers in Romans a negative political theology which in essence is a critique of law, and an exaltation of love as the revolutionary messianic principle that is both the meaning and the end of history. But he argues that Paul does this in such a way as to refuse to negate the universal and ultimately mediatorial and theurgic meaning of the Jewish covenant. Instead, Paul recovers the meaning of the covenant as a universal covenant, just as had the prophets in Exile. The new covenant overcomes the division between Jew and Gentile. But it sets up a new division, between *ecclesia* and empire, in which *ecclesia* is 'an autarchic polis that separates itself militantly from other communities but as a new universal world order', which brings it into conflict with the universal claim of empire.[49] The political implication of all this is a radical transvaluation of the politics of empire. The world is 'turned upside down', the mighty are put down from their thrones, and the lowly are lifted up because the only disburdenment from the guilt which weighs down Gentiles as well as the Jews is the radical decision to love the neighbour as one's self, which is the

46. Taubes, *The Political Theology of Paul,* 88.

47. Taubes, *The Political Theology of Paul,* 94.

48. Taubes, *The Political Theology of Paul,* 110.

49. Wolf-Daniel Hartwich, Aleida Assmann, and Jan Assmann, 'Afterword', in Taubes, *The Political Theology of Paul,* 130.

source of a radical new equality in history, an equality that comes from outside history, through the *logos* who is also the Messiah.[50] The true fulfilment of humanity therefore comes from outside history, from the end of history; but it is fulfilled in an *ecclesia* which stays true to the suffering Messiah, and not the *ecclesia* which becomes a state or an empire.

The Apocalypse of Albion Christ

As we have seen, the beginning of the era of the Anthropocene is often dated at the dawn of the industrial revolution in England with the invention of Watt's steam engine. But the Anthropocene, like the industrial revolution, seems to fly in the face of the Copernican claim that humanity is no longer in charge of history. Prometheanism is the perverse outcome of the Joachimite and Copernican claim that history is not in the hands of humanity. Only when there is no longer a steady correspondence between heaven and earth, which the Church underwrites through the Holy Roman Empire and its partnerships with Christian kings and emperors, is humanity free to take charge of her own history. But in taking charge, humanity achieves in the Anthropocene a new and unintended kind of correspondence between the heavens and the earth. The emissions from myriad chimneys, cars, burning forests, cement factories, planes, and smoke stacks are in the sky; when we look at the sky, when we experience the weather coming from the sky, we increasingly sense an unsought-for correspondence between the sky and what has been going up to the sky from earth.

The first 'modern' nation in the Christian West is England. The land-grab and state-making projects of Henry Tudor and Elizabeth I created a new society where labour was sundered increasingly from land and exiled to workhouses and mills in what became the first industrial cities, fueled first by water as well as by muscle power. Conditions in these new settlements were life shortening, but became far worse when coal became the principal fuel source.

That the smoke-filled cities of England were the first to presage a new and unsought-for era in human and earth history was first realised not by scientists or philosophers but by novelists and poets. And the poet above all others who saw the apocalypse coming in the dark satanic mills of industrial England was the working-class artisan, lithographer, and poet William Blake.

50. Hartwich et al., 'Afterword', 135.

Blake was also a Joachimite and deeply suspicious of the established church, which he saw as an agent of empire. Blake is 'the prophet-poet par excellence of English literature', whose poetry combines the ethical seriousness of the Old Testament with the ecstatic visionary power and nature-imagery of its prophetic prose and poetry.[51] For Blake saw more clearly than any other the social and environmental catastrophes that Promethean Albion visited on her own land and people, and, through empire, on other continents and peoples.

Most interpreters of Blake see him as Romantic in his exaltation of vision over reason and poetry over philosophy, though Northrop Frye describes Blake as anti-Romantic in his attitude to *nature*.[52] In a new ecocritical reading of Blake, Kevin Hutchings argues that the view of Blake as anti-natural, even anti-ecological, is a serious misrepresentation. The best evidence for Hutchings's revisionary reading is Blake's account of the 'Vision of Light' that defined his sense of call as an artist and an apocalyptic dissenter. Blake describes standing on the edge of the sea at Felpham, Sussex, and falling into an ecstasy in which he felt himself become one with the divine.[53] He describes his awe at the 'glorious beams' of the sun which beckoned to him,

> Saying: 'each grain of Sand,
> every Stone on the Land,
> each rock and each hill,
> each fountain and rill,
> each herb and each tree,
> Mountain, hill, earth and sea,
> Cloud, Meteor, and Star,
> Are Men Seen Afar.'[54]

He continues a few lines later,

> My eyes more and more,
> Like a sea without a shore
> Continue expanding,

51. Murray Roston, *Prophet and Poet: The Bible and the Growth of Romanticism* (Evanston, IL: Northwestern University Press, 1965), 100-101.

52. On the interpretation of Blake as anti-natural see Kevin Hutchings, *Imagining Nature: Blake's Environmental Poetics* (Quebec: McGill-Queen's University Press, 2002), 37-75; Hutchings's insightful ecological reading of Blake was influential in the account that follows here.

53. William Blake, 'To Thomas Butts,' line 2.

54. Blake, 'To Thomas Butts,' lines 25-32.

The heavens commanding;
Till the Jewels of Light,
Heavenly Men beaming bright,
Appear'd as One Man
Who Complacent began
My limbs to enfold
In his beams of bright gold.[55]

Blake's vision produces ecstasy: he feels himself touched by the gods, not only seeing their form but being embraced and raised up by the 'heavenly man'. But if the vision is of the heavenly man, the place where he has the vision is the seashore, the horizon of sea and sky, the refracted beams of the sun on ocean, shore, and sky. As Chris Rowland comments, Blake at Felpham attests that it is 'the natural world which launches him into a mystical experience.'[56]

The heavenly man at Felpham is inspired by Ezekiel's divine image of the heavenly man who moves through wheels within wheels, and hence Blake's vision has particular significance for the climate and ecological crisis.[57] For Ezekiel, the Babylonian destruction of Solomon's Temple meant that the God of the Hebrews had taken on a new divine-human form beyond the buildings of men, and he depicts God as a heavenly and wheeled man with four faces who, like the Tabernacle, moves around on the face of the earth and is no longer confined to a building. Blake draws Ezekiel's vision in his lithograph *Ezekiel's Wheels,* in which the wheeled divinity is clearly depicted as Christ, whose four-fold face and body (the fourth face and body are back of the painted figures and so not shown) is the core of the joined bodies of the four living creatures and the wheels that surround them in Ezekiel's vision.[58]

In *The Four Zoas* Blake gives an extensive interpretation of Ezekiel's vision. The Four Zoas represent the body, reason, emotions, and imagination; the four compass points; and the four elements. The Four Zoas are crucial to Blake's critique of the regnant ideologists of industrial England: Bacon, Newton, and Locke. In *Jerusalem,* these three are named as having drugged Albion, the divine 'Four-fold Man' 'in deadly sleep', over whom the terrors unleashed by Bacon and Newton hang 'sheathed in deadly steel'. 'Awake

55. Blake, 'To Thomas Butts,' lines 45-54.

56. Chris Rowland, *Blake and the Bible* (New Haven, CT: Yale University Press, 2010), 135.

57. Harold Bloom, 'Blake's *Jerusalem:* The Bard of Sensibility and the Form of Prophecy', cited in Rowland, *Blake and the Bible,* 66.

58. Rowland, *Blake and the Bible,* 142.

Albion', cries the prophet Blake, from the drudgery of the 'Looms of Locke' and the 'Water-wheels of Newton', whose 'wheel[s] without wheel' have tyrannical cogs which are 'not as those in Eden, which Wheel within Wheel, in freedom revolve in harmony & peace'.[59] As Rowland comments, for Blake 'the water-wheels of Newton are mechanical and utterly predictable, and contrast with those in Eden. In Eden, the chariot of life surrounded by the four living creatures offers true humanity and the entry into the world of the imagination.'[60] In *Milton,* Los calls Satan 'Newton's Pantocrator' since the satanic mills of Newton supplant the restorative rule over creation of the divine man Christ, and especially over the land on which Christ may once have walked.[61] Instead of the restored peace of Eden, Newton's rule promotes war, which permeated most of Blake's lifetime, first in the American Revolution and then in the Europe-encompassing Napoleonic Wars, both of which had political as well as ideological repercussions in England, from riots on the streets of London to new coastal defences.

Key to the demise, and the envisaged redemption, of Albion are the idolatrous dreams of Bacon, the mechanistic optics of Newton, and the monetised politics of Locke. These three great dreamers of modern England had delivered up both land and working people to the coercive service of mechanical governance. Newton's water wheels and satanic mills symbolise not only the ideological drift toward atomist mechanism but, as Hutchings notes, a 'mode of governmental *discourse*' which frustrates the creative and participative rule of Albion-Christ and subjects nature to a predictive scientism whose end is control and domination of nature's ends.[62] But Blake does not argue for a back-to-the-land primitivism. Instead, he conjures a new harmony between rationalism and feeling, mechanism and nature, property and equity, science and religion. Hence in the conclusion of *Jerusalem,* Blake has Bacon, Newton, and Locke meet with Chaucer, Shakespeare, and Milton. In his vision Albion is released from her captivity and, awakened from her slumber, leads humanity and Paradise from exile toward eternity.[63]

Far more than the Romantic nature poets, Blake looks into the dark heart of industrial Albion and sees there the multiple causes of the stripping bare of the hills and the silencing of bird and beast. Blake's account of England's fall is of a double tragedy: not only the fall of humanity and nature,

59. William Blake, *Jerusalem: The Emanation of the Giant Albion* 15.15-20.

60. Rowland, *Blake and the Bible,* 143.

61. Hutchings, *Imagining Nature,* 137.

62. Hutchings, *Imagining Nature,* 138.

63. Blake, *Jerusalem* 98.8-28, cited in Rowland, *Blake and the Bible,* 144.

but the spiritual fall from the promise of the new creation, the new Jerusalem, that the disparate tribes and shires of England, and then Wales and Scotland, had made their own unitive story. For Blake, Albion — England, Scotland, and Wales — is the New Jerusalem: the shires and counties of all three nations are counted off in *Jerusalem* with the twelve tribes of Israel. The divine man indwells Albion's people and the love of God flows through the creatures in her vales:

> Awake, awake O sleeper of the land of shadows, wake, expand!
> I am in you and you in me, mutual in love divine;
> Fibres of love from man to man through Albion's pleasant land.
> In all the dark Atlantic vale, down from the hills of Surrey,
> A black water accumulates. Return, Albion, return![64]

Albion's fall is fourfold: the enslavement of the energies of her people to mechanical labour and war as cogs in Newtonian machines; the stripping of life and wealth from the wooded hills and vales the people had once used for energy, grazing, hunting, and gathering; the monetisation of value and wealth and its concentration in the hands of merchants; the imperial export of these same processes to the nations. All four stages are present in *Jerusalem:*

> The banks of the Thames are clouded, the ancient porches of Albion are
> Darkened; they are drawn through unbounded (absolute Newtonian)
> space, scattered upon
> the void in incoherent despair. (5.1-2)

> Albion's mountains run with blood, the cries of war & of tumult
> Resound into the unbounded night, every Human perfection
> Of mountain & river & city, are small and withered and darkened.
> (5.6-8)

Meanwhile the fallen Sons of Los

> Devour the sleeping Humanity of Albion in rage & hunger
> They revolve into the furnaces southward & are driven forth northward
> Divided into Male and Female forms time after time
> From these Twelve all the families of England spread abroad (5.30-34)

64. Blake, *Jerusalem* ch. 1, ll. 6-10.

> Wash'd by the Water-wheels of Newton: black the cloth
> In heavy wreathes folds over every Nation: cruel Works
> Of many Wheels I view, wheel without wheel, with cogs tyrannic
> Moving by compulsion each other: not as those in Eden, which
> Wheel within Wheel in freedom revolve in harmony & peace.
>
> I see in deadly fear in London Los raging round his Anvil
> Of death: forming an Ax of gold; the Four Sons of Los
> Stand round him cutting the Fibres from Albion's hills,
> That Albion's Sons may roll apart over the Nations (15.15-24)

As humanity is enslaved to furnaces and looms, the love of human souls for one another is turned by Newtonian alchemy to a love for gold and manufactured luxuries. At the same time the fibres — the trees — are cut from Albion's hills, and as Albion is enslaved and stripped bare her devastation is then exported to the nations. The enslavement of Albion to machines and war takes her children, body and soul, and turns their energies, erotic and creative, into gold:

> The violent Man-slaughter, these are the Spectre's double Cave:
> The Sexual Death living on accusation of Sin & Judgment,
> To freeze Love & Innocence into the gold & silver of the Merchant.
> Without Forgiveness of Sin Love is Itself Eternal Death. (64.21-24)

For Blake, the betrayal is like a second fall, where sex, love, and innocence are again spoiled by sin and judgment and death. Forgiveness of sin had redeemed love in Albion-Christ, but in industrial enslavement she enters a second death.

Blake's vision of redemption is equally multilayered. One of the fullest accounts is the ninth chapter of the incomplete but extensive poem *The Four Zoas*. This chapter rests on the Revelation of Saint John of Patmos.[65] It commences with a reference to Matthew 24:29-31: the sun and moon are darkened as the 'vegetable hands' of Los crack the heavens at the death of Christ. But then the voice of the risen Christ 'awakes the dead' and

> The heavens are shaken and the earth removed from its place,
> The foundations of the eternal hills discovered.

65. W. H. Stevenson, *Blake: The Complete Poems* (London: Longman, 1971), 441.

> The thrones of kings are shaken; they have lost their robes and crowns
> The poor smite their oppressors, they wake up to the harvest,
> The naked warriors rush together down to the sea-shore
> Trembling before the multitude of slaves now set at liberty;
> They are become like wintry flocks, like forests stripped of leaves.
> The oppressed pursue like the wind; there is no room for escape.
> (9.16-23)

Drawing on the books of Ezekiel and Revelation, Blake describes the salvific rising of Albion between the Resurrection of Christ and the Last Judgment. Once redeemed by the Lamb of God, then again enslaved, Albion will be redeemed again as the New Jerusalem rises once more in England's green and pleasant lands, and at the heart of this new rising will be the forgiveness of sins as the merchant's measure of money debt gives way to love divine.

Christ the redeemer, as in the Book of Revelation, is the Lamb of God, communion with whom is the way out of alienation and back to redemption. As Rowland comments, 'The lamb offers an ever-present possibility for renewal, though the inhabitants of Britain are prevented from reaching that goal by their inability to recognise their true destiny.'[66] This destined salvation is not confined to the Church. Albion-Christ is a supra-ecclesial community whose messianic calling is to take the message to Europe, the Americas, and beyond.

Whereas in the biblical Book of Revelation there is a division at the Last Judgement between the faithful followers of the Lamb and those destined for the fire, in Blake's *Jerusalem* the refiner's fire of judgment overcomes the opposition of heaven and earth, good and evil, and leads to the ultimate restoration of all things. Blake's vision of God being 'all in all' (cf. 1 Corinthians 15:28) occurs when the trinity of Newton, Bacon, and Locke take their places in the chariots of the Almighty (*Jerusalem* 98.9). The universality of *Jerusalem*'s readers ('To Jews, Deists, Christians, and the Public') demonstrates that Blake includes in his audience a group much wider than the Church, for his vision concerns universal salvation and identification of the various obstacles in its way. The way of the Lamb emerged in his own labours and above in *Lamb*eth (*Jerusalem* 12.41). In this radical universalisation of the Christian apocalypse Blake goes back behind Revelation to the book of Ezekiel, which is also John of Patmos's key source. Ezekiel's vision, described in the first chap-

66. Rowland, *Blake and the Bible*, 148.

ter, is of the presence of God moving out from the Temple back through the Tabernacle to be realised universally, throughout creation.

Blake is not alone among the Romantics in being inspired by Ezekiel. Ezekiel's description of the four-headed heavenly man and the sapphire throne fired the eighteenth-century Romantics' imagination above all other biblical or Classical texts. For Burke, Ezekiel is the most sublime of the Hebrew prophets, and this distinctive sublimity, and Ezekiel's unique symbolic imagery, inspired poems by William Collins, Thomas Grey, William Wordsworth, and John Keats.[67] Ezekiel's appeal is that he goes back in his imagination before the Temple and the Tabernacle, before even the call of Moses, to envisage a time when the *Shekinah* of God dwelt in all the elements of creation. For Ezekiel, there would be no second temple. Instead, the divine man — whom the fathers of the Church read as Jesus Christ — would restore humanity and the cosmos to the time of Paradise when the presence of God was known in all creation.[68] The Temple is for the Hebrews a microcosm of the whole creation. After its destruction and the Exile of the Hebrews to Babylon, Ezekiel's message is that the Hebrews could 'sing the Lord's song' in a strange land, even by the rivers of Babylon. For those self-same rivers had once flowed out from the Garden of Eden, which were precisely the waters of the Tigris and the Euphrates which flowed through Babylon. Thus, for Ezekiel, the geography of exile is not the end of the divine plan of redemption. On the contrary, the new exile is the beginning of the end of the old exile from Eden: creation itself has become the temple again, the divine presence is returning, and the river of life will flow out freely from the redeemed city to the lands of the Gentiles as well as the Jews.

By the mid-eighteenth century Christianity was no longer at the heart of English intellectual culture. The message of Ezekiel — that the end of institutional religion is not the end of the sacred — was taken up by the Romantics, who found in nature the new sublime. The spiritual in Europe would outlive the demise of Christianity if people would recognise the divine presence in nature. This is not to say that the English or the German Romantics were pagan nature-worshippers. Rather, after the scientific, industrial, and French revolutions the Romantics wished to underline that God or the Spirit had not *left* Europe. For Blake in particular, the Spirit had

67. Brian Hepworth, *The Rise of Romanticism: Essential Texts* (London: Carcanet New Press, 1978), 183; and Gordon Strachan, *Prophets of Nature: Green Spirituality in Romantic Poetry and Painting* (Edinburgh: Floris Books, 2008), 36.

68. Robert W. Jenson, *Ezekiel*, Brazos Theological Commentary on the Bible (Grand Rapids: Brazos Press, 2009).

been set free by these revolutions from the confining structures of Christian monarchs and priests 'that all the people might prophesy', and free from prescientific metaphysics to be fully present in living nature so eternity might be glimpsed in a grain of sand.

This sense of the freeing of the Spirit from the confines of Church, patriarchy, and tradition is also central to Blake's politics. And hence Blake does something more with Ezekiel's vision than the other English nature Romantics, and the 'more' is at once more apocalyptic, messianic, and politically radical. For Blake, the key to the political is compassion for the suffering of the poor, combined with a sense that the time when their suffering will end is near. The coming of Christ had released the presence of God into humanity in a new way, and the practice that is the redemptive moment for those who are open to this presence is the forgiveness of sins:

> To Mercy, Pity, Peace, and Love,
> All pray in their distress:
> And to these virtues of delight
> Return their thankfulness.

Love is 'the human form divine' to whom everyman prays and therefore

> All must love the human form,
> In heathen, Turk, or Jew.
> Where Mercy, Love, & Pity dwell,
> There God is dwelling too.
>
> ('The Divine Image',
> *Songs of Innocence* XX, 1-4, 11, 16-20)

'The Divine Image' is key to Blake's theology, which, as the late *Everlasting Gospel* indicates more fully, turns centrally on the image of God in humanity, which had been sullied and obscured by sin and Sinai's law. In Christ, forgiveness and love are revealed again in their fullest flowering. As these divine behaviours are expressed by the giant man Albion-Christ — who as a country, a nation, and a people represents all humanity — the image is potentially restored. In the *Everlasting Gospel* the 'shadowy man' represents the fallen 'selfhood' which Christ's death on the cross 'roll'd away'. On the death of Christ the sinful self is nailed to the cross and the divine image is restored again in humanity's freedom to love for love's sake and not because of law. As Rowland comments, for Blake, Christ is the 'archetypal antinomian' who 'taught that God loved all Men' and 'forbad all contention for Worldly prosperity'.[69] The worldly are therefore restored to life and love when they find fellow feeling with the poor and respond to innocent suffering with mercy, pity, peace, and love. For Blake, the response of feeling — pity, compassion, love — to innocent suffering is the essence of Christian redemption and its moral import for soul and society.[70]

Blake might be said to be at one with the mainstream Romantics in their exaltation of heroic feeling as the point of resistance to the instrumentalising and controlling tendencies of rationalism. But for Blake this redemptive feeling is not for nature alone, and nature's release is not about feeling alone; instead, it requires that Albion rediscover the manual crafts and the more equitable harvests in which she once engaged nature. The release of the poor from enslavement to the furnaces, looms, and smoke-filled cities of drugged and sleeping Albion will come when the valleys and shires are re-peopled, with vegetable gardens and small farms, and the harvest of myriad small producers is again enjoyed up and down the land. As Christ longs for the peace of Jerusalem, Blake longs for the awakening of Albion and the freeing of her children, both at home and in her empire.[71] In 'Holy Thursday', the poor children of London troop into Saint Paul's; their voices joined in song are 'like a mighty wind', and the poem concludes, 'Then cherish pity, lest you drive an angel from your door'.[72] The moral virtues of Christ the heavenly man are not the Classical virtues of moral philosophers but the 'virtues of delight', which are 'mercy, pity, peace, and love'.[73] The human capacity for these delightful virtues emanates from the binding of Satan and the forgiveness of debts. Christ restores the divine image by freeing humankind from religion and debt, satanic power and mercantilism. The 'God of this World' is cast out just as the 'Merchant Canaanite' is evicted from 'the Temple of His [Christ's] Mind'.[74] Hence the goods of this world are again to be shared in common, as Blake indicates in his paraphrase of the Lord's Prayer:

69. Rowland, *Blake and the Bible,* 191, and citing Blake, 'Annotations to Watson's *Apology*', 387.

70. Analogously Thomas Nagel sees compassion and the alleviation of suffering as the central ethical sentiments and actions; Nagel, *The View from Nowhere* (Oxford: Oxford University Press, 1986).

71. Cf. 'The Little Black Boy', *Songs of Innocence* VI.

72. 'Holy Thursday', *Songs of Experience* IX.

73. Blake, 'Divine Image'; also Rowland, *Blake and the Bible,* 192.

74. Blake, *The Everlasting Gospel,* cited in Rowland, *Blake and the Bible,* 192.

> Give us This Eternal Day our own right Bread by taking away Money or Debt or Tax & Value or Price as we have all Things Common among us. Every Thing has as much right to Eternal Life as God, who is the Servant of Man.[75]

Forgiveness of sins is the spiritual prism through which Blake reads the Gospels and the message he prophetically addresses to the mercantilism and spreading immiseration of the London of his day: 'Jehovah's Salvation Is without Money & without Price, in the Continual Forgiveness of sins In the Perpetual Mutual Sacrifice in Great Eternity!' and hence the Covenant of Jehovah is that 'If you Forgive one-another, so shall Jehovah Forgive You'.[76] Forgiveness, pity, love, and mercy — these are the relational practices through which Blake interprets the Bible, as well as the apocalyptic miasma of smoke-filled industrial London and Albion's imperial sinews. In this way, as Rowland observes, Blake reads the present through the past, seeking to 'galvanise a complacent society to see the folly of religion hid in war and take up the practice of the forgiveness of sins' again.[77] Pity at the innocent suffering of the poor can find political recourse only by rejecting the appropriation of property by merchants, and reimagining the poor not as debtors but as proprietors again of their own productive *common* domain.

Blake's antinomianism, like that of Taubes, finds its core meaning in a morality of love which is not reactive but spontaneous gift emanating from justifying faith in Christ and unmerited forgiveness, but it is less negative in its outcome. For Blake, the fruit of forgiveness is freedom instead of slavery, plenty instead of scarcity. Blake imagines the restoration of a great harvest feast in which is ended the 'painful partnership of coke and ore in the smelting furnace', and harmony is restored between humanity and nature.[78] 'You shall forget your former state. Return O Love in peace, Into your place, the place of seed, not in the brain or heart'.[79] Harvest also represents the end of empire, the slave trade, and hierarchy: the 'New Song' to celebrate the great harvest feast is composed by a black African and sung not by heavenly an-

75. Blake, 'Annotations to Thornton's *The Lord's Prayer Newly Translated*', reconstructed in Rowland, *Blake and the Bible*, 193.

76. Blake, *Jerusalem*, cited in Rowland, *Blake and the Bible*, 197.

77. Rowland, *Blake and the Bible*, 241.

78. David V. Erdman, *Blake: Prophet Against Empire* (Princeton, NJ: Princeton University Press, 1954), 327.

79. Blake, *The Four Zoas*, lines 362-63.

gels but by 'all the Slaves from every Earth'. As Erdman comments, nothing in his oeuvre more deeply 'reveals Blake's thoroughgoing democracy'.[80]

For Blake, the aesthetic turn of the Romantic nature poets — and especially 'Lord' Wordsworth — left the deeper dimensions of the apocalyptic birth of scientific modernity untouched. What is now called the ecological crisis is for Blake part of a much larger crisis in the human psyche, society, and spirit, and Albion will return to building Jerusalem only when she ends 'Corporeal War' and takes up again '*mental* strife' against 'the Class of Men whose whole delight is in Destroying'.[81] At the root of war and destruction are Baconian mechanism and the Newtonian view of time and space. In *Milton* Blake opposes the Newtonian idea of absolute time, since it excludes the human sensual apprehension of time: 'Los is by mortals nam'd Time [and] Enitharmon is nam'd Space'.[82] The alienation of absolute Newtonian time and space from humanity is indicative of the fall of humanity, and it is to redeem humanity from this alienation that Christ had come and for which the apostles after him strove. But Blake does not take the millenarian path and posit fallen time over against a redeemed eternity. Instead, the redemption of time and space are already being revealed within fallen history before the final apocalypse, and he sees a restored Albion uniting in her limbs temporality and eternity.[83]

Hutchings describes Blake as the original human ecologist, but Blake differs from most who own that label by placing Christianity, and more especially the doctrine of the Incarnation and the restoration of the divine image in Christ, at the centre of human ecology. So in *Milton,* when men 'wildly' follow Los and abandon at least for a moment the Harrow and the Mill there will be a 'blank in Nature', but a fecund and potentially transformative blank as nature awaits the shaping of humanity redeemed from the drugged sleep of mechanistic servitude.[84] For Blake, mind and matter interpenetrate one another in a way that both Baconian mechanism and Lockean rationalism deny. This mutual interpenetration of humanity and nature through the senses, and more especially through images and words, results in a vision of the restoration of humanity and nature in the resurrection of the 'Vegetable Body' as described in *Milton.* As drugged Albion is awakened to a resurrection in the Vale, Blake sees himself in a

80. Erdman, *Blake,* 329.

81. Blake, *Milton,* Preface, 14, 19.

82. Blake, *Milton* 24.68, cited in Hutchings, *Imagining Nature,* 133.

83. Hutchings, *Imagining Nature,* 133.

84. Blake, *Milton* 8.12, cited in Hutchings, *Imagining Nature,* 139.

vegetable garden, his wife at his side, a lark singing, purple thyme moving in the wind.[85]

> Terror struck in the Vale: I stood at that immortal sound
> My bones trembled. I fell outstretchd upon the path
> A moment, & my Soul return'd into its mortal state
> To Resurrection & Judgment in the Vegetable Body
> And my sweet Shadow of Delight stood trembling by my side
> Immediately the Lark mounted with a loud trill from Felphams Vale
> And the Wild Thyme from Wimbletons green & impurpled Hills
> And Los & Enitharmon rose over the Hills of Surrey
> Their clouds roll over London with a south wind, soft Oothoon
> Pants in the Vales of Lambeth weeping o'er her Human Harvest
> Los listens to the Cry of the Poor Man: his Cloud
> Over London in volume terrific, low bended in anger.
> Rintrah & Palamabron view the Human Harvest beneath
> Their Wine-presses & Barns stand open; the Ovens are prepar'd
> The Waggons ready: terrific Lions & Tygers sport & play
> All Animals upon the Earth, are prepar'd in all their strength
> To go forth to the Great Harvest & Vintage of the Nations.[86]

Sometimes dismissed as a 'Muggletonian', Blake is no dualist. For Blake, the salvation of humanity also involves the redemption of the earth, for both are tarnished by the fall of Albion. Christ for Blake comes to nature as her preexisting Lord to recover dominion from the usurper Satan. As Christ gives embodied form to the divine Spirit on earth, Blake gives form to the spirits in the embodied particulars of the earth, seeing 'Heaven in a Wild Flower', a white cloud as the creator Spirit and not merely a combination of water droplets.

For Blake, consciousness is embodied and mediated through the sensory and imaginative apprehension of the material world, and this includes consciousness of divinity as well as nature and humanity. This is also indicated by Blake's geographical location of Christ-Albion and the New Jerusalem, which he situates in a real earthly geography, in intimate landscapes and urban and rural villages. In so doing he gives poetic and spiritual force to the struggle of men and women to free their locales in London, Surrey,

85. Hutchings, *Imagining Nature*, 149.

86. Blake, *Milton* 42.24–43.1.

and the rest of Britain from the dirt, noise, and smoke of Albion's fall and build Jerusalem again in a green land in which Christ had once walked.[87] In *Jerusalem* a restored and *sensual* harmony between humanity and nature is Christological in meaning:

> We live as One Man; for contracting our infinite senses
> We behold multitude; or expanding: we behold as one,
> As One Man all the Universal Family; and that One Man
> We call Jesus Christ: and he in us, and we in him,
> Live in perfect harmony in Eden the land of life.[88]

Blake's Christological spiritualised geography is combined with a deep sense of the interconnectedness of all things. Nothing 'lives alone' or 'for itself';[89] all things are deeply interrelated. For Blake, there is a profound analogy of being which sees all elements, living individuals, and life processes as the bones, limbs, organs, and sinews of the body of Christ-Albion. The body of Albion-Christ is both human and more than human, and the 'more than' includes not only divine life but also sensory and non-sensate life. Blake resists the tendency of the biologists of his time to eschew teleological language in their empiricist quest for pure cause-effect relations as means to explain the evolution of life. Blake also resists, as Whitehead did later, the tendency of scientists to dismiss the qualities of life such as the redness of the sky as 'secondary characteristics' which, because non-quantifiable, are not deemed scientifically significant.[90]

For many interpreters, Blake's identification of nature with Christ is seen as proof of Blake's alleged anti-naturalism. In this perspective, the Christocentric character of Blake's nature mysticism is problematic because it is 'anthropocentric'.[91] If cities, rivers, and mountains are human — if 'everything is Human', as Blake puts it in *Jerusalem* — and the Ten Commandments attend only to human relations, then the natural world is at risk of being excluded from the morally considerable in Blake's thought apart from its value to humans. This reading of Blake is underwritten by secular humanist history, according to which the first 1,500 years of the Christian era were a dark, superstitious interlude between the Classical age of reason and the scientific revolution.

87. Hutchings, *Imagining Nature,* 60.
88. *Jerusalem* 34.17-21, cited in Hutchings, *Imagining Nature,* 36.
89. Blake, *The Book of Thel* 3.26-27, cited in Hutchings, *Imagining Nature,* 62.
90. Hutchings, *Imagining Nature,* 63.
91. Hutchings, *Imagining Nature,* 74.

Christianity, on this view, was responsible for religious wars, superstition, and slavery; Classical philosophy, poetry, and rationalism enabled Western Europeans to embrace liberty, democracy, science, and progress and to leave their religious chains behind. Given the charge sheet, it is entirely logical that ecocide should be added to it. According to Aldo Leopold, Christians and Jews do not love nature because animals, land, soil, and water are not part of the Abrahamic moral world, not subjects for their ethical consideration.[92] For environmental philosopher J. Baird Callicott, Christianity is a dualistic religion that teaches domination of nature and the superiority of soul to body.[93]

For environmental philosophers such as Callicott, and even the Calvinist Holmes Rolston III, the quest to ascribe value to nature is a modern project which requires the deepening and extension of Enlightenment rationalism beyond the human. The derivation of transcendental values about the treatment of nature *from* nature takes its rise from the Kantian identification of intrinsic value with the reasoning capacities of persons.[94] The extension of Kantian intrinsic value to animals and non-sentient species forms the central philosophical plank of environmental ethics as elaborated by Holmes Rolston.[95] Lemurs may not be reasoners, and trees may not be conscious, but the behaviours of such sentient and non-sentient individuals represent, if not an intention, a 'goal-directed' orientation to continued *being*. They may therefore be said to be valuing subjects with intrinsic value worthy of respect; or, as Callicott puts it, they may be regarded as *values* if not valuers.[96] In Rolston's account the concept of intrinsic value can be extended beyond individuals to species as a whole, and to ecosystems,[97]

92. Aldo Leopold, 'The Land Ethic', in *A Sand County Almanac and Sketches Here and There* (New York: Oxford University Press, 1949), 167-89.

93. J. Baird Callicott, *Earth's Insights: A Multicultural Survey of Ecological Ethics from the Mediterranean Basin to the Australian Outback* (Berkeley: University of California Press, 1994).

94. It also rests upon Kant's account of the aesthetic as a sphere of human judgment independent of scientific and moral judgement, and his related account of the poetic: Immanuel Kant, *The Critique of Judgement* (1790). See also Allan Megull, *Prophets of Extremity: Nietzsche, Heidegger, Foucault, Derrida* (Berkeley: University of California Press, 1985), 11-12; and Max Oelschlaeger, *The Idea of Wilderness: From Prehistory to the Age of Ecology* (New Haven: Yale University Press, 1991), 115.

95. Holmes Rolston III, *Environmental Ethics: Duties to and Values in the Natural World* (Philadelphia: Temple University Press, 1988).

96. J. Baird Callicott, 'The Pragmatic Power and Promise of Theoretical Environmental Ethics: Forging a New Discourse', *Environmental Values* 11 (2002): 3-25.

97. Holmes Rolston III, 'Duties to Endangered Species', *BioScience* 35 (1985): 718-26.

and this approach has been so influential that it now appears in national environmental law, including the United States' Endangered Species Act and New Zealand's Resource Management Act.[98]

If mainstream environmental ethics is a coda in post-Enlightenment philosophy — a corrective within the modern tradition of rationalism rather than its rejection — this explains in part why the move to endow nature with moral value, even with legal rights, has not been more successful in challenging the larger destructive tenor and drift of industrial societies with respect to the natural world. Indeed, the failure of this approach is now clearly manifest as the industrial assault on the last stands of wilderness and the last redoubts of endangered species continues apace, while environmental law — such as the 'REDD' treaty procedures mandated under the UNFCCC to protect forests — accelerates the incorporation of the last forest wildernesses into a property regime of forest accounting which permits their conversion into plantations for more 'efficient' carbon capture than that of old growth forests.[99]

As Adorno and Horkheimer argued a century after Blake, the attitude of domination to nature is not a mere side effect of Enlightenment rationalism but front and centre to the science-informed culture which births the Enlightenment: 'in thought human beings distance themselves from nature in order to arrange it in such a way that it can be mastered.'[100] It was precisely the Baconian project to control and reorder nature that lent an empirical tenor to enclosures and 'agricultural improvement' that made peasants vagrants, and the subsequent Enlightenment project, taken up first by David Hume and then by Kant, to locate morality empirically in categories of mind and in societal expressions of taste and legal prescriptions rather than in the history of life and of human customs and traditions in sharing life.

Blake's antinomianism, like that of Taubes and Saint Paul, is particularly instructive in relation to the failure of environmental law or climate treaty to prevent the headlong rush toward ecological and climate catastrophe. Blake's critique of the homogenising tendencies of 'One Law' suggests,

98. See further Michael Northcott, 'Artificial Persons against Nature: Environmental Governmentality, Economic Corporations and Ecological Ethics', *Annals of the New York Academy of Sciences: The Year in Ecology and Conservation Biology* 1249 (2011): 104-17.

99. Alain Karsenty, 'The Architecture of Proposed REDD Schemes after Bali: Facing Critical Choices', *International Forestry Review* 10 (2008): 443-57.

100. Max Horkheimer and Theodor W. Adorno, 'The Concept of Enlightenment', in Horkheimer and Adorno, *Dialectic of Enlightenment: Philosophical Fragments*, ed. Gunzelin Noerr, trans. Edmund Jephcott (Stanford, CA: Stanford University Press, 2002), 1-34.

as Hutchings puts it, 'that the law's main problem is its inability to acknowledge and respect the particular otherness of individual creatures and contexts, human or otherwise.'[101]

The Blakean resolution to social and ecological alienation is not law, but emotional and sensory recognition of the interrelatedness of all things and pity, mercy, peace, and love toward all creatures who are oppressed, from the 'robin red-breast in the cage' to the slave in Africa. This is because for Blake 'every thing that lives is Holy', and everything lives not alone to itself but in interconnection, spiritual and material, with all others. If human government is oppressive, this oppression will extend to the animal world; if humans are enslaved in coal-besmirched cities, not even the countryside is safe from pollution's tyranny. This implies an environmental philosophy that acknowledges particularity and difference, while at the same time recognising an underlying organic and spiritual interrelatedness and unity in all things. It also implies that no amount of environmental legislation will restore the human-nature relationship when human beings are in large numbers excluded from a proper share of the fruits of the earth *and* cut off from poetic, sensory engagement with the delights of creaturely life by oppressive ugliness in polluted industrial cities, or by penury and unemployment as empire replaces them in a global mechanised economy.

Blake's vision of the interpenetration of human and biological life involves a politics. For Blake, there is no problem in recognising that humanity and nature are interpenetrated. The answer to ecological alienation is not to fence off field and forest from human beings in protected wildernesses but to remove the maw of empire from the heart of humanity. The contest for a just and peaceable harvest is not between humans and nature but between the masters of empire and the ordinary people of the cities and the shires.

Blake's poetic challenge for Albion-Christ to awake from the stupor of industrial empire to her quasi-messianic role in advancing compassion and justice for the poor, and the restoration of the fruits of the land to all the people, provides powerful resources for reimagining the moral and spiritual calling of the nations in facing the climate crisis in the present century. But after almost two centuries of wars between the new nations of Europe, the idea of a nation as a moral community with a shared purpose to sustain the common good of its people is still in retreat to a dream of global community and trade that is a tragic simulacrum of genuine community.

101. Hutchings, *Imagining Nature*, 74.

Among social scientists, the nation is represented as a social construct, an *imagined* entity, rather than as something that is intrinsic to the progress of humankind from tribalism to cosmopolitanism. Among economists, the nation, and more especially the nation-state, is viewed as a morally dubious enterprise which is best excluded from oversight of the principal mechanisms by which power and wealth are accumulated from land and sea and acquired and stored up by the wealthy arbiters of a global capitalist empire.

Transition from the Imperial Economy to the Distributed Economy of the Shires

The gap between the climate that imperial capitalism promotes and that which climate scientists and climate activists advocate is wide and getting wider. In its Fifth Assessment Report, which was available in prepublication draft at time of writing, the IPCC predicts that 'business as usual' growth in economic activities and fossil fuels will see average global temperatures rise by seven degrees Celsius over pre-industrial temperatures by 2100, with temperatures in the Arctic eleven degrees Celsius above historic norms, while oceans will likely be rising by one foot per *decade* by 2100.[102] This will be a new world. And it will be a world in which global capitalism will collapse, since the deep seaports of the world's major trading cities will be under water, along with all coastal or low lying airports. At the same time, stronger storms will make global ocean shipping more difficult and costly. It will also be a world in which many nations will be uninhabitable, and many others reduced to bare subsistence; and it may mean a war of all against all for the remaining areas of viable farmland and for potable water sources. These worst extremes of planetary change may still be avoided if corporations stop digging and drilling for fossil fuels and burning forests. But the IPCC continues, even in its Fifth Assessment Report, to advocate market-based approaches to greenhouse gas emission restraint which, under the preference-maximising 'laws' of supply and demand of global markets, stand no prospect of restraining private and public corporations from extracting sufficient fossil fuels to destabilise the climate.

In essence, there is a conflict between the corporate and imperial agen-

102. IPCC, *Fifth Assessment Report: Working Group 1: Summary for Policymakers,* advance draft at http://www.stopgreensuicide.com/SummaryForPolicymakers_WG1AR5-SPM_FOD_Final.pdf (accessed 17 December 2012).

cies of global industrial capitalism and the health of planet, people, and species.[103] As Blake saw, industrial capitalism is an empire, directed in Blake's time by the City of London and spreading its oppressive and extractive tentacles to every continent. But the imperial beast had a home, in the city, and it was in London that Blake conducted his spiritual war against the imperial antichrist through his apocalyptic art. The choice for Blake is between an imperial antichrist and the messianic Albion-Christ of the shires. Building Jerusalem field by field, village by village, town by town, involves spiritual strife in exposing and dethroning empire, and physical and mental struggle in creating and crafting the alternative.

For Blake, as for Taubes, Jewish and Christian apocalyptic resists the power of empire with the subversive messianic challenge of love of neighbour and the formation of equitable communities where spiritual struggle and mutual interdependence challenge the will to power and systems of domination. This messianic communitarianism is not only a feature of the politics of the early Church and Joachimite millenarians. It may also be found in the contemporary growth of dissenting citizens, communities, and networks of individuals, households, and businesses who are challenging the claim of the fossil fuelled empire to continue to provision human life without regard for planetary limits. Alongside the failure of the top-down approach of the UNFCCC to mitigate climate change by reducing fossil fuel extraction and deforestation is the emergence of myriad bottom-up, small-scale responses to climate change in the home, the church, the village or neighbourhood, the workplace, the town, and the city. If climate change is coming — and it is coming, even if in the future human beings find ways to slow it down — there is an urgent need to find ways to live in the new world of the Anthropocene which do not involve the continued large-scale sacrifice of distant ecosystems and the stable climate of future generations. These initiatives are increasingly focused on ways that recover the ability of households and communities locally to procure food, fuel, and fibre, and to foster alternative sources of wealth in new local economies which are less fragile, as well as more participative, than the global grids and networks that currently provision the majority of residents of the earth's still growing cities.

The low carbon energy technology which is already well adapted to

103. See also Brett Clark and Richard York, 'Carbon Metabolism: Global Capitalism, Climate Change, and the Biospheric Rift', *Theory and Society* 34 (2005): 391-428; it is also the premise of Naomi Klein's forthcoming book *Capitalism and Climate Change*.

a local and distributed energy economy is solar power. My first graduate supervisor, Professor Stephen Sykes, devoted some of his retirement pension to putting solar panels on the roof of his home in Durham, England, to reduce its carbon footprint. A number of churches which are part of the Eco-Congregations movement in the UK are installing renewable power generation plants in their church buildings. Alban Thurston, a member of an Eco-Congregation, Christchurch, Wimbledon, has worked with solar energy suppliers to facilitate church members' group purchase of solar installations for their homes, so extending the Eco-Congregation ideal of conserving energy and promoting renewable energy for churches by enabling households to do the same.[104] Solar energy is rising so fast as a source of household power in Australia, Germany, and the United States that the large corporate providers of fossil fuel–produced electricity are already having to adapt to declining rather than expanding householder demand for power. The head of the nuclear energy company NRG in the United States said at a recent conference that 'consumers are realising they don't need the power industry at all', while Jim Rogers of Duke Energy, the largest utility in the United States, said that 'if the cost of solar panels keeps coming down, installation costs come down, and if they combine solar with battery technology and power management system, then we have someone just using us for backup.'[105] Solar energy is also becoming the default electricity source for off-grid rural villages in the developing world.[106] The old model of inefficient and fossil fuelled centralised electricity power generation and distribution has for more than a hundred years promoted householder dependency on large corporations, but its time is coming to an end. An alternative vision of human development using locally harvested biopower is already becoming a reality in both the developed and developing worlds.

Paradoxically, and despite the global threat of extreme climate change,

104. Published research on Eco-congregations is unavailable at time of writing; an undergraduate dissertation by Catherine M. Harmer, 'Eco-Congregation: A Successful Blend of Theology and Ecology?' for the University of Chichester provides a useful ethnographic portrait, at http://ew.ecocongregation.org/downloads/CatHarmerDissertationUnivChichesterJun07.pdf (accessed 17 May 2013).

105. Giles Parkinson, 'Gone Solar', *Inside Story: Current Affairs from Australia and Beyond*, at www.inside.org.au/gone/solar (accessed 17 May 2013). See also David Roberts, 'Solar Panels Could Destroy US Utilities, according to US Utilities', *Grist Magazine*, 10 April 2013.

106. Paul Cook, 'Rural Electrification and Rural Development', in Subhes Bhattachayra, ed., *Rural Electrification Through Decentralised Off-Grid Systems in Developing Countries* (London: Springer, 2013), 13-38.

the city-based fossil fuel corporations, and public agencies such as the World Bank and the United States federal government, and publicly owned banks such as the Royal Bank of Scotland, continue to pour investment funds into the increasingly polluting business of nonconventional oil and gas and opencast coal extraction to sustain the profits of the centralising fossil fuel industry and its networks of pipelines and grids. But the climate crisis is already revealing the power of individual householders and community energy projects to challenge empire and return the energy economy to the naturally distributed light of the sun as it warms the earth and moves the winds over the earth's surface. Available sunlight was the past of human energy provision. With what are increasingly low-cost and simple technologies, in the hands of ordinary people, it will replace stored sunlight and become the future of energy again.[107]

It is not from a city but from a county town of the shire that a globally influential vision of a locally crafted, self-sufficient, and post–fossil fuel economy is emerging in England. Totnes in Devon is only three hundred miles from the City of London, but in 2006 it became the first Transition Town in the world. Rob Hopkins, the founder of the Transition movement, describes Transition as a response to peak oil and climate change. It began as a community education and permaculture initiative in a village in Ireland and then in Totnes, England. It has in less than ten years become a loose network of community and neighbourhood groups which has spread to more than eighty countries.[108] Transition groups are informal voluntary groups which plan actions to localise sustainable energy and food procurement and to promote vibrant local economies. Less a political movement than a network of groups facilitating community organising for adaptation to less energy-intense and more local forms of economic life, the Transition Towns movement is sometimes described as a 'new social movement'. Cato and Hillier argue that Transition's eschewal of fossil fuel dependence and of statist political economy, and movement toward a decentralised form of associationalism, is reminiscent of the project of Paulo Freire's liberative philosophy of education and Gilles Deleuze's philosophy of an emancipatory micropolitics.[109] The Transition

107. Jeremy K. Leggett, *The Solar Century: The Past, Present and World Changing Future of Solar Energy* (London: Green Profile Books, 2009).

108. Rob Hopkins, *The Transition Handbook: From Oil Dependency to Local Resilience* (Totnes: Green Books, 2008).

109. Molly Scott-Cato and Jean Hillier, 'How Could We Study Climate-Related Social Innovation? Applying Deleuzian Philosophy to Transition Towns', *Environmental Politics* 18 (2010): 869-87.

ideal of a bottom-up, household driven and small business driven economy, where food, energy, building materials, and other essentials are procured from local biomass rather than imported fossil fuels and other materials, recalls in some respects the premodern economy of self-sufficient squirarchy and peasant farmers and craftsmakers.

Perhaps the best analogy for the form of economy advocated by Transition is that of Catholic distributism, as commended by Hilaire Belloc and G. K. Chesterton. For Belloc, the central evil of the modern world was the loss of the liberty and self-possession of the householder by the accumulation of land and wealth by the modern state and modern private corporations. Belloc saw capitalism and communism as two aspects of the same problem: in depriving men and women of ownership of land, and hence livelihood, both systems, or hybrids of them, had ended what for Belloc were the greatest political achievements of Christianity: the ending of slavery in the late Middle Ages and the creation of a free peasantry of smallholders with independent use rights to common grazing and woodlands. For Belloc, a society in which global warming really did begin at home, where home is a small holding with an independent supply of energy and food, and whose owners are therefore able to reduce the impact of their lives on the climate, would be an ideal, even a Christian, society. Belloc's vision is underwritten by the Catholic account of subsidiarity as developed in response to the rise of industrial capitalism by Pope Leo XIII in *Rerum Novarum*. Leo XIII argued that in a complex social system such as that of industrial capitalism, deliberative moral power ought to be located at the lowest permissible point in the social hierarchy. In this perspective, concern about global warming really ought to begin at home, and hence Leo XIII's successor, Pope Benedict XVI, sought to reduce the carbon footprint of his own papal apartments by fitting solar panels to the roof of the Vatican.

Some argue that the transition that is required 'from oil dependency to local resilience' is a necessity not only for mitigating climate change but also for adapting to climate change and diminishing reserves of energy.[110] In the midst of extreme weather events, such as Hurricane Sandy in New York and New Jersey, or Typhoon Bopha in the Philippines, both in 2012, communities and households are cut off in some cases for weeks or even months from life's necessities — including food, light, water, and warmth. This is because flooding and storms destroy national and international power, food, and water supply delivery systems. The Transition vision is

110. Hopkins, *Transition Handbook*.

that householders and small businesses take back responsibility for energy, food, and water supply from the extended but geophysically fragile supply chain delivery systems of the modern state to local and community-level provision. In so doing, Transition is sponsoring a transformation of community infrastructure and ownership patterns which will enable greater local community resilience in the face of climate shocks.

Transition is not, however, simply a technical adaptation to business and household consumption and energy supply systems, which would be radical enough in itself; it is also a root and branch social change strategy. The aim is not only to create communities which supply their own power, food, and other essentials, but also to create really functioning residential neighbourhoods where people cease to be strangers and learn to depend on one another for the necessities of life instead of on anonymous and distant market and statist providers. The Transition Towns movement, while it has no explicit politics, promotes cooperation, friendship, and mutual assistance among neighbouring households and businesses as the first step to designing a society of sustainable and self-sufficient economies which are no longer dependent on large corporations or big government for the necessities of life. The clear intent, as with the geographic parish-based model of Christian *ecclesia,* and Blake's vision of a recovered harvest in the shires, is to re-create a society where economic exchange is remapped onto local geography and local relationships. While not anti-capitalist, the Transition movement involves a radical reform of energy, food, and other kinds of procurement away from the global market and nation-state organisation of capitalism and remaps most economic exchanges onto geographically contiguous social relationships in a manner envisaged by Karl Polanyi in his *The Great Transformation.*

The vision of a post–oil dependent society promoted by Transition Towns and other ecological change movements exemplifies the communitarian ethics and politics of Alasdair MacIntyre and the messianic politics of Taubes. It also seems to offer at least the possibility that Albion may awake from her empire-induced stupor and take up the messianic role Blake envisages for her in demonstrating to the nations what the New Jerusalem looks like. Transition began in Britain and in less than ten years has spawned thousands of Transition initiatives in more than sixty nations.[111] From the capture by the nation-state and private corporations of land, production, and economic and political power, Transition reenacts a kind of

111. See further Rob Hopkins, *The Power of Just Doing Stuff* (Totnes: Transition Publishers, 2013).

politics where people take back the capacity to act together for the common good of a stable climate. At the same time, Transition groups are finding more fulfilling ways of creating and sharing common *wealth* through local economies in which building, craft, horticulture, and power production are all remapped onto local communities. Hardt and Negri argue that such local organising by groups of 'the multitude', when seen together, represent a global and historical movement toward a post-republican social order in which commonwealth is recovered from the institution of private property and the control of states and corporations:

> The multitude produces efficiently, and moreover develops new productive forces, only when it is granted the freedom to do so on its own terms, in its own way, with its own mechanisms of cooperation and communication. This freedom requires an exodus from the republic of property as an apparatus of control in both its private and public guises.[112]

Freedom from the coerciveness of market and state-enforced contracts depends upon the abolition of extreme individualism which the market and the state together promote. Instead, the recovery of wealth as held in common will require that people act cooperatively together, for 'only a multitude can produce the common'.[113] Here they recall the claim of Alasdair MacIntyre that the nation-state cannot promote or serve the common good, precisely because the nation-state rests upon the accumulation of the local powers and wealth of a multitude of local associations who are over time dispossessed by the modern and related processes of nation-state and capital formation, and the seemingly impregnable domain of private property:

> The shared public goods of the modern nation-state are not the common goods of a genuine nation-wide community and, when the nation-state masquerades as the guardian of such a common good, the outcome is bound to be either ludicrous or disastrous or both.[114]

The micropolitical response to climate change of the Transition movement rejects the claim of the nation-state to serve the common good of a stable

112. Michael Hardt and Antonio Negri, *Commonwealth* (Cambridge, MA: Belknap Press, 2009), 302-3.

113. Hardt and Negri, *Commonwealth*, 303.

114. Alasdair MacIntyre, *Dependent Rational Animals* (Chicago and La Salle, IL: Open Court, 1999), 37.

climate precisely because nation-states, while masquerading as a defender of the climate behind participation in the UNFCCC and, in some cases, the Kyoto Protocol, for the most part refuse to acknowledge a duty to serve the common good which might conflict with the general political goal of promoting the interests of private property and growth in wealth creation and accumulation.

The answer to the coming emergency of climate change and extreme weather is no different from the answer to organising human affairs in such a way as to reduce, and hence mitigate, the dependence of modern political economy on the pollution of the climate by fossil fuels. That these two responses are related is confirmed by a Scottish story of recent provenance. The village of Comrie in Perthshire was inundated by flooding in November 2012 from a river that is changing its course in response to growing sudden rain incidents in the region in the last decade. While the villagers blamed the local authority for not putting in place flood defences despite earlier heavy flooding in the same location, they also acted together in ways that considerably reduced suffering and dislocation by those affected. Many householders in the village whose homes were not flooded offered hospitality for weeks, and in some cases months, to those whose homes were flooded so that the community could stay together, children could continue to attend school, and people would not be dislocated from their home surroundings. Alastair McIntosh argues that this surprising instance of friendship and hospitality in a flooded village is linked to the earlier training in community responses to climate change in which he took part three years earlier.[115] Training for adaptation to a post-oil world had generated sufficient community bonds and friendships that when the community was faced with an imminent, rather than a long-term, emergency, residents acted cooperatively together to share what it had with its own.

This is a heartwarming story and indicates that the proof of the Transition pudding is in the eating. But we are still left with the structural problem presented by the Jevons paradox. If exemplary householders, local communities, and even cities take action to collectively move away from fossil fuels to renewable energy, and from global to local procurement of everyday essentials from food and clothing to household artefacts and construction materials, there remains the global market reality that these reductions in

115. Alastair McIntosh, 'Radio Scotland Thought for the Day on Flooding in Comrie', at http://ecocongregationscotland.blogspot.co.uk/2012/11/radio-scotland-thought-for-day-on.html (accessed 10 December 2012).

demand for fossil fuel energy will translate into a lower market price for energy only if the extraction of fossil fuels from existing reserves is unabated. Only the nations individually within their territories are in a position to produce an authoritative inventory of greenhouse gas emissions from coal, gas, and oil extraction sites, electric power production, the built environment, vehicles, waste facilities, and land use. Only the nations individually are capable of encouraging or 'nudging' their citizens through appropriate pricing of energy and taxation, through building and transport regulations and infrastructure provision, toward lower fossil fuel consumption. Only the nations acting together in a concerted fashion can restrain large multinational fossil fuel corporations, power companies, and advertising and marketing corporations from producing ongoing streams of fossil fuels into global markets and enticing consumers to ongoing growth in fossil fuel consumption and use.

Many modern theorists of the nation-state argue that it is a social construct which arose in the service of, and for the purposes of promoting, the accumulation of capital and the capture of land and natural resource wealth from the common ownership of the people to the private ownership of landowners and public and private corporations.[116] Hence, as MacIntyre, Hardt and Negri, and others have it, the nations cannot act for the common good or commonwealth. On this account, mitigation of the climate emergency by the nations is neither feasible nor even *desirable.* But as we have seen, for Schmitt, as for Blake, the nations do have a providential purpose to sustain the fertility of the earth, and hence the flourishing of their citizens. Transition is already showing signs of its capacity to stimulate government and the state to recover their duties to foster national and sustainable economies rather than collude with the corporate agents of empire. The local authority of the county of Devon, in which is situated the first Transition Town of Totnes, has begun to incorporate Transition ideas about local economy and sustainability through community resilience in its planning and organisational procedures.[117]

116. This is a view widely shared by contemporary social scientists and theologians. The best known social scientific treatment is Benedict Anderson, *Imagined Communities: Reflections on the Origin and Spread of Nationalism,* rev. ed. (London: Verso, 1991). For a theological argument for a similar position see William T. Cavanaugh, '"Killing for the Telephone Company": Why the Nation-State Is Not the Keeper of the Common Good', in Cavanaugh, *Migrations of the Holy: God, State, and the Political Meaning of the Church* (Grand Rapids: Eerdmans, 2011), pp. 7-45.

117. Sarah Neal, 'Transition Culture: Politics, Localities and Ruralities', *Journal of Rural Studies* 32 (2013): 60-69.

The nations may have failed internationally to construct a global politics to defend the climate. But outlier nations, and especially Germany, have taken up the need for transition to a post–fossil fuel economy and distributed power as a national priority. There are twelve hundred community energy cooperatives in Germany, and their promotion by the German government is at the cost of the power and profits of electricity companies. But the problem, as I argue more fully elsewhere, is that fossil fuelled grids, networks, and pipelines have constructed a mode of power delivery which continues to shape most governments and large private companies in their relations to householders and small producers.[118] Governments and corporations are addicted to the centralisation of power and control over land and resources which coal, oil, and gas have made possible across much of the habitable surface of the earth in the last century. This shaping is only enhanced by technologies such as metal shipping containers and digital money flows.

Despite the corrosive effects on political community of a borderless global economy, it remains the case that no coal, oil, or gas can be dug, drilled, or extracted without governmental authority; *extraction* is still a bordered activity even if emissions from the burning of the fuels are extra-territorial. If we try to envisage a messianic, anti-imperial solution to the problem of fossil fuel overuse, it may be by analogy with the Transition model of a post–fossil fuel economy. On this approach, climate activists, post–fossil fuel communities of the kind Transition fosters, faith communities, and environmental organisations might form national networks which work together to put beyond use fossil fuel reserves by exposing and resisting government licensing of the remaining reserves. Something like this is envisaged in the fossil fuel divestment movement begun in the United States and Canada by Bill McKibben.[119] However, the movement's primary focus has been on investments in corporate fossil fuel stocks and shares. But if governments license extraction, fossil fuels will be extracted and burned, and it is hard to see how disinvestment by an enlightened few will change this.

An analogy for the approach I am proposing is already taking shape as a way of addressing deforestation and biodiversity loss. 'Distributed conservation' involves an approach to forest conservation that, instead of relying on state forestry and conservation agencies, facilitates the management and

118. Michael S. Northcott, *A Moral Climate: The Ethics of Global Warming* (London: Darton, Longman and Todd, 2007).

119. 'The Fossil Fuel Free Divestment Movement', at www.GoFossilFree.org (accessed 20 March 2013).

ownership of forests by local communities or charitable agencies which may draw on overseas as well as domestic resources.[120] This approach has been pioneered by, among others, North Face founder Douglas Tompkins, who established the Conservation Land Trust in 1992 with the aim of protecting wildlands in Chile and Argentina. The trust bought more than a million hectares of some of the most biodiverse wild land in South America, part of which is now constituted as Pumalin Park and designated a Nature Sanctuary by the government of Chile. The land is put beyond exploitation for natural resource use and is now owned by the Fundación Pumalin, which is a Chilean foundation. There are many related distributed conservation initiatives now emerging on every continent which, while they have engaged government and scientific organisations, represent a new model of nature conservation based on community ownership and private trust law rather than on top-down state management and ownership.[121] If a similar approach was adopted in Malaysia, Indonesia, Brazil, and the Congo, areas of rainforest whose current destruction is creating more CO_2 emissions than were ever intended to be 'saved' by the Kyoto Protocol could be put beyond use. But just as the UNFCCC has failed to focus minds on slowing fossil fuel extraction to address the chief cause of climate change, it has also failed to promote efforts to prevent deforestation, let alone to reforest warming cities and rural areas as ways to at least slow the impacts of rising global temperatures with the cooling effects of trees and their oxygenating and water-fixing properties in atmosphere and soil.[122]

In 2011, climate activist Tim DeChristopher was sentenced to two years in prison at a United States federal court in Salt Lake City. His 'crime' was to participate in a state government auction of oil and gas leases on land in the South Utah wilderness area, which is a National Park. DeChristopher intended to prevent the government from selling the rights to develop rare and wild land for fossil fuel production.[123] Auctions of coal, oil, and gas

120. Janet Silbernagel, Jessica Price, Randy Swaty, et al., 'The Next Frontier: Projecting the Effectiveness of Broad-scale Forest Conservation Strategies', in C. Li, R. Lafortezza, J. Chen, eds., *Landscape Ecology in Forest Management and Conservation: Challenges and Solutions for Global Change* (New York: Springer, 2011), 209-30.

121. Here I am sympathetic to Scruton's claim that English trust law is a unique contribution to the law of peoples and to the practices of ecological conservation; Roger Scruton, *Green Philosophy: How to Think Seriously about the Planet* (London: Atlantic Books, 2012).

122. Thomas Jones, 'How Can We Live with It?' *London Review of Books* 35 (23 May 2013): 3-7.

123. DeChristopher's actions were a key inspiration for the campaign Unburnable

reserves go on every day on every continent. Governments, when they do not own fossil fuel agencies themselves, commission and license fossil fuel companies quoted on American, Asian, and European stock exchanges to seek out, drill, dig up, refine, and transport fossil fuels from every conceivable tract of land and the ocean floor, including most recently parts of the now ice-free Arctic Ocean. If publicly and privately owned fossil fuel corporations are not stopped, six degrees of climate change by 2070 is entirely feasible, and the earth will no longer be the habitable place it now is for seven billion humans. DeChristopher's actions show one potentially very effective way forward. That he served time in jail for his efforts, though these efforts harmed no one and were intended to prevent future harms to many millions of people, perhaps indicates just how effective his action has the potential to be.

In 2012, the Harper government passed a law in the Canadian Parliament, ominously entitled the Omnibus Bill, which altered land use rights granted under earlier treaty to native peoples, and changed the nation's Navigable Water Protection Act, which since 1882 had made it illegal 'to block, alter, or destroy any water deep enough to float a canoe without federal approval.'[124] In response, Chief Allan Adam of the Athabasca Chipewyan First Nation in Canada formally announced the intention of his people to oppose the bill and the permission it gave to Shell Oil and the Canadian government to build a network of roads and pipelines across aboriginal land to enable the extraction of tar sands oil from Northern Alberta:

> Today, history was made. The Harper government passed Bill C-45, a bill that strips down First Nations Treaty rights and protection and diminishes the democracy of this country. We have seen the erosion of our people's Treaty rights throughout various forms of legislation in the past, but this bill is proof the government holds little stock in our rights and title and is attempting to create more loopholes for industry to continue annihilating our lands.[125]

Carbon, which is linked to the Carbon Tracker Initiative at http://www.carbontracker.org/unburnable-carbon (accessed 1 January 2013), and has prompted student protests calling for disinvestment by universities from fossil fuel companies.

124. Brooke Jarvis, 'Idle No More: Native-led Protest Movement Takes on Canadian Government', *Rolling Stone,* 4 February 2013.

125. Press statement from Chief Allan Adam in reaction to the passing of Bill C-45 at http://acfnchallenge.wordpress.com/2012/12/14/for-immediate-release-press-statement-from-chief-allan-adam-in-reaction-to-the-passing-of-bill-c-45/ (accessed 15 March 2013).

Adam subsequently announced plans to engage in nonviolent protest and blockade the existing road — Highway 63 — into their territory.

Jews and Christians recite the following words from what Christians call the *Venite* in daily prayers at synagogues, churches, and chapels around the world:

> For the Lord is a great God
> And a great King above all gods.
> In his hands are the depths of the earth;
> The heights of the mountains are his.
> The sea is his, for he made it himself,
> And his right hand formed the dry land.
> Come let us bow down
> And kneel before the Lord, our maker. (Psalm 95:3-6)

According to the *Venite,* the fossil fuels that remain in the depths of the earth are in the hands of God. The climate crisis indicates that, to honour the God who rules over earth and heaven, local and national communities should find ways to conserve their own fossil fuels in the depths of the earth, while at the same time creating and commissioning a new energy economy dependent on sunlight, wind, and biomass, and so re-create the historic and customary connections between nature and culture, land and life, love for neighbour and nature which are central to the Jewish and Christian messianism of empire-challenging love.

Index